Test Bank

James V. Frugale
Wesleyan University

CALCULUS
ONE AND SEVERAL VARIABLES
9th Edition

Saturnino Salas

Einar Hille

Garret Etgen
University of Houston

WILEY

John Wiley & Sons, Inc.

To order books or for customer service call 1-800-CALL-WILEY (225-5945).

Copyright © 2003 by John Wiley & Sons, Inc.

Excerpts from this work may be reproduced by instructors for distribution on a not-for-profit basis for testing or instructional purposes only to students enrolled in courses for which the textbook has been adopted. *Any other reproduction or translation of this work beyond that permitted by Sections 107 or 108 of the 1976 United States Copyright Act without the permission of the copyright owner is unlawful. Requests for permission or further information should be addressed to the Permissions Department, John Wiley & Sons, Inc., 111 River Street, Hoboken, NJ 07030.*

ISBN 0-471-27523-9

Printed in the United States of America

10 9 8 7 6 5 4 3 2 1

Printed and bound by Bradford & Bigelow, Inc.

TO THE INSTRUCTOR

This test bank has been prepared for the ninth edition of Salas, Hille and Etgen's Calculus. Each of chapters 1–11 is constructed with three sections: A, B, C. Chapters 12–18 are constructed using two sections: B and C.

In Section A the questions are designed to see if the student knows the basic content of the chapter. The questions are about definitions, theorems and "straight forward" applications. In this section you will be able to choose T-F, fill ins, and easier calculation questions. Most of the questions do not require a substantial amount of time to answer.

The questions in Section B are from the 8th edition test bank prepared by Deborah Betthauser Britt. These vary in the amount of time required to solve them. The questions are primarily focused on the students ability to apply the concepts, in the chapter, in a variety of situations.

The questions in Section C are primarily from two sources, some of them are the even numbered problems in the text and the others are a collection of problems that I have used in the past. The problems in Section C are generally more theoretical and frequently extend the concepts in that chapter. Many of these are both classical and historic problems in mathematics.

I wish to thank Sharon Prendergast, production editor, at Wiley; and Marie Vanisko, who reviewed and made many suggestions and corrections to the test bank.

James V. Frugale

CONTENTS

Chapter	1	Introduction	1
		Part A	1
		Part B	7
		Part C	17
Chapter	2	Limits and Continuity	21
		Part A	21
		Part B	25
		Part C	40
Chapter	3	Differentiation	43
		Part A	43
		Part B	46
		Part C	59
Chapter	4	The Mean-Value Theorem and Applications	65
		Part A	65
		Part B	68
		Part C	83
Chapter	5	Integration	89
		Part A	89
		Part B	95
		Part C	109
Chapter	6	Some Applications of the Interval	113
		Part A	113
		Part B	116
		Part C	126
Chapter	7	The Transcendental Functions	129
		Part A	129
		Part B	132
		Part C	146
Chapter	8	Techniques of Integration	151
		Part A	151
		Part B	154
		Part C	164

Chapter 9	Conic Sections; Polar Coordinates; Parametric Equations....	169	
	Part A	169	
	Part B	172	
	Part C	186	
Chapter 10	Sequences; Intermediate Forms; Improper Integrals	191	
	Part A	191	
	Part B	195	
	Part C	203	
Chapter 11	Infinite Series	207	
	Part A	207	
	Part B	213	
	Part C	228	
Chapter 12	Vectors	233	
	Part B	233	
	Part C	241	
Chapter 13	Vector Calculus	243	
	Part B	243	
	Part C	252	
Chapter 14	Functions of Several Variables	255	
	Part B	255	
	Part C	265	
Chapter 15	Gradients; Extreme Values; Differentials	269	
	Part B	269	
	Part C	281	
Chapter 16	Double and Triple Integrals	285	
	Part B	285	
	Part C	296	
Chapter 17	Line Integrals and Surface Integrals	301	
	Part B	301	
	Part C	314	
Chapter 18	Elementary Differential Equations	319	
	Part B	319	
	Part C	325	

CHAPTER 1 Introduction

Part A

1. a) Define rational numbers:
   b) Given the set $A = \{-\pi, 2, 0, 1, \sqrt{2}, 17/2, \sqrt[3]{1001}\}$. (1.2)
      Indentify all:
      (i) Integers _____.
      (ii) Rational numbers _____.
      (iii) Irrational numbers _____.

2. Write $2.\overline{21}$ as a fraction, $\frac{a}{b}$, a, b are integers. (1.2)

3. Write $.\overline{714}$ as a fraction, $\frac{a}{b}$, a, b are integers. (1.3)

4. for $|a + b| \leq |a| + |b|$ find real numbers a, b
   So that a) $|a + b| = |a| + |b|$
   b) $|a + b| < |a| + |b|$

   Complete:

5. $\frac{x-2}{x+1} > 0$ iff _____. (1.3)

6. Express the following graphs with interval notation

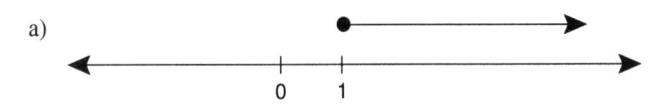

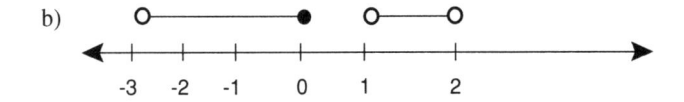

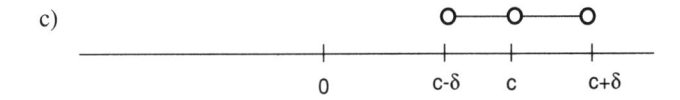

   a)

   b)

   c)

For Questions 7–13 Determine if the given set is bounded, and if it is suggest suitable upper and lower bounds (1.2)

7. $\{1, 3, 5, 9\}$.

8. $\{1, 3, 5, \ldots\}$.

9. $\{x : x \leq 3, x \in R\}$.

10. $\{x : x < 3, x \in R\}$.

11. $\{x : x < 3, x \in I^+\}$.

2 Calculus: One and Several Variables

12. $\{x : x < \sqrt{3}, x \in I^+\}$.

13. $\{x : x = \dfrac{1}{n}, n \in I^+\}$.

For Questions 14–20 True or False (1.2)

14. if $x < 4$ then $-x < -4$.

15. if $x < 2$ then $x - 4 < -2$.

16. $|x| > 2$ iff $-2 < x < 2$. (1.2)

17. $|x - 1| < 3$ iff $-2 < x < 4$.

18. $|x| = 0$ iff $x = 0$.

19. $|-3|^2 = -3^2$.

20. a) $\dfrac{x-2}{x+1} < 0$ iff $(x-2)(x+1) < 0$. (1.3)
    b) $|x - 2| < 7$ is equivalent to $0 < |x - 2| < 7$.

21. Find 2 real numbers a, b such that
    a) $|a + b| = |a| + |b|$ (1.3)
    b) $|a + b| < |a| + |b|$

For Questions 22–25 Given the points $(2, 4)$ and $(-3, 0)$ (1.4)

Find

22. The distance between them.

23. The midpoint of the line segment joining them.

24. The slope of the line joining them. (1.4)

25. Find the slopes of any line perpendicular to the line joining the two points.

26. Given two sets D, R Define a function from D into R. (1.5)

For Questions 27–34 True or False (1.5)

27. $f(x) = x^2$ is an even function.

28. $f(x) = x^3 + 6x^2 - 10$ is an odd function.

29. $f(x) = 4 \sin x$ is an even function.

30. $f(x) = x^2, g(x) = \cos x \quad f + g$ is an even function.

31. $f(x) = \dfrac{1}{x}$ is an even function.

32. The domain $f(x) = \sqrt{9 - x^2}$ is $D_f = 3 < x < -3$.

33. The domain $f(x) = \dfrac{(x+2)(x-1)}{(x+1)(x^2-9)}$ is D_f all R except $x = -3, -2, -1, 1, 3$.

34. The range of $f(x) = x^2 - 4$ $R_s = [-4, \infty)$.

35. a) Give an example of a polynomial (1.6)
    b) Give an example that is not a polynomial

36. Define a Rational function.

For Questions 37–44 True or False

37. $y = 2$ is a function.

38. $x = -4$ is a function.

39. $\pi/3 = 45°$.

40. $\sin(\pi/4) + \cos\left(\dfrac{3\pi}{4}\right) = 0$. (1.6)

41. $\sin(\pi) + \cos(\pi) = 0$.

42. $\sin^2(2) + \cos^2(2) = 1$.

43. $\sin\theta = \cos(\pi/2 - \theta)$.

44. The domain of tangent function is all R.

For Questions 45–50 $f(x) = -\dfrac{1}{2}(x) + 2$ $g(x) = \dfrac{1}{2}x - 2$ (1.7)

Find:

45. $f + g =$

46. $f \cdot g =$

47. $f/g =$

48. $f \circ g =$

49. $g \circ f =$

50. Are $f_1 g$ inverses of each other?

For Questions 51–53 Find the inverse of (1.7)

51. $f(x) = \dfrac{2}{5}x + 1$.

52. $f(x) = x^3 - 1$.

53. $f(x) = \dfrac{1}{x}$.

54. If f, g are odd functions, what can you conclude about $f \cdot g$? Give an example justifying your answer. (1.7)

55. State the Axiom of Induction
    Let S be a set of positive integers (1.8)

        If (A) _____.

        (B) _____.

    Then _____.

Answers to Chapter 1, Part A Questions

1. a) a/b | a,b ∈ I and b ≠ 0}
   (i)   {2, 0, 1}
   (ii) {2, 0, 1, 17/2}
   (iii) $\{-\pi, \sqrt{2}, \sqrt[3]{1001}\}$

2. 219/99

3. $\dfrac{714}{999} = \dfrac{238}{333}$

4. ans vary
   a) 2 positive or 2 negative numbers.
   b) one negative and one positive number

5. $(x-2)(x+1) > 0$

6. a) $[1, \infty)$
   b) $(-3, 0] \cup (1, 2)$
   c) $(c - \delta, c) \cup (c, c + \delta)$

7. bounded any $LB \leq 1$, $UB \geq 9$

8. not bounded

9. not bounded

10. not bounded

11. bounded $LB \leq 1$ $UB \geq 2$

12. bounded $LB \leq 1$ $UB \geq 2$

13. bounded $LB \leq 0$ $UB \geq 1$

14. F

15. T

16. F

17. T

18. T

19. F

20. a) T   b) F

21. ans vary
    a) two positive numbers or two negative numbers
    b) use one positive and one negative number

22. $\sqrt{41}$

23. $(-1/2, 2)$

24. $4/5$

25. $-5/4$

26. Def

27. T

28. F

29. F

30. T

31. F

32. F

33. F

34. T

35. any vary all exponents natural numbers

36. $R(x) = \dfrac{P(x)}{Q(x)}$ $P(x), Q(x)$ polynomials
    $Q(x) \neq 0$

37. T

38. F

39. F

40. T

41. F

42. T

43. T

44. F

45. 0

46. $-\dfrac{1}{4}x^2 + 2x - 4$

47. $\dfrac{-\dfrac{1}{2}x + 2}{\dfrac{1}{2}x - 2}$ or $\dfrac{-x + 4}{x - 4}$ or -1 $x \neq 4$

48. $-\dfrac{1}{4}x + 3$

49. $-\dfrac{1}{4}x - 1$

50. No

51. $f(x) = \dfrac{2}{5}x + 1$

   $f^{-1}(x) = \dfrac{5}{2}x - \dfrac{5}{2}$

52. $f(x) = x^3 - 1$
    $f^{-1}(x) = \sqrt[3]{x+1}$

53. $f^{-1}(x) = \dfrac{1}{x}$

54. $f \cdot g = $ even function
    $f(x) = x \quad g(x) = x^3 \quad f \cdot g(x) = x^y$

55. (A) $l \in S$
    (B) $k \in S$ implies $(k+1) \in S$
    then all the positive integers are in S

Part B

1.2 Review of Elementary Mathematics

1. Is the number $\sqrt{13^2 - 12^2}$ rational or irrational?

2. Is the number $5.121122111222\ldots$ rational or irrational?

3. Write $6.27272727\ldots$ in rational form p/q.

4. Find, if any, upper and lower bounds for the set $S = \{x : x^3 > 1\}$.

5. Find, if any, upper and lower bounds for the set $S = \left\{\dfrac{2n-1}{n} : n = 1, 2, 3, \ldots\right\}$.

6. Rewrite $27 - 8x^3$ in factored form.

7. Rewrite $x^4 - 18x^2 + 81$ in factored form.

8. Evaluate $\dfrac{4!}{6!}$

9. Evaluate $\dfrac{7!}{4!3!}$

10. What is the ratio of the surface area of a cube of side x to the surface area of a sphere of diameter x?

1.3 Review of Inequalities

11. Solve $x + 3 < 2x - 8$.

12. Solve $\dfrac{-2}{3}x \geq \dfrac{1}{6} - \dfrac{3}{4}x$.

13. Solve $8 - x^2 > 7x$.

14. Solve $x^2 - 5x + 5 \geq 1$.

15. Solve $3x^2 - 1 \geq \dfrac{1}{2}(3 + x^2)$.

16. Solve $\dfrac{2x+3}{4x-1} < 1$.

17. Solve $\dfrac{1}{x} < \dfrac{1}{x+1}$.

18. Solve $\dfrac{1}{2}(3x + 2) < 2 - \dfrac{2}{3}(5 - 3x)$.

19. Solve $\dfrac{3}{x+2} < \dfrac{2}{x-1}$.

20. Solve $\dfrac{2x-3}{x+2} \geq 1$.

8   Calculus: One and Several Variables

21. Solve $x(2x - 1)(3x + 2) < 0$.

22. Solve $\dfrac{x^2}{x^2 - 9} < 0$.

23. Solve $x^2(x - 1)(x + 2)^2 > 0$.

24. Solve $\dfrac{x + 2}{x^2(x + 3)} > 0$.

25. Solve $|x| > 2$.

26. Solve $|x + 1| \leq \dfrac{1}{3}$.

27. Solve $|x - 1| \leq \dfrac{1}{2}$.

28. Solve $0 < \left| x - \dfrac{1}{4} \right| < 1$.

29. Solve $|2x - 1| < 5$.

30. Solve $|3x - 5| \geq \dfrac{2}{3}$.

31. Find the inequality of the form $|x - c| < \delta$ whose solution is the open interval $(-1, 5)$.

32. Find the inequality of the form $|x - c| < \delta$ whose solution is the open interval $(-2, 3)$.

33. Determine all values of $A > 0$ for which the following statement is true. If $|2x - 5| < 1$, then $|4x - 10| < A$.

34. Determine all values of $A > 0$ for which the following statement is true. If $|2x - 3| < A$, then $|6x - 9| < 4$.

1.4   Coordinate Plane; Analytic Geometry

35. Find the distance between the points $P_0(2, -4)$ and $P_1(1, 5)$.

36. Find the midpoint of the line segment from $P_0(a, 2b)$ to $P_1(3a, 5b)$.

37. Find the slope of the line through $P_0(-4, 2)$ and $P_1(3, 5)$.

38. Find the slope of the line through $P_0(-2, -4)$ and $P_1(3, 5)$.

39. Find the slope and the y-intercept for the line $2x + y - 10 = 0$.

40. Find the slope and the y-intercept for the line $8x + 3y = 6$.

41. Write an equation for the line with the slope -3 and y-intercept -4.

42. Write an equation for the horizontal line 4 units below the x-axis.

43. Write an equation for the vertical line 2 units to the left of the y-axis.

44. Find an equation for the line that passes through the point $P(2, -1)$ and is parallel to the line $3y + 5x - 6 = 0$.

Introduction 9

45. Find an equation for the line that passes through the point $P(1, -1)$ and is perpendicular to the line $2x - 3y - 8 = 0$.

46. Find an equation for the line that passes through the point $P(2, 2)$ and is parallel to the line $2x + 3y = 18$.

47. Find an equation for the line that passes through the point $P(3, 5)$ and is perpendicular to the line $6x - 7y + 17 = 0$.

48. Determine the point(s) where the line $y = 2x$ intersects the circle $x^2 + y^2 = 4$.

49. Find the point where the lines l_1 and l_2 intersect. $l_1: x + y - 2 = 0$; $l_2: 3x + y = 0$.

50. Find the area of the triangle with vertices $(-1, 1), (4, -2), (3, 6)$.

51. Find the area of the triangle with vertices $(-1, -1), (-4, 3), (3, 3)$.

52. Find an equation for the line tangent to the circle $x^2 + y^2 - 4x - 2y = 0$ at the point $P(4, 2)$.

1.5 Functions

53. If $f(x) = \dfrac{|x + 2|}{x^2 + 4}$, calculate (a) $f(0)$, (b) $f(1)$, (c) $f(-2)$, (d) $f(-5/2)$.

54. If $f(x) = \dfrac{x^3}{x^2 + 2}$, calculate (a) $f(-x)$, (b) $f(1/x)$, (c) $f(a + b)$.

55. $f(x) = |1 - 2x|$ Find the number(s), if any, where f takes on the value 1.

56. $f(x) = 1 - \cos x$ Find the number(s), if any, where f takes on the value 1.

57. Find the exact value(s) of x in the interval $[0, 2\pi]$ which satisfy $\cos 2x = -\dfrac{\sqrt{3}}{2}$.

58. Find the domain and range for $f(x) = 2 - x - x^2$.

59. Find the domain and range for $f(x) = \sqrt{3x + 4}$.

60. Find the domain and range for $h(x) = -\sqrt{4 - x^2}$.

61. Find the domain and range for $f(x) = \dfrac{1}{\sqrt{x} - 1}$.

62. Find the domain and range for $f(x) = \dfrac{1}{(x + 3)^2}$.

63. Find the domain and range for $f(x) = \dfrac{1}{x^2 + 3}$.

64. Find the domain and range for $f(x) = \left|\cos x - \dfrac{1}{2}\right|$.

65. Sketch the graph of $f(x) = 3 - 4x$.

66. Sketch the graph of $f(x) = -\sqrt{6x - x^2}$.

10 Calculus: One and Several Variables

67. Sketch the graph of $f(x) = x - \dfrac{4}{x}$.

68. Sketch the graph of $g(x) = 2 + \sin x$.

69. Sketch the graph of $g(x) = \begin{cases} 2 & \text{if } x < 1 \\ 3 & \text{if } x = 1 \\ 2x - 1 & \text{if } x > 1 \end{cases}$ and give its domain and range.

70. Sketch the graph of $f(x) = \begin{cases} x^2 & \text{if } x < 1 \\ 2x & \text{if } x \geq 1 \end{cases}$ and give its domain and range.

71. Is an ellipse the graph of a function?

72. Determine whether $f(x) = x^4 - x^2 + 1$ is odd, even, or neither.

73. Determine whether $f(x) = x^5 + x^3 - 3x$ is odd, even, or neither.

74. Determine whether $f(x) = \dfrac{x^3 - 2x^2 + 5x + 1}{x}$ is odd, even, or neither.

75. Determine whether $f(x) = \cos(x + \pi/6)$ is odd, even, or neither.

76. Determine whether $f(x) = 3x - 2\sin x$ is odd, even, or neither.

77. Determine whether $f(x) = \cos x + \sec x$ is odd, even, or neither.

78. A given rectangle is twice as long as it is wide. Express the area of the rectangle as a function of the (a) width, (b) length, (c) diagonal.

1.6 The Elementary Functions

79. Find all real numbers x for which $R(x) = \dfrac{3x^2 + 2x + 5}{x - x^2}$ is undefined.

80. Find all real numbers x for which $R(x) = \dfrac{x^3 - 4x^2 + 3x}{x^2 + x - 2}$ is zero.

81. Find the inclination of the line $x - \sqrt{3}y + 2\sqrt{3} = 0$.

82. Write an equation for the line with inclination $45°$ and y-intercept -2.

83. Find the distance between the line $4x + 3y + 4 = 0$ and (a) the origin (b) the point $P(1, 3)$.

84. Find the distance between the line $2x - 5y - 10 = 0$ and (a) the origin (b) the point $P(-2, -1)$.

85. In the triangle with vertices $(0, 0)$, $(2, 6)$, $(7, 0)$, which vertex is farthest from the centroid?

1.7 Combinations of Functions

86. Given that $f(x) = \dfrac{x^2 - x - 6}{x}$ and $g(x) = x - 3$, find (a) $f + g$ (b) $f - g$ (c) $\dfrac{f}{g}$.

Introduction　　11

87. Sketch the graphs of the following functions with f and g as shown in the figure.
    (a) $\frac{1}{2}f$　　(b) $-g$　　(c) $g - f$

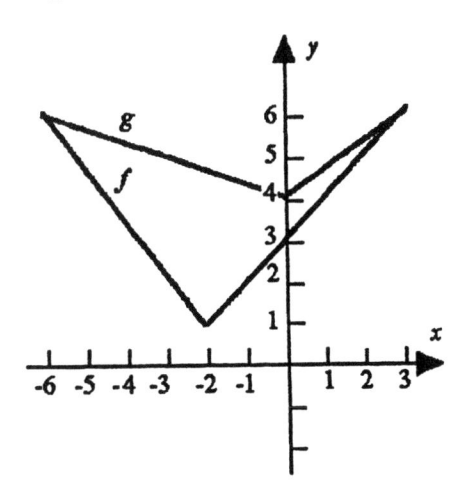

88. $f(x) = \sqrt{x+5}, g(x) = \sqrt{x}$.
    (a) Form the composition of $f \circ g$.　　(b) Form the composition of $g \circ f$.

89. $f(x) = x^2 + 1, g(x) = \dfrac{2}{\sqrt{x}}$.
    (a) Form the composition of $f \circ g$.　　(b) Form the composition of $g \circ f$.

90. $f(x) = \dfrac{1}{x} + 1, g(x) = 3x^2 + 2$.
    (a) Form the composition of $f \circ g$.　　(b) Form the composition of $g \circ f$.

91. $f(x) = \sqrt{4 + x^2}, g(x) = \dfrac{2}{x}$.
    (a) Form the composition of $f \circ g$.　　(b) Form the composition of $g \circ f$.

92. $f(x) = |x|, g(x) = x^3 + 1$.
    (a) Form the composition of $f \circ g$.　　(b) Form the composition of $g \circ f$.

93. Form the composition of $f \circ g \circ h$ if $f(x) = \dfrac{1}{4}x$, $g(x) = 2x - 1$, and $h(x) = 3x^2$.

94. Form the composition of $f \circ g \circ h$ if $f(x) = x^2$, $g(x) = 2x + 1$, and $h(x) = 2x^2$.

95. Form the composition of $f \circ g \circ h$ if $f(x) = \dfrac{2}{x}$, $g(x) = \dfrac{2}{3x+2}$, and $h(x) = x^2$.

96. Find f such that $f \circ g = F$ given that $g(x) = 2x^2$ and $F(x) = x + 2x^2 + 3$.

97. Find f such that $f \circ g = F$ given that $g(x) = \sqrt{x} + 1$ and $F(x) = x + 2\sqrt{x}$.

98. Find g such that $f \circ g = F$ given that $f(x) = x^2 - 1$ for all real x and $F(x) = 3x - 1$ for $x \geq 0$.

99. Find g such that $f \circ g = F$ given that $f(x) = x^2$ and $F(x) = (2x + 5)^2$.

100. Form $f \circ g$ and $g \circ f$ given that $f(x) = 4x + 1$ and $g(x) = 4x^2$.

101. Form $f \circ g$ and $g \circ f$ given that $f(x) = \begin{cases} \dfrac{1}{x} & \text{if } x < 0 \\ 2x - 1 & \text{if } x \geq 0 \end{cases}$ and that $g(x) = \begin{cases} 2x & \text{if } x < 1 \\ x^2 & \text{if } x \geq 1 \end{cases}$.

102. Decide whether $f(x) = 4x + 3$ and $g(x) = \dfrac{1}{4}x - 3$ are inverses of each other.

103. Decide whether $f(x) = (x - 1)^5 + 1$ and $g(x) = (x - 1)^{1/5} + 1$ are inverses of each other.

1.8 A Note on Mathematical Proof; Mathematical Induction

104. Show that $3n \leq 3^n$ for all positive integers n.

105. Show that $n(n + 1)(n + 2)(n + 3)$ is divisible by 8 for all positive integers n.

106. Show that $1 + 5 + 9 + \cdots + (4n - 3) = 2n^2 - n$ for all positive integers n.

Answers to Chapter 1, Part B Questions

1. rational
2. irrational
3. 69/11
4. lower bound 1, no upper bound
5. lower bound 1, upper bound 2
6. $(3 - 2x)(9 + 6x + 4x^2)$
7. $(x - 3)^2(x + 3)^2$
8. 1/30
9. 35
10. $\dfrac{6}{\pi}$
11. $(11, \infty)$
12. $(2, \infty)$
13. $(-8, 1)$
14. $(-\infty, 1] \cup [4, \infty)$
15. $(-\infty, -1] \cup [1, \infty)$
16. $(-\infty, 1/4] \cup (2, \infty)$
17. $(-1, 0)$
18. $(14/3, \infty)$
19. $(-\infty, -2) \cup (1, 7)$
20. $(-\infty, -2) \cup [5, \infty)$
21. $(-\infty, -2/3) \cup (0, 1/2)$
22. $(-3, 0) \cup (0, 3)$
23. $(1, \infty)$
24. $(-\infty, -3) \cup (-2, 0) \cup (0, \infty)$
25. $(-\infty, -2) \cup (2, \infty)$
26. $(-\infty, -4/3] \cup [-2/3, \infty)$
27. $[1/2, 3/2]$
28. $(-3/4, 1/4) \cup (1/4, 5/4)$
29. $(-2, 3)$
30. $(-\infty, 13/9) \cup (17/9, \infty)$
31. $|x - 2| < 3$
32. $|x - 1/2| < 5/2$
33. $A \geq 2$
34. $0 \leq A \leq 4/3$
35. $\sqrt{82}$
36. $\left(2a, \dfrac{7}{2}b\right)$
37. $m = 3/7$
38. $m = 9/5$
39. $m = -2$; y-intercept: 10
40. $m = -8/3$; y-intercept: 2
41. $y = -3x - 4$
42. $y = -4$
43. $x = -2$
44. $y = \dfrac{-5}{3}x + \dfrac{7}{3}$
45. $y = \dfrac{-3}{2}x + \dfrac{1}{2}$
46. $y = \dfrac{-2}{3}x + \dfrac{10}{3}$
47. $y = \dfrac{-7}{6}x + \dfrac{17}{2}$
48. $\left(\dfrac{2\sqrt{5}}{5}, \dfrac{4\sqrt{5}}{5}\right), \left(\dfrac{-2\sqrt{5}}{5}, \dfrac{-4\sqrt{5}}{5}\right)$
49. The point of intersection is $P(-1, 3)$.
50. 37/2
51. 14
52. $y = -2x + 10$

14   Calculus: One and Several Variables

53. (a) 1/2   (b) 3/5   (c) 0   (d) 2/41

54. (a) $\dfrac{-x^3}{x^2+2}$ (b) $\dfrac{1}{x+2x^3}$ (c) $\dfrac{(a+b)^3}{(a+b)^2+2}$

55. $x = 0, 1$

56. $(2n+1)\dfrac{\pi}{2}, n = \text{integer}$

57. $5\pi/12, 7\pi/12, 17\pi/12, 19\pi/12$

58. domain: $(f) = (-\infty, \infty)$
    range: $(f) = (-\infty, 9/4)$

59. domain: $(f) = [-4/3, \infty)$
    range: $(f) = [0, \infty)$

60. domain: $(f) = [-2, 2]$
    range: $(f) = [-2, 0]$

61. domain: $(f) = [0, 1) \cup (1, \infty)$
    range: $(f) = [-1, 0) \cup (0, \infty)$

62. domain: $(f) = (-\infty, -3) \cup (-3, \infty)$
    range: $(f) = (0, \infty)$

63. domain: $(f) = (-\infty, \infty)$
    range: $(f) = (0, 1/3]$

64. domain: $(f) = (-\infty, \infty)$
    range: $(f) = [0, 3/2]$

65. $y = 3 - 4x$

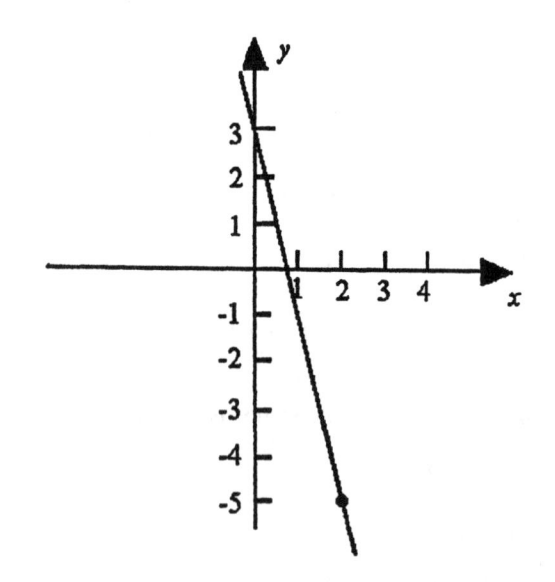

66. $y = -\sqrt{6x - x^2}$

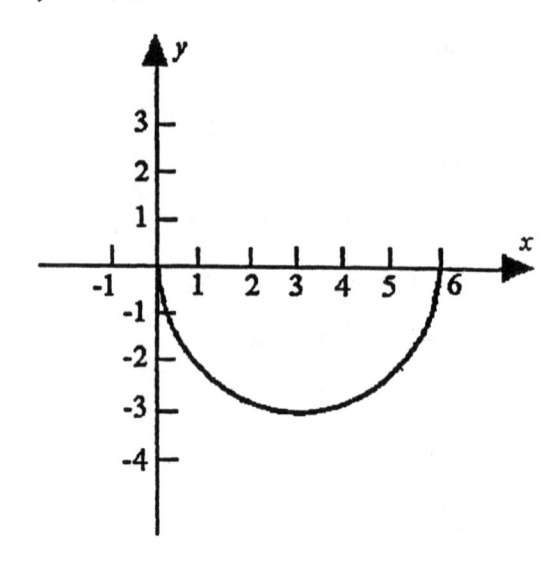

67. $y = x - \dfrac{4}{x}$

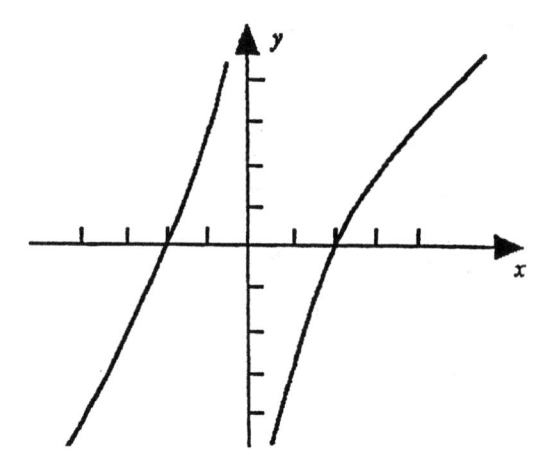

68. $y = 2 + \sin x$

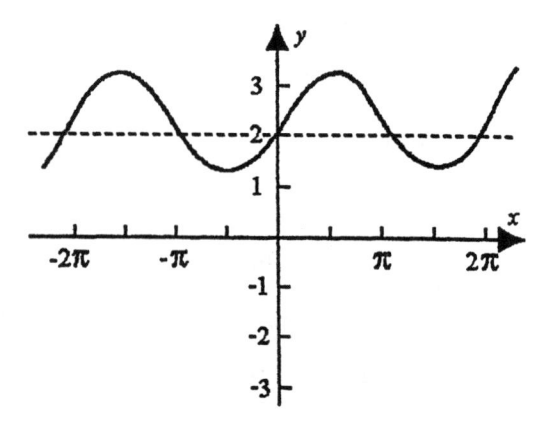

69. domain: $(f) = (-\infty, +\infty)$
    range: $(f) = [2, +\infty)$

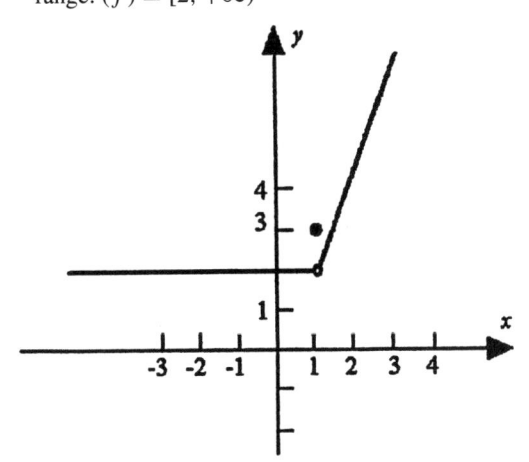

70. domain: $(f) = (-\infty, +\infty)$
    range: $(f) = [0, +\infty)$

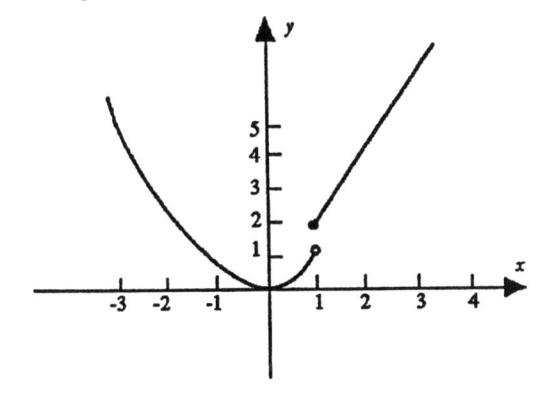

71. no

72. even

73. odd

74. neither

75. neither

76. odd

77. even

78. (a) $A = 2w^2$ (b) $A = \dfrac{l^2}{2}$ (c) $A = \dfrac{2}{5}d^2$

79. 0, 1

80. 0, 3

81. $\theta = \pi/6$

82. $x - y - 2 = 0$

83. (a) 4/5 (b) 17/5

84. (a) $\dfrac{10}{\sqrt{29}}$ (b) $\dfrac{9}{\sqrt{29}}$

85. $(7, 0)$

86. (a) $\dfrac{2x^2 - 4x - 6}{x}$ (c) $\dfrac{x+2}{x}, x \neq 3$
    (b) $\dfrac{2x - 6}{x}$

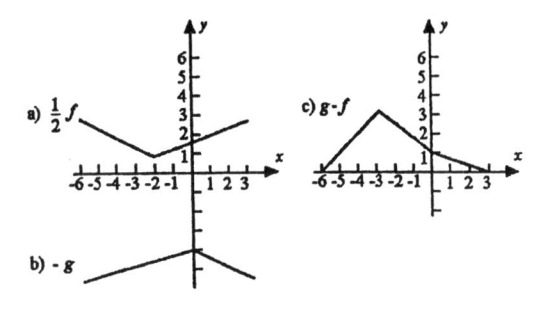

87. a) $\frac{1}{2}f$ b) $-g$ c) $g-f$

88. (a) $\sqrt{\sqrt{x} + 5}$ (b) $\sqrt[4]{x + 5}$

89. (a) $\dfrac{4}{x} + 1, x > 0$ (b) $\dfrac{2}{\sqrt{x^2 + 1}}$

90. (a) $\dfrac{1}{3x^2 + 2} + 1$ (b) $3\left(\dfrac{1}{x} + 1\right)^2 + 2$

91. (a) $\sqrt{4 + \dfrac{4}{x^2}}$ (b) $\dfrac{2}{\sqrt{4 + x^2}}$

92. (a) $|x^3 + 1|$ (b) $|x|^3 + 1$

93. $\dfrac{1}{4}(6x^2 - 1)$

94. $(4x^2 + 1)^2$

95. $3x^2 + 2$

96. $f(x) = x + 3$

97. $f(x) = x^2 - 1$

98. $g(x) = \sqrt{3x}$

99. $g(x) = 2x + 5$

100. $(f \circ g)(x) = 16x^2 + 1$
     $(g \circ f)(x) = 4(4x + 1)^2$

101.
$$(f \circ g)(x) = \begin{cases} \dfrac{1}{2x} & \text{if } x < 0 \\ 4x - 1 & \text{if } 0 \le x < 1 \\ 2x^2 - 1 & \text{if } x \ge 1 \end{cases}$$

$$(g \circ f)(x) = \begin{cases} 2/x & \text{if } x < 0 \\ 2(2x - 1) & \text{if } 0 \le x < 1 \\ (2x - 1)^2 & \text{if } x \ge 1 \end{cases}$$

102. not inverses

103. inverses

104. True for $n = 1 : 3 \le 3$. Assume true for n. Then $3(n + 1) = 3n + 3 \le 3^n + 3 \le 3^n + 3^n = 2(3^n) < 3(3^n) = 3^{n+1}$, so the inequality is true for $n + 1$. Therefore, by induction, it is true for $n \ge 1$.

105. True for $n = 1 : 1 \cdot 2 \cdot 3 \cdot 4 = 3 \cdot 8$. Assume true for n. Then $(n + 1)(n + 2)(n + 3)(n + 4) = n(n + 1)(n + 2)(n + 3) + 4(n + 1)(n + 2)(n + 3)$. The first term is divisible by 8 by the induction hypothesis, and the second term is divisible by 8 since at least one of $(n + 1), (n + 2), (n + 3)$ is even. Hence the result is true for $n + 1$. Therefore, by induction, it is true for all $n \ge 1$.

106. True for $n = 1 : 1 = 2(1)^2 - 1$. Assume true for n. Then $1 + 5 + 9 + \cdots + [4(n + 1) - 3] = 1 + 5 + 9 + \cdots + (4n - 3) + [4(n + 1) - 3] = 2n^2 - n + (4n + 1) = 2(n + 1)^2 - (n + 1)$, so the result is true for $n + 1$. Therefore, by induction, it is true for all $n \ge 1$.

Part C

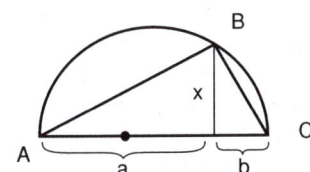

1. $a, b > 0$, ABC are of semicircle, $\overline{AC}$ diameter prove $\sqrt{ab} \leq \frac{1}{2}(a+b)$. (1.2)

If a is a nonzero rational number and b is an irrational number.

2. Prove $a + b$ is an irrational number. (1.2)

3. Is $a \cdot b$ is an irrational number? (1.2)

4. Prove $\dfrac{a}{b}$ is an irrational number. (1.2)

5. Find the values of b for which the line $y = 3x + b$ intersects the circle $x^2 + y^2 = 4$. (1.4)

6. If the line $y = mx + b$ is tangent to the circle $x^2 + y^2 = r^2$ find an equation relating m, b, and r. (1.4)

7. Find the equations of the lines that pass through $(1, 3)$ and are tangent to the circle $x^2 + y^2 = 2$. (1.4)

8. $f(x) = \dfrac{1+x}{1-x}$ find a) $f\left(\dfrac{1}{x}\right)$ b) $f(f(x))$. (1.5)

9. Given $\triangle ABC$ sides a, b, c (1.6)

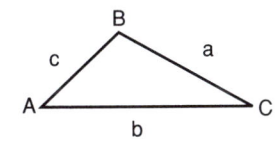

   Prove $\dfrac{a}{\sin A} = \dfrac{b}{\sin B} = \dfrac{c}{\sin C}$

   and that the area of

   $\triangle ABC = \dfrac{1}{2}ab\sin C = \dfrac{1}{2}ac\sin B = \dfrac{1}{2}bc\sin A$.

Use mathematical induction to prove (10–15).

10. Prove $\dfrac{1}{1 \cdot 2} + \dfrac{1}{2 \cdot 3} + \dfrac{1}{3 \cdot 4} + \cdots + \dfrac{1}{n(n+1)} = \dfrac{n}{n+1}$ (1.8)

11. Prove $1 \cdot 2 + 2 \cdot 3 + 3 \cdot 4 + \cdots + n(n+1) = \dfrac{n(n+1)(n+2)}{3}$. (1.8)

12. Prove $1 + \dfrac{1}{2} + \dfrac{1}{4} + \cdots + \dfrac{1}{2^n} = 2 - \dfrac{1}{2^n}$. (1.8)

13. Prove $1 + r + r^2 + \cdots + r^n = \dfrac{1 - r^{n+1}}{1 - r}$ $r \neq 1$. (1.8)

14. Prove $1^3 + 2^3 + 3^3 + \cdots + n^3 = \left[\dfrac{1}{2}n(n+1)\right]^2$. (1.8)

15. Prove $1^4 + 2^4 + 3^4 + \cdots + n^4 = \dfrac{1}{30}n(n+1)(6n^3 + 9n^2 + n - 1)$. (1.8)

16. Find a formula for $3 + 8 + 13 + \cdots + (5n - 2)$. (1.8)

17. Proofs (1.6)

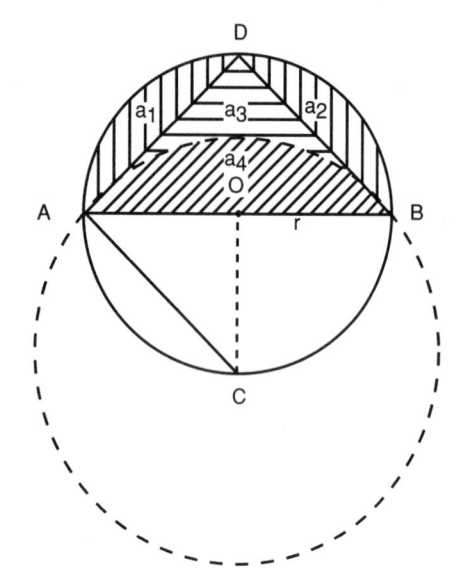

Given: circle O, radius r, circle C radius $\overline{AC}$ $\overline{AD}$, $\overline{DB}$ are chords of circle O and are tangent to circle C
a_1, a_2, a_3, a_4 are areas

Prove $a_1 + a_2 + a_3 = r^2$

Hint: Find $\overline{AC}$ first then $\dfrac{a_1}{a_4} = *$

* Segments of two circles with equal central angles are to each other as the squares of the radii.

18. Given Points $(6, 2), (8, 6), (4, 8), (2, 4)$ Prove they are the vertices of a rectangle. (1.6)

19. Prove that a graph that is symmetric with respect to both coordinate axes is also symmetric to the origin. (1.6)

20. If $a, b \in R$, denote the larger $\max(a, b)$ the smaller $\min(a, b)$ show that (1.6)
$$\max(a, b) = \frac{1}{2}(a + b + |a - b|)$$
and $\min(a, b) = \dfrac{1}{2}(a + b - |a - b|)$.

*The lune of Hippocrates.

Answers to Chapter 1, Part C Questions

1. $\dfrac{a}{x} = \dfrac{x}{b} \longleftrightarrow x^2 = ab$

   $x < \sqrt{ab}$

   $a + b = $ AC diameter

   $\dfrac{1}{2}(a+b) = $ radius

   $x \leq $ radius for all x

   so $\dfrac{1}{2}(a+b) \leq \sqrt{ab}$

   assume $(a+b)$ is rational

2. $(a+b) - a = b$ is rational but b is irrational so $(a+b)$ is not rational

3. Assume $a \cdot b$ is rational $a \neq 0$

   Thus $\dfrac{a \cdot b}{a}$ is rational

   but $\dfrac{a \cdot b}{a} = b$ is irrational

4. Again assume $\dfrac{a}{b}$ is rational

   but this $\dfrac{a}{b} \div a$ is rational

   hence $\dfrac{1}{b}$ is rational

   and b is rational contradicts hypothesis

5. $b^2 \leq 40$ so $-40 \leq b \leq 40$

6. $b^2 = r^2(m^2 + 1)$ or $|b| = r\sqrt{m^2+1}$

7. $y = x + 2$
   $y = -7x + 10$

8. (a) $\dfrac{x+1}{x-1}$ (b) $\dfrac{-1}{x}$

9.

   (triangle ABC with altitude h from B to AC, sides c, a, b, right angle at foot)

   $\sin A = \dfrac{h}{c}$ $c \sin A = h$

   $\sin C = \dfrac{h}{a}$ $a \sin C = h$

   $A = \dfrac{1}{2}(\text{base})(\text{height})$

   $= \dfrac{1}{2}bc \sin A = \dfrac{1}{2}ba \sin C = \dfrac{1}{2}ac \sin B$

   $\dfrac{1}{1}$ by $\dfrac{1}{2}abc = \dfrac{\sin A}{a} = \dfrac{\sin B}{b} = \dfrac{\sin C}{c}$

   $\dfrac{a}{\sin A} = \dfrac{b}{\sin B} = \dfrac{c}{\sin C}$

10. $P_1 = \dfrac{1}{1 \cdot 2} = \dfrac{1}{1+1} = \dfrac{1}{2}$ term

    $P_{k+1} = \dfrac{k+1}{k+2}$ $(k+1)^{th}$ term $= \dfrac{1}{(k+1)(k+2)}$

    $P_k + (k+1)^{th}$ term $= \dfrac{k}{k+1} + \dfrac{1}{(k+1)(k+2)}$

    $= \dfrac{k(k+2)+1}{(k+1)(k+2)} = \dfrac{k^2+2k+1}{(k+1)(k+2)} = \dfrac{(k+1)}{k+2}$

11. $P_1 = \dfrac{1(1+1)(1+2)}{3} = \dfrac{2(3)}{3} = 2 = 1 \cdot 2$ True

    $P_k = \dfrac{k(k+1)(k+2)}{3} \Rightarrow P_{k+1}$

    $= \dfrac{(k+1)(k+2)(k+3)}{3}$

    $P_k + (k+1)^{th}$ term $(k+1)^{th}$ term $(k+1)(k+2)$

    $\dfrac{k(k+1)(k+2)}{3} + (k+1)(k+2)$

    $= \dfrac{k(k+1)(k+2) + 3(k+1)(k+2)}{3}$

    $= \dfrac{(k+1)(k+2)(k+3)}{3}$

12. $P_1 = 2 - \dfrac{1}{2^1} = 2 - 1 = 1$ True

    $P_k = 2 - \dfrac{1}{2^k} \Rightarrow P_{k+1} = 2 - \dfrac{1}{2^{k+1}} = \dfrac{2^{k+2}-1}{2^{k+1}}$

    $(k+1)^{th}$ term $\dfrac{1}{2^{k+1}}$

    $P_k(k+1)^{th}$ term $= 2 - \dfrac{1}{2^k} + \dfrac{1}{2^{k+1}} = \dfrac{2^{k+2}-2+1}{2^{k+1}}$

    $= \dfrac{2^{k+2}-1}{2^{k+1}}$ $P_k \Rightarrow P_{k+1}$

13. $P_k = 1 = \dfrac{1-r^1}{1-r} = 1$

    $P_k = \dfrac{1-r^{k+1}}{1-r} \Rightarrow P_{k+1} \dfrac{1-r^{k+2}}{1-r}$

    now $(k+1)^{th}$ term $= r^{k+1}$

    $P_k(k+1)^{th}$ term $= \dfrac{1+r^{k+1}}{1-r} + r^{k+1} = \dfrac{1-r^{k+2}}{1-r}$

    $P_k \Rightarrow P_{k+1}$

14. $P_1 = $ Term

$P_k = \left[\dfrac{1}{2}k(k+1)\right]^2 \Rightarrow P_{k+1}$

$= \left[\dfrac{1}{2}(k+1)(k+2)\right]^2$ $(k+1)^{th}$ term $(k+1)^3$

$P_k + (k+1)^{th}$ term $= \left[\dfrac{1}{2}(k)(k+1)\right]^2 + (k+1)^3$

$\qquad = (k+1)^2 \left[-\dfrac{1}{4}k^2 + k + 1\right]$

$\qquad = (k+1)^2 \left[\dfrac{k^2 + 4k + 4}{4}\right]$

$\qquad = \left[\dfrac{1}{2}(k+1)(k+2)\right]^2$

$\qquad P_k \Rightarrow P_{k+1}$

15. $P_1 = \dfrac{1}{30}(1)(2)(6+9+1-1) = 1$

$P_k = \dfrac{1}{30}k(k+1)(6k^3 + 9k^2 + k - 1)$

$\Rightarrow P_{k+1} = \dfrac{1}{30}(k+1)(k+2)(6(k+1)^3$
$\qquad\qquad + 9(k+1)^2 + k)$

$P_k + (k+1)^{th}$ term

$= \dfrac{1}{30}k(k+1)(6k^3 + 9k^2 + k - 1) + (k+1)^4$

$= \dfrac{1}{30}(k+1)[k(6k^3 + 9k^2 + k - 1) + 3(k+1)^3]$

$= \dfrac{1}{30}(k+1)[6k^4 + 39k^3 + 91k^2 + 89k + 30]$

$= \dfrac{1}{30}(k+1)(k+2)(6k^3 + 27k^2 + 37k - 15)$

$= \dfrac{1}{30}(k+1)(k+2)[6(k+1)^3$
$\qquad + 9(k+1)^2 + k + 1 - 1]$

16. $\dfrac{n(5n+1)}{2} \qquad P_k \Rightarrow P_{k+1}$

1) $3 + 8 + 13 + \cdots + (5n - 2)$
2) $(3+2) + (8+2) + (13+2) + \cdots + 5n$
3) $5 + 10 + 15 + \cdots + 5n - 2n$
4) $5(1 + 2 + 3 + \cdots n) - 2n$
5) $5\left[\dfrac{n(n+1)}{2}\right] - 2n$
6) $\dfrac{5n^2 + 5n}{2} - \dfrac{4n}{2}$
7) $\dfrac{5n^2 + n}{2} = n\left(\dfrac{5n+1}{2}\right)$

17. Proof

18. Slopes are $2, -1/2, 2, -1/2$ Hence 2 pairs of parallel lines that are perpendicular

19. Proof

20. $a \geq b \rightarrow |a - b| = a - b$ so

$\dfrac{1}{2}(a + b + a - b) = a$

$a < b \rightarrow |a - b| = b - a$ so

$\dfrac{1}{2}(a + b + b - a) = b$

min similar to the above

CHAPTER 2 Limits and Continuity

Part A

Complete:

1. $\lim_{x \to c} f(x) = L$ iff _____. (2.1)

2. $\lim_{x \to c^+} f(x) = L$ is called _____. (2.2)

3. $\lim_{x \to c^-} f(x) = L$ is called _____. (2.2)

4. If $f(x) \to \infty$ as $x \to 2$ then $x = 2$ is a _____. (2.1)

5. If $f(x) \to \infty$ as $x \to 2^-$ and $f(x) \to -\infty$ as $x \to 2^+$ then $\lim_{x \to 2} f(x)$ _____. (2.1)

6. Draw a graph of a function so that the $\lim_{x \to c} f(x)$ exists and $f(c)$ exists and $f(c) \neq \lim_{x \to c} f(x)$. (2.2)

7. When does the $\lim_{x \to c} f(x)$ fail to exist? (2.2)

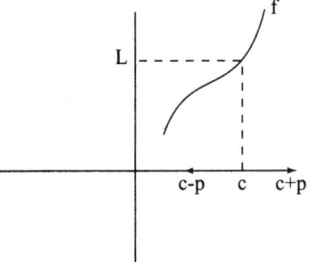

8. Given f a function defined on $(c-p, c) \cup (c, c+p)$ with $p > 0$
   Then $\lim_{x \to c} f(x) = L$ iff for each $\epsilon > 0$ (2.2)

   _____

   _____

9. For any constant function, $f(x) = k$
   $\lim_{x \to c} f(x) =$ (2.2)

10. For $f(x) = x$ the $\lim_{x \to c} f(x) =$ (2.2)

11. True or false
    (a) $\lim_{x \to c} f(x) = L$ is equivalent to $\lim_{h \to 0} f(c + h) = L$ (2.2)
    (b) $\lim_{x \to c} f(x) = L$ is equivalent to $\lim_{x \to c}(f(x) - c) = 0$
    (c) $\lim_{x \to c}(f(x) - L) = 0$ is equivalent to $\lim_{x \to c}(L - f(x)) = 0$

12. Complete:
    If $\lim_{x \to 3} f(x) = 10$ and $\lim_{x \to 3} f(x) = z$ (2.3)
    then _____.

13. If $\lim_{x \to 4} f(x) = 2$ and $\lim_{x \to 4} g(x) = -2$ (2.3)

    Find
    (a) $\lim_{x \to 4}[f(x) + g(x)] =$
    (b) $\lim_{x \to 4}[f(x) - g(x)] =$
    (c) $\lim_{x \to 4}[f(x) \cdot g(x)] =$
    (d) $\lim_{x \to 4}[-7f(x)] =$
    (e) $\lim_{x \to 4}[4f(x) - 3g(x)] =$
    (f) $\lim_{x \to 4} \dfrac{[f(x)]^3}{g(x)} =$
    (g) $\lim_{x \to 4} \dfrac{1}{g(x)} =$

14. (a) If $p(x)$ is a polynomial the $\lim_{x \to c} p(x) =$ _____. (2.3)
    (b) If $p(x) = x^3 - 2x^2 + \sqrt{2}$
    $\lim_{x \to 1} p(x) =$ _____.

15. True or false
    $\lim_{x \to 4} x^2 = (\lim_{x \to 4} x)(\lim_{x \to 4} x)$ (2.3)

16. True or false
    If $g(x) = x$
    then $\lim_{x \to 0} g(x)$ exists (2.3)

17. True or false
    if $f(x) = x^2 - 5x + \pi$ and $g(x) = x - 2\pi$
    $\lim_{x \to 0} \dfrac{f(x)}{g(x)} = -1/2$ (2.3)

18. Complete:
    f is continuous at c iff $\lim_{x \to c} f(x)$ exists and $f(c)$ exists and _____. (2.4)

19. Give an example of a function that has a removable discontinuity at $x = 1$. (2.4)

20. Give an example of a function that has a jump discontinuity at $x = 1$. (2.4)

21. Give an example of a function that has infinite discontinuities at $x = 1$. (2.4)

22. Define: a function is continuous from the right at c iff _____. (2.4)

23. $f(x) = \sin x$ is continuous on all R and $g(x) = \cos x$ is continuous on all R. Which of the following are not continuous on R and if any are not continuous list the points of discontinuity. (2.4)
    (a) $f + g$
    (b) $f - g$
    (c) $f \cdot g$
    (d) f/g
    (e) $f(g(x))$

24. A function f is continuous on $[a, b]$ means that f is continuous on _____ and f is _____ and _____. (2.4)

25. $f(x) = \sqrt{x - 1}$ is continuous on _____. (2.4)

26. Evaluate (2.5)
    (a) $\lim_{x \to 0} \sin x =$
    (b) $\lim_{x \to 0} \cos x =$
    (c) $\lim_{x \to 0} \dfrac{\sin x}{x} =$
    (d) $\lim_{x \to 0} \dfrac{1 - \cos x}{x} =$
    (e) $\lim_{x \to c} \sin x =$
    (f) $\lim_{x \to c} \cos x =$
    (g) $\lim_{x \to \pi/2} \tan x =$

27. For $f(x) = x + 2$, $f(0) = 2$, $f(1) = 3$ (2.6)
    then there exists $c \in [0, 1]$ such that $f(c) = 2,009$
    (a) What theorem is illustrated by the above statement?_____.
    (b) Find c so that $f(c) = 2,009$

28. For $f(x) = x^3 - 100$ on $[0, 10]$, $f(x)$ will take on all values between _____
    and _____. (2.6)

29. For $f(x) = x^3 - 100$ on $[0, 10]$, f is continuous on $[0, 10]$ hence we can conclude that f will attain
    a _____ and a _____ by the extreme value theorem. (2.6)

30. Find m, M, the minimum and maximum values for
    (a) $f(x) = x$ on $[1, 5]$
    (b) $f(x) = x^2$ on $[1, 5]$

Answers to Chapter 2, Part A Questions

1. $\lim_{x \to c^-} f(x) = L$ and $\lim_{x \to c^+} f(x) = L$
2. Right-hand limit
3. Left-hand limit
4. Vertical asymptote
5. DNE
6. ans vary
7. $\lim_{x \to c^-} f(x) \neq \lim_{x \to c^+} f(x)$
8. there exist a $\delta > 0$ such that if $0 < |x - c| < \delta$ then $|f(x) - L| < \epsilon$
9. k
10. c
11. (a) T (c) T
    (b) F
12. $Z = 10$
13. (a) 0   (e) 14
    (b) 4   (f) -4
    (c) -4 (g) $-1/2$
    (d) -14
14. (a) $p(c)$ (b) $-1 + \sqrt{2}$
15. T
16. T
17. T
18. $\lim_{x \to c} f(x) = f(c)$
19. ans vary $f(x) = \dfrac{x^2 - 1}{x - 1}$
20. ans vary $f(x) = \begin{cases} x & x < 1 \\ 5 & x \geq 1 \end{cases}$
21. ans vary $f(x) = \dfrac{1}{x - 1}$
22. iff $\lim_{x \to c^+} f(x) = f(c)$
23. (a) continuous
    (b) continuous
    (c) continuous
    (d) continuous except at $\cos \pi/2$ and $(2n + 1)\pi/2 \quad n \in I$
    (e) continuous
24. f is continuous from the right as $x \to a$ and from the left $x \to b$
25. $x \in \lim_{x \to 1^-} (1, \infty)$
26. (a) 0   (e) $\sin c$
    (b) 1   (f) $\cos c$
    (c) 1   (g) does not exist
    (d) 0
27. (a) Intermediate Value Theorem
    (b) .009
28. $[-100, 900]$
29. maximum value
    minimum value
30. (a) $m = 1 \quad M = 5$
    (b) $m = 1 \quad M = 25$

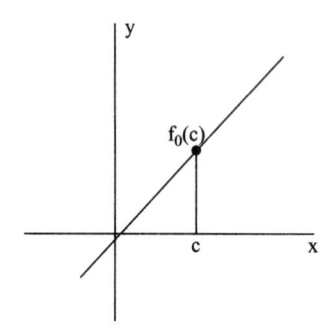

Limits and Continuity   25

Part B

2.1 The Idea of Limit

1. For the function f graphed below, $c = 0$. Use the graph of f to find
   (a) $\lim_{x \to c^-} f(x)$ (b) $\lim_{x \to c^+} f(x)$ (c) $\lim_{x \to c} f(x)$ (d) $f(c)$

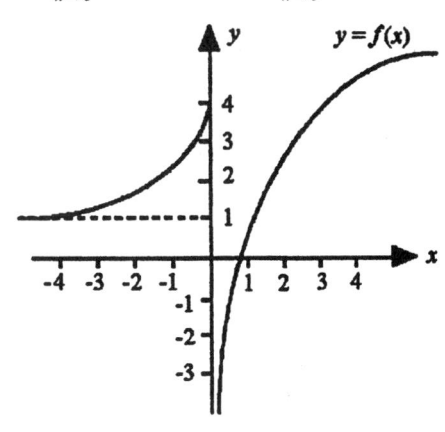

2. For the function f graphed below, $c = 2$. Use the graph of f to find
   (a) $\lim_{x \to c^-} f(x)$ (b) $\lim_{x \to c^+} f(x)$ (c) $\lim_{x \to c} f(x)$ (d) $f(c)$

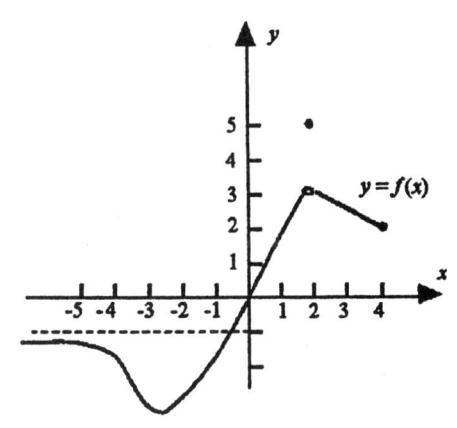

3. For the function f graphed below, $c = -3$. Use the graph of f to find
   (a) $\lim_{x \to c^-} f(x)$ (b) $\lim_{x \to c^+} f(x)$ (c) $\lim_{x \to c} f(x)$ (d) $f(c)$

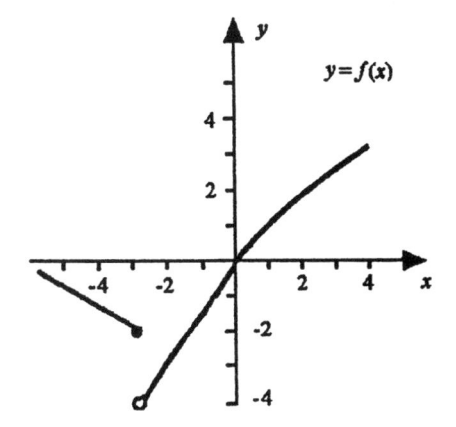

26 Calculus: One and Several Variables

4. For the function f graphed below, $c = 2$. Use the graph of f to find
   (a) $\lim_{x \to c^-} f(x)$
   (b) $\lim_{x \to c^+} f(x)$
   (c) $\lim_{x \to c} f(x)$
   (d) $f(c)$

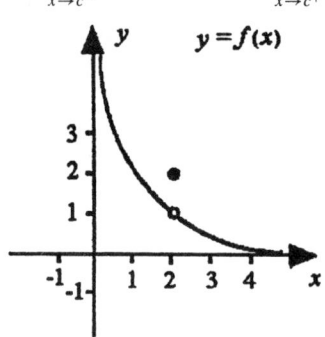

5. For the function g graphed below, $c = -2$. Use the graph of g to find
   (a) $\lim_{x \to c^-} f(x)$
   (b) $\lim_{x \to c^+} f(x)$
   (c) $\lim_{x \to c} f(x)$
   (d) $f(c)$

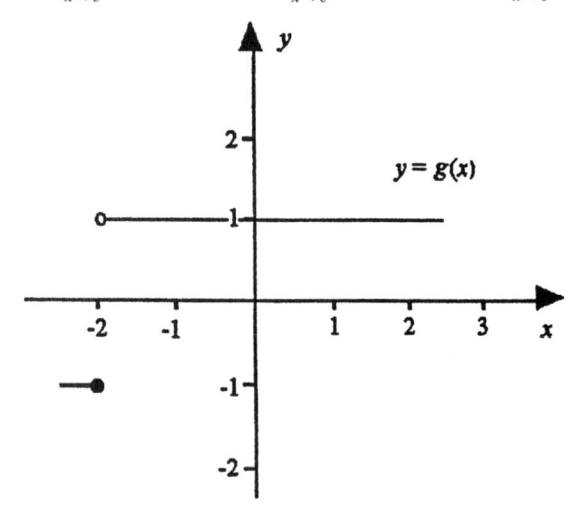

6. For the function f graphed below, $c = -1$. Use the graph of f to find
   (a) $\lim_{x \to c^-} f(x)$
   (b) $\lim_{x \to c^+} f(x)$
   (c) $\lim_{x \to c} f(x)$
   (d) $f(c)$

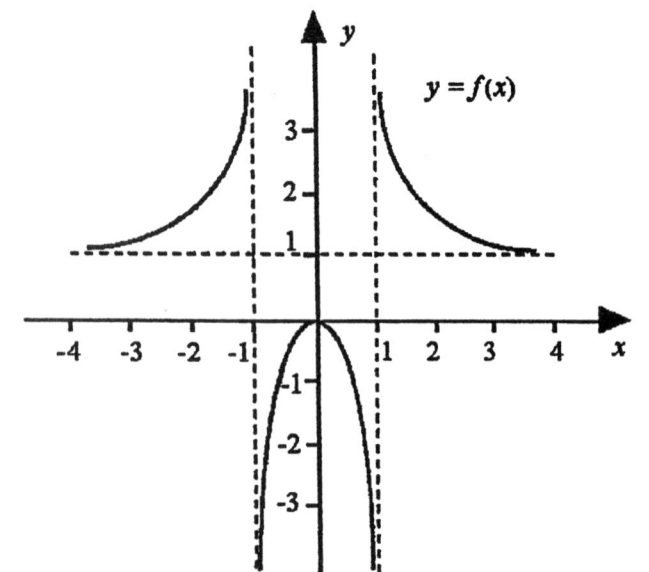

Limits and Continuity   27

7. For the function f graphed below, $c = 0$. Use the graph of f to find
    (a) $\lim_{x \to c^-} f(x)$ (b) $\lim_{x \to c^+} f(x)$ (c) $\lim_{x \to c} f(x)$ (d) $f(c)$

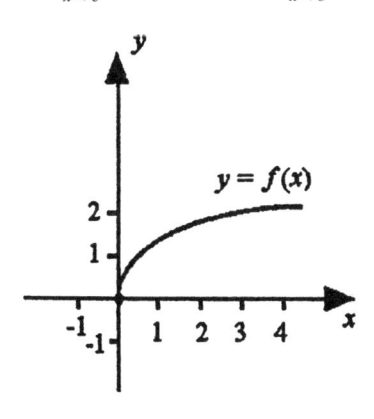

8. For the function f graphed below, $c = 2$. Use the graph of f to find
    (a) $\lim_{x \to c^-} f(x)$ (b) $\lim_{x \to c^+} f(x)$ (c) $\lim_{x \to c} f(x)$ (d) $f(c)$

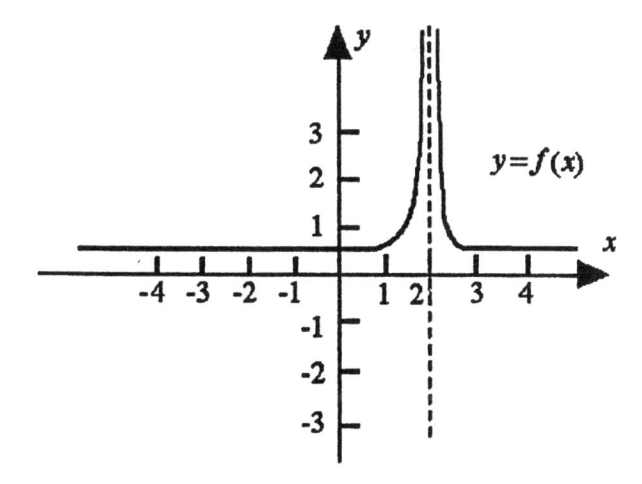

9. For the function f graphed below, $c = -1$. Use the graph of f to find
    (a) $\lim_{x \to c^-} f(x)$ (b) $\lim_{x \to c^+} f(x)$ (c) $\lim_{x \to c} f(x)$ (d) $f(c)$

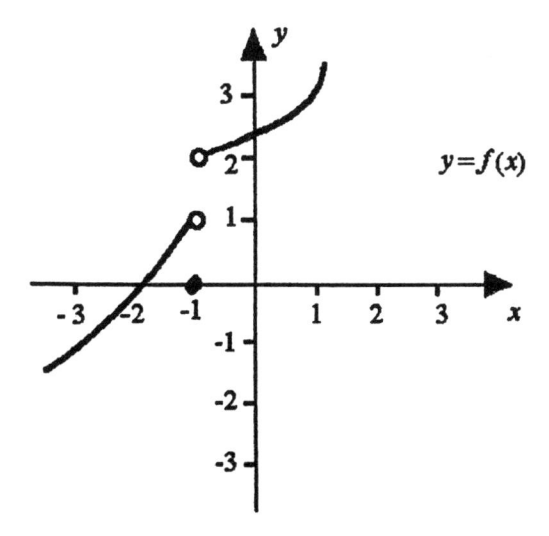

Calculus: One and Several Variables

10. Consider the function f graphed below. State the values of c for which $\lim_{x \to c} f(x)$ does not exist.

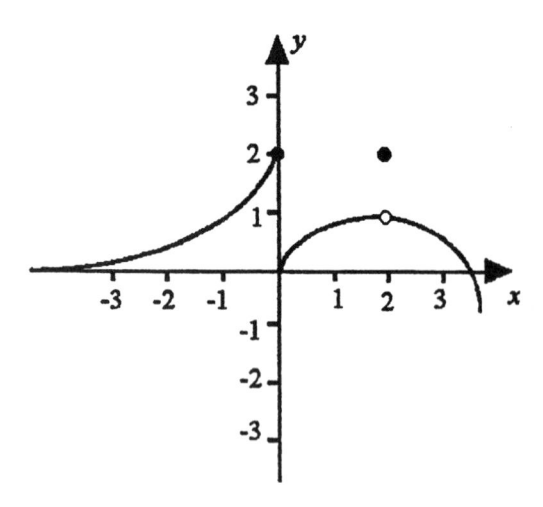

11. Consider the function f graphed below. State the values of c for which $\lim_{x \to c} f(x)$ does not exist.

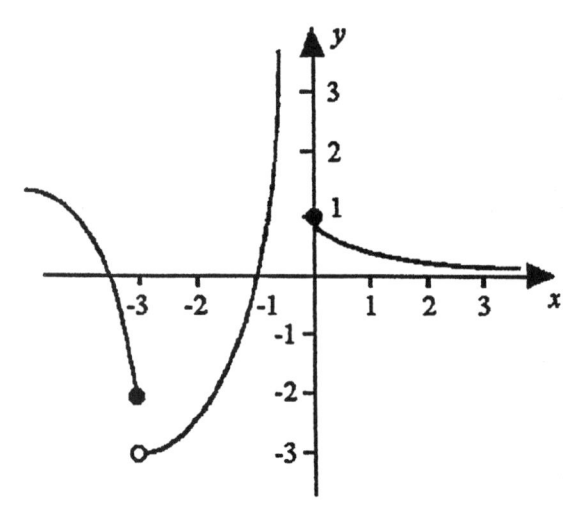

12. Consider the function f graphed below. State the values of c for which $\lim_{x \to c} f(x)$ does not exist.

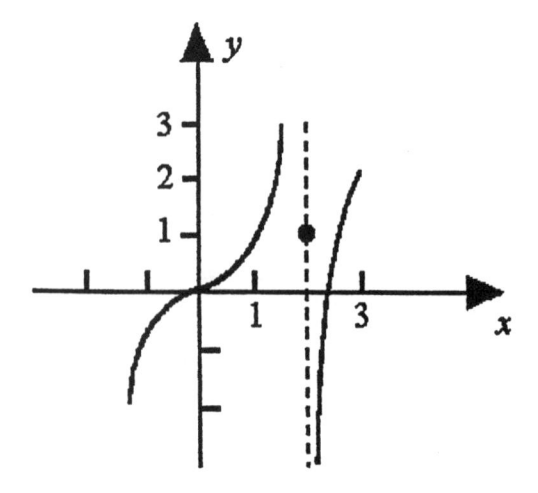

Limits and Continuity   29

13. Consider the function f graphed below. State the values of c for which $\lim_{x \to c} f(x)$ does not exist.

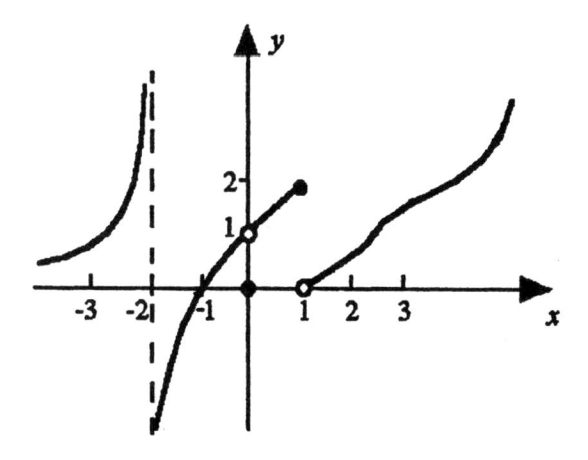

14. Evaluate $\lim_{x \to 2}(2x - 5)$, if it exists.

15. Evaluate $\lim_{x \to 0} \pi^2$, if it exists.

16. Evaluate $\lim_{x \to -3} x^3$, if it exists.

17. Evaluate $\lim_{x \to -2} \dfrac{5}{x - 2}$, if it exists.

18. Evaluate $\lim_{x \to -2} (x^3 + 6x^2 - 16)$, if it exists.

19. Evaluate $\lim_{x \to 4} \dfrac{x^2 + 9}{x^2 - 1}$, if it exists.

20. Evaluate $\lim_{x \to 2} \dfrac{3x - 6}{2x - 4}$, if it exists.

21. Evaluate $\lim_{x \to 0} \dfrac{7x - 5x^2}{x}$, if it exists.

22. Evaluate $\lim_{x \to 2} \dfrac{x^3 - 8}{x - 2}$, if it exists.

23. Evaluate $\lim_{x \to -1} \dfrac{x^2 - 3}{x^2 + 1}$, if it exists.

24. Evaluate $\lim_{x \to a} \dfrac{x^2 - a^2}{x - a}$, if it exists.

25. Evaluate $\lim_{x \to 4} \dfrac{x^2 - 16}{x^2 + x - 20}$, if it exists.

26. Evaluate $\lim_{x \to 0} \dfrac{x^2 + 2x}{x^2 - 2x^2}$, if it exists.

27. Evaluate $\lim_{x \to 2} f(x)$, if it exists. $f(x) = \begin{cases} 3, x \neq 2 \\ 2, x = 2 \end{cases}$

28. Evaluate $\lim_{x \to -1} f(x)$, if it exists. $f(x) = \begin{cases} 2x - 1, & x < -1 \\ 3x, & x \geq -1 \end{cases}$

29. Evaluate $\lim_{x \to 1} f(x)$, if it exists. $f(x) = \begin{cases} \dfrac{1}{x+2}, & x < 1 \\ 1 - 2x, & x > 1 \end{cases}$

30. Evaluate $\lim_{x \to 1} \dfrac{\sqrt{x^2 + 4} - \sqrt{5}}{x - 1}$, if it exists.

2.2 Definition of Limit

31. Evaluate $\lim_{x \to 1} \dfrac{2x}{3x + 1}$, if it exists.

32. Evaluate $\lim_{x \to 0} \dfrac{x^3(4x + 1)}{5x^2}$, if it exists.

33. Evaluate $\lim_{x \to 2} \dfrac{\sqrt{9x}}{\sqrt{x + 3}}$, if it exists.

34. Evaluate $\lim_{x \to 1} \dfrac{1 - x^2}{x^2 + 5x - 6}$, if it exists.

35. Evaluate $\lim_{x \to 1} \dfrac{x^2 + x - 2}{x^2 - 4x + 3}$, if it exists.

36. Evaluate $\lim_{x \to 1} \dfrac{x^3 - 3x^2 + 2x}{x - 2}$, if it exists.

37. Evaluate $\lim_{h \to 1} \dfrac{|h - 2| - 2}{h}$, if it exists.

38. Evaluate $\lim_{x \to 2-} \dfrac{|x - 2|}{x - 2}$, if it exists.

39. Evaluate $\lim_{x \to 1+} \dfrac{x - 1}{|x - 1|}$, if it exists.

40. Evaluate $\lim_{x \to 3-} f(x)$, if it exists, if $f(x) = \begin{cases} \dfrac{|x - 3|}{x - 3}, & x < 3 \\ x, & x > 3 \end{cases}$

41. Evaluate the right-hand limit at $x = 1$, if it exists, for $f(x) = \begin{cases} 1 + x, & x < 1 \\ 6, & x = 1 \\ 1 - x, & x > 0 \end{cases}$

42. Evaluate the right-hand limit at $x = 0$, if it exists, for $f(x) = \begin{cases} x + 1, & x > 0 \\ x^3 - 1, & x \geq 0 \end{cases}$

43. Evaluate $\lim_{x \to 1} f(x)$, if it exists, if $f(x) = \begin{cases} 3x^2, & x < 1 \\ 5, & x = 1 \\ 2x^2 - 1, & x > 1 \end{cases}$

44. Evaluate the largest δ that "works" for a given arbitrary ε. $\lim_{x \to 1} 5x = 5$.

45. Evaluate the largest δ that "works" for a given arbitrary ε. $\lim\limits_{x \to \frac{10}{3}} \frac{3}{5}x = 2$.

46. Give an ε, σ proof for $\lim\limits_{x \to 2}(3x - 2) = 4$.

47. Give an ε, σ proof for $\lim\limits_{x \to 1}(5x - 2) = 3$.

48. Give an ε, σ proof for $\lim\limits_{x \to 5} |x - 3| = 2$.

49. Give the four equivalent limit statements displayed in (2.2.5), taking $f(x) = \dfrac{4}{2x+1}$, $c = 2$.

50. Give an ε, σ proof for $\lim\limits_{x \to 3} x^2 = 9$.

51. Evaluate $\lim\limits_{x \to 3} f(x)$, if it exists, if $f(x) = \begin{cases} x^2 - 1, & x < 3 \\ (x-1)^3, & x > 3 \end{cases}$

2.3 Some Limit Theorems

52. Given that $\lim\limits_{x \to c} f(x) = 0$, $\lim\limits_{x \to c} g(x) = 3$, $\lim\limits_{x \to c} h(x) = -4$, evaluate the limits that exist.

   (a) $\lim\limits_{x \to c}[f(x) - g(x)]$
   (b) $\lim\limits_{x \to c}[h(x)]^2$
   (c) $\lim\limits_{x \to c} \dfrac{g(x)}{h(x)}$
   (d) $\lim\limits_{x \to c} \dfrac{f(x)}{g(x)}$
   (e) $\lim\limits_{x \to c} \dfrac{g(x)}{f(x)}$
   (f) $\lim\limits_{x \to c} \dfrac{1}{g(x) - h(x)}$
   (g) $\lim\limits_{x \to c}[3f(x) - 2g(x) - h(x)]$

53. Evaluate $\lim\limits_{x \to 1} 5$, if it exists.

54. Evaluate $\lim\limits_{x \to 2}(2 - 3x)^2$, if it exists.

55. Evaluate $\lim\limits_{x \to 2}(2x^3 - 3x^2 + 2)$, if it exists.

56. Evaluate $\lim\limits_{x \to -1} 2|2x - 1|$, if it exists.

57. Evaluate $\lim\limits_{x \to 1} \dfrac{2x}{3x - 1}$, if it exists.

58. Evaluate $\lim\limits_{x \to 0} 2x - \dfrac{5}{x}$, if it exists.

59. Evaluate $\lim\limits_{h \to 2} \dfrac{h^3 - 4h}{h^3 - 2h^2}$, if it exists.

60. Evaluate $\lim\limits_{x \to 0} x\left(2x + \dfrac{3}{x}\right)$, if it exists.

61. Evaluate $\lim\limits_{x \to 2} \dfrac{x - 2}{x^3 - 8}$, if it exists.

62. Evaluate $\lim\limits_{x \to 2} \dfrac{x^3 - 8}{x - 2}$, if it exists.

32   Calculus: One and Several Variables

63. Evaluate $\lim\limits_{x \to 1} \dfrac{2x - 3x^2}{(2x - 1)(x^2 - 4)}$, if it exists.

64. Evaluate $\lim\limits_{x \to -5} \dfrac{(x^2 + 2x - 15)^2}{x + 5}$, if it exists.

65. Evaluate $\lim\limits_{x \to 3} \dfrac{x^2 + 2x - 15}{(x + 5)^2}$, if it exists.

66. Evaluate $\lim\limits_{x \to -5} \dfrac{x^2 + 2x - 15}{(x + 5)^2}$, if it exists.

67. Evaluate $\lim\limits_{x \to -3} \dfrac{(x + 5)^2}{x^2 + 2x - 15}$, if it exists.

68. Evaluate $\lim\limits_{x \to a} \dfrac{\frac{1}{x} - \frac{1}{a}}{x - a}$, if it exists.

69. Evaluate $\lim\limits_{h \to 0} \dfrac{\frac{1}{2 + h} - \frac{1}{2}}{h}$, if it exists.

71. Evaluate $\lim\limits_{x \to 2} \dfrac{1 - \frac{4}{x^2}}{1 - \frac{2}{x}}$, if it exists.

72. Evaluate $\lim\limits_{x \to 0} \dfrac{x + \frac{2}{x}}{x - \frac{3}{x}}$, if it exists.

73. Evaluate $\lim\limits_{x \to 3} \left(\dfrac{2x}{x - 3} + \dfrac{5}{x - 3} \right)$, if it exists.

74. Evaluate $\lim\limits_{x \to 3} \left(\dfrac{2x}{x - 3} + \dfrac{-6}{x - 3} \right)$, if it exists.

75. Evaluate the following limits that exist.
   (a) $\lim\limits_{x \to 2} \left(\dfrac{3}{x} - \dfrac{1}{2} \right)$
   (b) $\lim\limits_{x \to 2} \left[\left(\dfrac{3}{x} - \dfrac{1}{2} \right) \left(\dfrac{1}{x - 2} \right) \right]$
   (c) $\lim\limits_{x \to 2} \left[\left(\dfrac{3}{x} - \dfrac{1}{2} \right) (x - 2) \right]$
   (d) $\lim\limits_{x \to 2} \left[\left(\dfrac{3}{x} - \dfrac{1}{2} \right)^2 \left(\dfrac{1}{x - 6} \right) \right]$

76. Given that $f(x) = x^2 - 3x$, evaluate the limits that exist.
   (a) $\lim\limits_{x \to 3} \dfrac{f(x) - f(3)}{x - 3}$
   (b) $\lim\limits_{x \to 5} \dfrac{f(x) - f(5)}{x - 5}$
   (c) $\lim\limits_{x \to -3} \dfrac{f(x) - f(5)}{x + 3}$
   (d) $\lim\limits_{x \to 1} \dfrac{f(x) - f(1)}{x - 1}$
   (e) $\lim\limits_{x \to 2} \dfrac{f(x) - f(1)}{x - 2}$
   (f) $\lim\limits_{x \to -2} \dfrac{f(x) - f(-2)}{x + 2}$

2.4 Continuity

77. Determine whether or not $f(x) = 2x^2 - 3x - 5$ is continuous at $x = 1$. If not, determine whether the discontinuity is a removable discontinuity, a jump discontinuity, or neither.

78. Determine whether or not $f(x) = \sqrt{(x-2)^2 + 2}$ is continuous at $x = 2$. If not, determine whether the discontinuity is a removable discontinuity, a jump discontinuity, or neither.

79. Determine whether or not $f(x) = |5 - 2x^2|$ is continuous at $x = 3$. If not, determine whether the discontinuity is removable discontinuity, a jump discontinuity, or neither.

80. Determine whether or not $f(x) = \dfrac{9x^2 - 4}{3x - 2}$ is continuous at $x = 2/3$. If not, determine whether the discontinuity is a removable discontinuity, a jump discontinuity, or neither.

81. Determine whether or not $f(x) = \begin{cases} -3, & x < -1 \\ 1, & x = -1 \\ 2, & x > -1 \end{cases}$ is continuous at $x = -1$. If not, determine whether the discontinuity is a removable discontinuity, a jump discontinuity, or neither.

82. Determine whether or not $f(x) = \begin{cases} x^2, & x < 1 \\ 3, & x = 1 \\ 2x + 1, & x > 1 \end{cases}$ is continuous at $x = 1$. If not, determine whether the discontinuity is a removable discontinuity, a jump discontinuity, or neither.

83. Determine whether or not $f(x) = \begin{cases} \frac{1}{2}x^3, & x < 2 \\ 1, & x = 2 \\ 2x, & x > 2 \end{cases}$ is continuous at $x = 2$. If not, determine whether the discontinuity is a removable discontinuity, a jump discontinuity, or neither.

84. Determine whether or not $f(x) = \begin{cases} \dfrac{1}{x-4}, & x \neq 1 \\ 1, & x = 4 \end{cases}$ is continuous at $x = 4$. If not, determine whether the discontinuity is a removable discontinuity, a jump discontinuity, or neither.

85. Determine whether or not $f(x) = \dfrac{x-1}{x(x+1)}$ is continuous at $x = 0$. If not, determine whether the discontinuity is a removable discontinuity, a jump discontinuity, or neither.

86. Determine whether or not $f(x) = \dfrac{1}{(x-1)^3}$ is continuous at $x = 1$. If not, determine whether the discontinuity is a removable discontinuity, a jump discontinuity, or neither.

87. Sketch the graph of $f(x) = \dfrac{x^2 - 9}{x + 3}$ and classify the discontinuities, if any.

88. Sketch the graph of $f(x) = \dfrac{x + 2}{x^2 - 4}$ and classify the discontinuities, if any.

89. Sketch the graph of $f(x) = |x - 3|$ and classify the discontinuities, if any.

90. Sketch the graph of $f(x) = \begin{cases} |x+1|, & x \leq -2 \\ x + 1, & -2 < x < 1 \\ 5, & x \geq 1 \end{cases}$ and classify the discontinuities, if any.

91. Sketch the graph of $f(x) = \begin{cases} 2x+1, & x < 1 \\ 1, & x = 1 \\ 2x-1, & x > 1 \end{cases}$ and classify the discontinuities, if any.

92. Sketch the graph of $f(x) = \begin{cases} x^2, & x < 1 \\ 0, & x = 1 \\ 2x, & x > 1 \end{cases}$ and classify the discontinuities, if any.

93. Define $f(x) = \dfrac{x^3+1}{x+1}$ at $x = -1$ so that it becomes continuous at $x = -1$.

94. Define $f(x) = \dfrac{x^2-9}{x+3}$ at $x = -3$ so that it becomes continuous at $x = -3$.

95. Define $f(x) = \dfrac{x^2+x-6}{x-2}$ at $x = 2$ so that it becomes continuous at 2.

96. Let $f(x) = \begin{cases} \dfrac{x^2-x-2}{x+1}, & x \geq -1 \\ A, & x < -1 \end{cases}$. Find A given that f is continuous at -1.

97. Prove that if $f(x)$ has a removable discontinuity at c, then $\lim_{x \to c}(x-c)f(x) = 0$.

2.5 The Pinching Theorem; Trigonometric Limits

98. Evaluate $\lim_{x \to 0} \dfrac{\sin 7x}{x}$, if it exists.

99. Evaluate $\lim_{x \to 0} \dfrac{x\sqrt{2}}{\sin \dfrac{x}{2}}$, if it exists.

100. Evaluate $\lim_{x \to 0} \dfrac{4x^2}{1-\cos 3x}$, if it exists.

101. Evaluate $\lim_{x \to 0} \dfrac{1-\cos 3x^2}{5x^2}$, if it exists.

102. Evaluate $\lim_{\theta \to 0} \dfrac{\tan \theta}{\theta}$, if it exists.

103. Evaluate $\lim_{\theta \to 0} \dfrac{\sin 2\theta}{\tan \theta}$, if it exists.

104. Evaluate $\lim_{\alpha \to 0} \dfrac{\sin \alpha - \tan \alpha}{\sin^3 \alpha}$, if it exists.

105. Evaluate $\lim_{\theta \to 0} \theta \cot 4\theta$, if it exists.

106. Evaluate $\lim_{\theta \to 0} \dfrac{\sin \sqrt{2\theta}}{\sqrt{\theta}}$, if it exists.

107. Evaluate $\lim_{\theta \to 0} \dfrac{\theta^2}{\sin 3\theta^2}$, if it exists.

108. Evaluate $\lim_{\theta \to 0} \dfrac{3}{\theta \csc \theta}$, if it exists.

109. Evaluate $\lim_{\theta \to 0} \dfrac{\sin 3\theta}{\sin 2\theta}$, if it exists.

110. Evaluate $\lim_{\alpha \to 0} \dfrac{\alpha}{\cos \alpha}$, if it exists.

111. Evaluate $\lim_{t \to 0} \dfrac{t^2}{1 - \cos^2 t}$, if it exists.

112. Evaluate $\lim_{\theta \to 0} \dfrac{3\theta}{\cos 2\theta}$, if it exists.

113. Evaluate $\lim_{\theta \to 0} \dfrac{\sin^2 \theta}{\tan \theta}$, if it exists.

114. For $f(x) = \sin x$ and $a = \pi/3$, find $\lim_{h \to 0} \dfrac{f(a+h) - f(a)}{h}$ and give an equation for the tangent line to the graph of f at $(a, f(a))$.

115. Use the pinching theorem to find $\lim_{x \to 0} \sqrt{x} \cos \dfrac{1}{x^2}$.

2.6 Some Basic Properties of Continuous Functions

116. Sketch the graph of a function f that is defined on [0, 1] and meets the following conditions (if possible): f is continuous on [0, 1], minimum value $1/2$, maximum value 1.

117. Sketch the graph of a function f that is defined on [0, 1] and meets the following conditions (if possible): f is continuous on (0, 1], no minimum value, maximum value $1/2$.

118. Sketch the graph of a function f that is defined on [0, 1] and meets the following conditions (if possible): f is continuous on (0, 1), takes on the values $1/2$ and 1 but does not take on the value 0.

119. Sketch the graph of a function f that is defined on [0, 1] and meets the following conditions (if possible): f is continuous on [0, 1], does not take on the value 0, minimum value -1, maximum value $1/2$.

120. Sketch the graph of a function f that is defined on [0, 1] and meets the following conditions (if possible): f is discontinuous at $x = 3/4$, but takes on both a minimum value and a maximum value.

121. Show the equation $x^3 - \cos^2 x = 0$ has a root in [0, 2].

122. Given that $f(x) = x^4 - x^2 + 5x + 2$, show that there exist at least two real numbers x such that $f(x) = 3$.

Answers to Chapter 2, Part B Questions

1. (a) 4 (c) does not exist
   (b) does not exist (−∞) (d) 0

2. (a) 3 (c) 3
   (b) 3 (d) 5

3. (a) −2 (c) does not exist
   (b) −4 (d) −2

4. (a) 1 (c) 1
   (b) 1 (d) 2

5. (a) −1 (c) does not exist
   (b) 1 (d) −1

6. (a) does not exist (+∞) (c) does not exist
   (b) does not exist (−∞) (d) does not exist

7. (a) does not exist (c) does not exist
   (b) 0 (d) 0

8. (a) does not exist (+∞) (c) does not exist (+∞)
   (b) does not exist (+∞) (d) does not exist

9. (a) 1 (c) does not exist
   (b) 2 (d) 0

10. $c = 0$

11. $c = -3$ and $c = 0$

12. $c = 2$

13. $c = -2$ and $c = 1$

14. -1

15. π^2

16. -27

17. $-5/4$

18. 0

19. 5/3

20. 3/2

21. 7

22. 12

23. -1

24. $2a$

25. 8/9

26. 2

27. 3

28. -3

29. does not exist

30. $\sqrt{5}/5$

31. ½

32. 0

33. $\dfrac{3\sqrt{10}}{5}$

34. $-2/7$

35. $-3/2$

36. 2

37. -1

38. -1

39. 1

40. -1

41. 0

42. 1

43. does not exist

44. $\dfrac{1}{5}\varepsilon$

45. $\dfrac{5}{3}\varepsilon$

46. Since $|(3x - 2) - 4| = |3x - 6| = 3|x - 2|$, we can take $\delta = \dfrac{1}{3}\varepsilon$: if $0 < |x - 2| < \dfrac{1}{3}\varepsilon$, then $|(3x - 2) - 4| = 3|x - 2| < \varepsilon$.

47. Since $|(5x - 2) - 3| = |5x - 5| = 5|x - 1|$, we can take $\sigma = \dfrac{1}{5}\varepsilon$: if $0 < |x - 1| < \dfrac{1}{5}\varepsilon$, then $|(5x - 2) - 3| = 5|x - 1| < \varepsilon$.

48. Since $|(x - 3) - 2| = |x - 5|$, we can take $\delta = \varepsilon$: if $0 < |x - 5| < \varepsilon$, then $|(x - 3) - 2| = |x - 5| < \varepsilon$.

Answers

49. (i) $\lim_{x \to 2} \dfrac{4}{2x+1} = \dfrac{4}{5}$

    (ii) $\lim_{h \to 0} \dfrac{4}{2(2+h)+1} = \dfrac{4}{5}$

    (iii) $\lim_{x \to 2} \left(\dfrac{4}{2x+1} - \dfrac{4}{5} \right) = 0$

    (iv) $\lim_{x \to 2} \left| \dfrac{4}{2x+1} - \dfrac{4}{5} \right| = 0$

50. If $|x-3| < 1$, then $-1 < x-3 < 1$, $2 < x < 4$, $5 < x+3 < 7$, and $|x+3| < 7$.
    Take $\delta =$ minimum of 1 and $\varepsilon/7$.
    If $0 < |x-3| < \sigma$, then $2 < x < 4$ and $|x-3| < \varepsilon/7$. Therefore,
    $|x^2 - 9| = |x+3||x-3| < 7|x-3| < 7(\varepsilon/7) = \varepsilon$.

51. 8

52. (a) -3 (e) does not exist
    (b) 16   (f) 1/7
    (c) $-3/4$ (g) -2
    (d) 0

53. 5

54. 16

55. 6

56. 6

57. 1

58. does not exist

59. 2

60. 3

61. 1/12

62. 12

63. 1/3

64. 0

65. 0

66. does not exist

67. does not exist

68. $\dfrac{-1}{a^2}$

69. $\dfrac{-1}{4}$

70. $3a^2$

71. 2

72. $\dfrac{-2}{3}$

73. does not exist

74. 2

75. (a) 1   (c) 0
    (b) does not exist   (d) $-1/4$

76. (a) 3   (d) $-1/4$
    (b) 7   (e) 1
    (c) does not exist   (f) -7

77. continuous

78. continuous

79. continuous

80. removable discontinuity at $x = 2/3$.

81. jump discontinuity

82. jump discontinuity

83. removable discontinuity

84. discontinuity of neither type

85. discontinuity of neither type

86. discontinuity of neither type

87. removable discontinuity at $x = -3$.

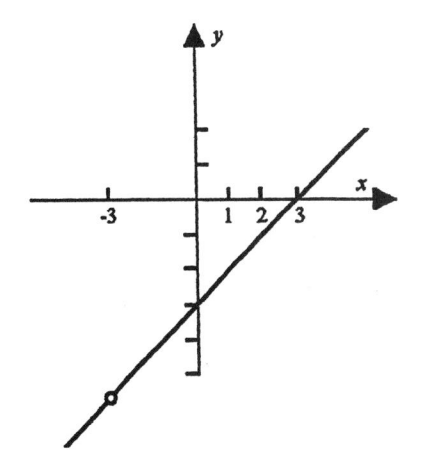

38   Calculus: One and Several Variables

88. nonremovable, nonjump discontinuity at $x = 2$.

91. jump discontinuity at $x = 1$.

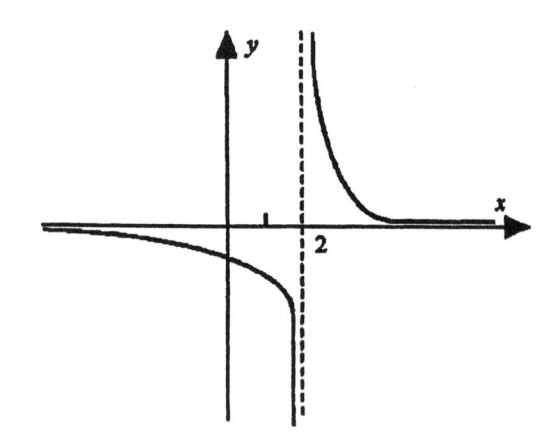

92. jump discontinuity at $x = 1$.

89. no discontinuities

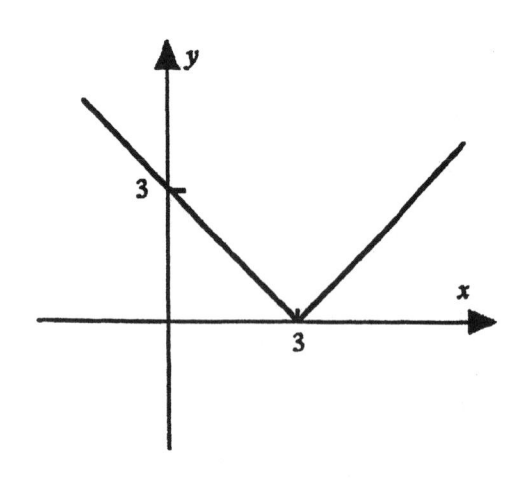

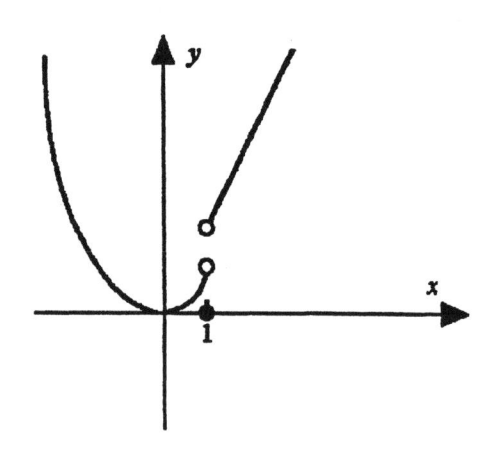

90. jump discontinuity at $x = -2$ and $x = 1$.

93. $f(-1) = 3$

94. $f(-3) = -6$

95. $f(2) = 5$

96. -3

97. Since f has a removable discontinuity at c, for $\lim_{x \to c} f(x) = L$ some real number L. Then $\lim_{x \to c}(x - c)f(x) = L \cdot \lim_{x \to c}(x - c) = L \cdot 0 = 0$.

98. 7

99. $2\sqrt{2}$

100. 8/9

101. 0

Answers

102. 1

103. 2

104. $-1/2$

105. 1/4

106. $\sqrt{2}$

107. 1/3

108. 3

109. 3/2

110. 0

111. 1

112. 0

113. 0

114. limit: $1/2$; tangent line: $y = \dfrac{1}{2}\left(x - \dfrac{\pi}{3}\right) + \dfrac{\sqrt{3}}{2}$

115. 0

116.

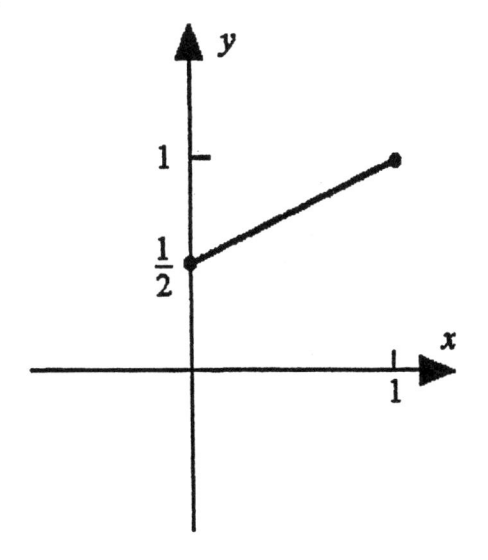

117. impossible

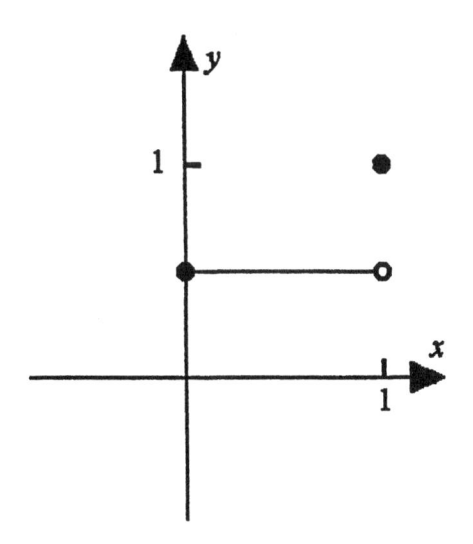

118.

119. impossible

120.

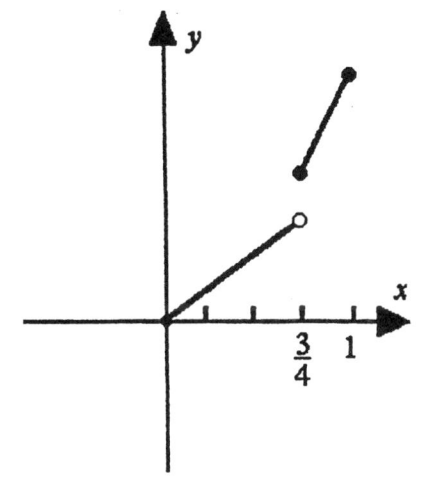

121. If $f(x) = x^3 - \cos^2 x$, then f is continuous, and $f(0) = -1 < 0$, $f(2) = 8 - \cos^2(2) \geq 7 > 0$, so by the intermediate-value theorem $f(c) = 0$ for some c in $[0, 2]$.

122. f is continuous, and $f(0) = 2$, $f(-2) = 4$, $f(1) = 7$, so by the intermediate-value theorem $f(x) = 3$ for some x in $[-2, 0]$ and for some x in $[0, 1]$.

Part C

1. Prove $\lim_{x \to c} = 0$ for $f(x) = |x|$. (2.2)

2. Find $\lim_{x \to 0} f(x)$, $f(x) = \begin{cases} 3x & x < 1 \\ x + 2 & x \geq 1 \end{cases}$. (2.2)

3. Give an ε, δ proof for $\lim_{x \to 1}(3x - 2) = 1$. (2.2)

4. Give an ε, δ proof for $\lim_{x \to 2} x^2 = 4$. (2.2)

5. Give an ε, δ proof for $\lim_{x \to 1}(x^3 - 10) = -9$. (2.2)

6. Prove that if $g(x) = \begin{cases} x & x \text{ rational} \\ 0 & x \text{ irrational} \end{cases}$ then $\lim_{x \to 0} g(x) = 0$. (2.2)

7. Fix a number x and evaluate $\lim_{h \to 0} \dfrac{f(x+h) - f(x)}{h}$ for
   (a) $f(x) = x$
   (b) $f(x) = x^2$
   (c) $f(x) = x^3$
   (d) $f(x) = x^4$
   (e) Guess the limit for $f(x) = x^n$, $n \in I^+$

8. For what points (if any) is the given function continuous?
$g(x) = \begin{cases} x & x \text{ is a rational number} \\ 0 & x \text{ is an irrational number} \end{cases}$ (2.4)

9. Let f and g be continuous at c. Prove that if $f(c) > 0$ then there exists $\delta > 0$ such that $f(x) > 0$ for all $x \in (c - \delta, c + \delta)$. (2.4)

10. Prove $\lim_{x \to \pi}(x - \pi)\cos^2\left(\dfrac{1}{x - \pi}\right) = 0$.

11. Prove that any polynomial of odd number degree has at least one root.

12. Given $f(x) = [|x|]$ the greatest integer function. Where $[|x|] = $ greatest integer n such that $n \leq x$
    (a) find $\lim_{x \to 0^-} [|x|]$
    (b) find $\lim_{x \to 0^+} [|x|]$

13. Find $\lim_{x \to \pi} \dfrac{\sin x}{x - \pi}$.

14. Determine the value of c so that (2.4)
$f(x) = \begin{cases} x + 3 & x \leq 2 \\ cx + 6 & x > 2 \end{cases}$ will be continuous for all x.

15. For $f(x) = \sqrt{x(x - 1)}$ (2.3) + (2.4)
    (a) find the domain of f
    (b) find $\lim_{x \to 1^+} f(x)$
    (c) find $\lim_{x \to 0^-} f(x)$

16. For $f(x) = \begin{cases} bx - 4 & x \leq 3 \\ x^2 - 6x + 9 & x > 3 \end{cases}$. What value of b makes $f(x)$ continuous? (2.4)

17. True or false

 If $f(x) = \dfrac{3x^2 + x}{x}$ and $0 < |x| < \epsilon/3$ then $|f(x) - 1| < \varepsilon$.

18. Find $\lim\limits_{x \to 0} \dfrac{1 - \cos^2 x}{x^2}$.

19. True or false

 If $\lim\limits_{x \to 0} f(x) = \infty$ and $\lim\limits_{x \to 0} g(x) = \infty$,
 then $\lim\limits_{x \to 0} \dfrac{f(x)}{g(x)}$ DNE. Give an example to support your answer.

20. Prove if $\lim\limits_{x \to c} f(x) = L$ then $\lim\limits_{x \to c} |f(x)| = |L|$
 Hint: $||a| - |b|| \le |a - b|$.

21. For $f(x) = \begin{cases} 0 & \text{if } x \text{ is an irrational number} \\ \dfrac{1}{b} & \text{if } x \text{ is a rational number } \dfrac{a}{b} \text{ in lowest terms } b > 0. \end{cases}$

 Show $f(x)$ is continuous at irrational points and discontinuous at rational points.

Answers to Chapter 2, Part C Questions

1. take $\delta = \varepsilon$

2. 0

3. take $\delta = \varepsilon/3$

4. take $\delta = \min(1, \varepsilon/5)$

5. take $\delta = \min(1, \varepsilon/7)$

6. take $\delta = \varepsilon$ suppose $0 < |x - 0| < \delta$, that is $0 < |x| < \delta$ then for x rational
$|g(x) - 0| = |x| < \delta = \varepsilon$ and for x irrational
$|g(x) - 0| = 0 < \varepsilon$
Thus, if $0 < |x - 0| < \delta$ then $|g(x) - 0| < \varepsilon$
hence $\lim_{x \to 0} g(x) = 0$

7. Fix a number x and evaluate $\lim_{h \to 0} \dfrac{f(x+h) - f(x)}{h}$ for
   (a) $f(x) = x \Rightarrow$ ans 1
   (b) $f(x) = x^2 \Rightarrow$ ans $2x$
   (c) $f(x) = x^3 \Rightarrow$ ans $3x^2$
   (d) $f(x) = x^4 \Rightarrow$ ans $4x^3$
   (e) $f(x) = x^n \Rightarrow$ ans nx^{n-1} $\quad n \in I^+$

8. Continuous only at 0

9. Let $\varepsilon = f(c) > 0$. By continuity of f at c, there exists $\delta > 0$ such that $|f(x) - f(c)| < f(c)$ for all $x \in (c - \delta, c + \delta)$ that implies $f(x) > 0$ for all $x \in (c - \delta, c + \delta)$

10. Since $0 \leq \cos^2\left(\dfrac{1}{x - \pi}\right) \leq 1$ for all $x \neq \pi$
    then $\left|(x - \pi)\cos^2\left[\dfrac{1}{x - \pi}\right]\right| \leq |x - \pi|$
    thus $-|x - \pi| \leq \left|(x - \pi)\cos^2\left[\dfrac{1}{x - \pi}\right]\right| \leq |x - \pi|$
    Since $\lim_{x \to \pi}(-|x - \pi|) = 0$
    and $\lim_{x \to \pi} |x - \pi| = 0$ by pinching then the result follows

11. $p(x) = a_n x^n + a_{n-1} x^{n-1} + \cdots + a_1 x + a_0$
    where n is an odd positive integer and $a_n \neq 0$
    $p(x)$ is continuous

    $p(x) = x^n\left(a_n + \dfrac{a_{n-1}}{x} + \dfrac{a_{n-2}}{x^2} + \cdots + \dfrac{a_1}{x^{n-1}} + \dfrac{a_0}{x_n}\right)$

    Case 1 $a_n > 0$
    as $x \to \infty$ $p(x) \to a_n x^n > 0$
    as $x \to -\infty$ $p(x) \to a_n x^n < 0$
    hence by the intermediate-value theorem there exists c such that $p(c) = 0$
    Case 2 $a_n < 0$
    as $x \to \infty$ $p(x) \to a_n x^n < 0$
    as $x \to -\infty$ $p(x) \to a_n x^n > 0$
    again by the intermediate-value theorem there exists c such that $p(c) = 0$

12. (a) -1 (b) 0

13. Evaluate $\lim_{x \to \pi} \dfrac{\sin x}{x - \pi} \Rightarrow \lim_{x \to \pi} -\dfrac{\sin(x - \pi)}{x - \pi} = -1$

14. $c = -1/2$

15. (a) $(-\infty, 0] \cup [1, \infty)$ (c) 0
    (b) 0

16. $4/3$

17. True

18. 1

19. False

20. use 2.2.5 $\lim_{x \to c} f(x) = L \Longleftrightarrow \lim_{x \to c} |\delta(x) - L| = 0$

21. Proof

CHAPTER 3 Differentiation

Part A

1. (a) If $\lim\limits_{x \to 0} \dfrac{f(x+h) - f(x)}{h}$ exists then f is said to be _____ at x. (3.1)
   (b) Given $f'(x)$ exists write the equation of the tangent line to f at $(x_0, f(x_0))$ _____ (3.1)

2. True or false (3.1)
   (a) If f is differentiable at x then f is continuous at x.
   (b) If f is continuous at x then f is differentiable at x.
   (c) If f is a polynomial then f is continuous and f is differentiable at all x.

3. Write the rules of differentiation for f, g continuous and differentiable at x (3.2)
   (a) $f(x) = x^n$, $n \in I'$ _____
   (b) $f(x) = \alpha x^n$, $n \in I'$ α a constant _____
   (c) $f(x) \cdot g(x)$ _____
   (d) $\dfrac{1}{g(x)}$, $g(x) \neq 0$ _____
   (e) $\dfrac{f(x)}{g(x)}$, $g(x) \neq 0$ _____ (3.3)

4. True or false
   If $f(x) = x^n$ $n \in I'$ then the nth derivative of f is $n!$.

5. True or false
   Given $x(t)$ is the position of an object moving in a straight line at time $t \geq 0$. (3.4)
   (a) speed at time $t = x'(t)$
   (b) the velocity at time t, $v(t) = x'(t)$
   (c) $a(t)$ the acceleration at time $t = x''(t) = v'(t)$

6. True or false
   The height, y, of an object in free fall at time t is given by $y = \dfrac{1}{2}gt^2 + v_0 t$. (3.4)

7. Given $y = f(u)$ and $u = g(x)$ y, u differentiable functions write the chain rule
   $\dfrac{dy}{dx} = $ _____. (3.5)

8. Given $y = f(u)$ and $u = g(x)$ y, u differentiable functions then $\dfrac{d}{dx}(u^n) = $ _____. (3.5)

9. True or false
   If g is differentiable at x and f is differentiable at $g(x)$ then the composition $f \circ g = f(g(x))$ is differentiable at x and $(f \circ g)'(x) = f'(g(x))$ (3.5)

10. Given $f(x) = \sin x$, $g(x) = \cos x$ (3.6)
    and $f'(x) = \cos x$, $g'(x) = -\sin x$
    $h(x) = \cot x$
    find $h'(x)$

11. If $x^4 + 4y^2 = -1$ (3.7)
    then $\dfrac{dy}{dx} = \dfrac{-x^3}{y}$
    Comment on the above result.

12. For $x^2y + 2xy = 7$ solving for y (3.7)
we have $y = \dfrac{7}{x^2 + 2x}$ and $\dfrac{dy}{dx} = \dfrac{-14x - 14}{(x^2 + 2x)^2}$. Now find $\dfrac{dy}{dx}$ using implicit differentiation and rectify the two answers.

13. Complete the following:
If Q is a quantity that varies with time then $\dfrac{dQ}{dt}$ gives _____. (3.8)

14. $\triangle ABC$ is a right triangle if at a given instant $\overline{AC} = 6$ $\overline{BC} = 8$ and $\dfrac{dx}{dt} = -1\dfrac{\text{in}}{\text{sec}}$ $\dfrac{dy}{dt} = 2\dfrac{\text{in}}{\text{sec}}$
find $\dfrac{dA}{dt}$ after 2 seconds

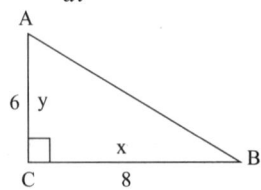

15. True or false
The formula used to approximate a root by Newton-Raphson method is $x_n = x_{n-1} + \dfrac{f(x_{n-1})}{f'(x_{n-1})}$. (3.9)

16. If we use Newton-Raphson method to approximate the roots of $f(x) = x^2 - 4$, what initial choice for x will fail? (3.9)

17. If $h \neq 0$ $f(x + h) - f(x) = \Delta f$
$f'(x) \cdot h$ is called the _____. (3.9)

Answers to Chapter 3, Part A Questions

1. (a) differentiable
   (b) $y - f(x_0) = f'(x_0)(x - x_0)$

2. (a) T (b) F (c) T

3. (a) $f'(x) = n\, x^{n-1}$
   (b) $f'(x) = \alpha\, n\, x^{n-1}$
   (c) $f(x)g'(x) + g(x)f'(x)$
   (d) $\dfrac{g'(x)}{[g(x)]^2}$
   (e) $\dfrac{g(x)\cdot f'(x) - f(x)g'(x)}{[g(x)]^2}$

4. T

5. (a) F (b) T (c) T

6. F

7. $\dfrac{dy}{dx} = \dfrac{dy}{du} \cdot \dfrac{du}{dx}$

8. $\dfrac{d}{dx}(u^n) = nu^{n-1}\dfrac{du}{dx}$

9. F

10. $h'(x) = -\csc^2 x$ (use quotient rule)

11. $x^4 + 4y^2 = -1$ Not possible

12. $y' = \dfrac{-2(xy + y)}{x(x+2)}$ use $y = \dfrac{7}{x^2 + 2x}$

    $y' = \dfrac{-2\left[x\left(\dfrac{7}{x^2+2x}\right) + \left(\dfrac{7}{x^2+2x}\right)\right]}{x^2 + 2x}$

    $y' = \dfrac{-14x - 14}{(x^2 + 2x)^2}$

13. the rate of change of that quantity with respect to time.

14. $1\,\dfrac{\text{in}^2}{\text{sec}}$

15. F

16. $x = 0$

17. differential of f at x with movement h

Part B

3.1 The Derivative

1. Differentiate $f(x) = 7$ by forming a difference quotient $\dfrac{f(x+h) - f(x)}{h}$ and taking the limit as h tends to 0.

2. Differentiate $f(x) = 3 - 4x$ by forming a difference quotient $\dfrac{f(x+h) - f(x)}{h}$ and taking the limit as h tends to 0.

3. Differentiate $f(x) = 2x^2 + x$ by forming a difference quotient $\dfrac{f(x+h) - f(x)}{h}$ and taking the limit as h tends to 0.

4. Differentiate $f(x) = x^3$ by forming a difference quotient $\dfrac{f(x+h) - f(x)}{h}$ and taking the limit as h tends to 0.

5. Differentiate $f(x) = \sqrt{x+2}$ by forming a difference quotient $\dfrac{f(x+h) - f(x)}{h}$ and taking the limit as h tends to 0.

6. Differentiate $f(x) = \dfrac{1}{x+1}$ by forming a difference quotient $\dfrac{f(x+h) - f(x)}{h}$ and taking the limit as h tends to 0.

7. Differentiate $f(x) = \dfrac{1}{2x^2}$ by forming a difference quotient $\dfrac{f(x+h) - f(x)}{h}$ and taking the limit as h tends to 0.

8. Find $f'(2)$ for $f(x) = (2x+3)^2$ by forming a difference quotient $\dfrac{f(2+h) - f(2)}{h}$ and taking the limit as $h \to 0$.

9. Find $f'(2)$ for $f(x) = (x^3 - 2x)$ by forming a difference quotient $\dfrac{f(2+h) - f(2)}{h}$ and taking the limit as $h \to 0$.

10. Find $f'(2)$ for $f(x) = 2x + \sqrt{x+2}$ by forming a difference quotient $\dfrac{f(2+h) - f(2)}{h}$ and taking the limit as $h \to 0$.

11. Find $f'(2)$ for $f(x) = \dfrac{3}{2x+1}$ by forming a difference quotient $\dfrac{f(2+h) - f(2)}{h}$ and taking the limit as $h \to 0$.

12. Find equations for the tangent and normal to the graph of $f(x) = 2x^3 + 1$ at the point $(1, \ f(1))$.

13. Find equations for the tangent and normal to the graph of $f(x) = x^3 - 3x$ at the point $(2, \ f(2))$.

14. Find equations for the tangent and normal to the graph of $f(x) = \sqrt{2x}$ at the point $(2, \ f(2))$.

15. Find equations for the tangent and normal to the graph of $f(x) = \sqrt{2x+1}$ at the point $(3/2, \ f(3/2))$.

16. Draw a graph of $f(x) = |2x - 1|$ and indicate where it is not differentiable.

17. Draw a graph of $f(x) = \begin{cases} x^2 + 1, & x \leq 1 \\ 2, & x > 1 \end{cases}$ and indicate where it is not differentiable.

18. Find $f'(c)$ if it exists. $f(x) = \begin{cases} 2x^2 + 2, & x \le 1 \\ x^4 + 3, & x > 1 \end{cases}$ $c = 1$.

19. Find $f'(c)$ if it exists. $f(x) = \begin{cases} 3x + 1, & x \le -1 \\ 2(x + 1)^2, & x > -1 \end{cases}$ $c = -2$.

20. Sketch the graph of the derivative of the function with the graph shown below.

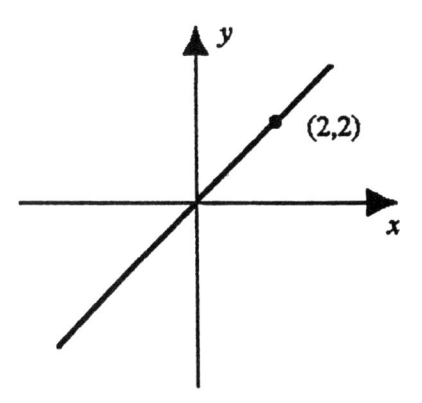

21. Sketch the graph of the derivative of the function with the graph shown below.

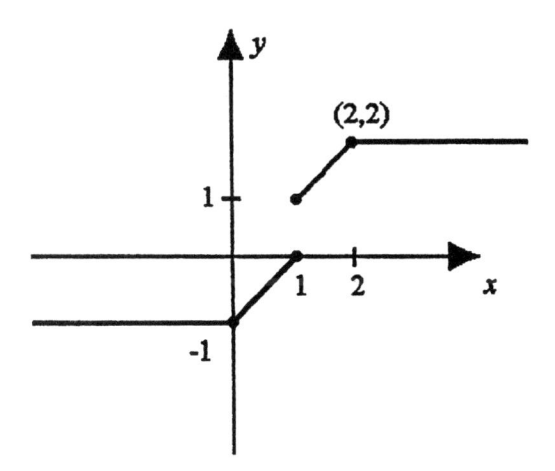

22. $\lim\limits_{h \to 0} \dfrac{(8 + h)^{2/3} - 4}{h}$ represents the derivative of a function f at a point c. Determine f and c.

23. $\lim\limits_{h \to 0} \dfrac{\sin h}{h}$ represents the derivative of a function f at a point c. Determine f and c.

3.2 Some Differentiation Formulas

24. Differentiate $F(x) = 1 - 3x$.

25. Differentiate $F(x) = 4x^5 - 8x^2 + 9x$.

26. Differentiate $F(x) = \dfrac{2}{x^4}$.

27. Differentiate $F(x) = (2x^2 - 1)(3x + 1)$.

28. Differentiate $F(x) = \dfrac{(2x^2 - 5)}{x^4}$.

29. Differentiate $F(x) = \dfrac{3x^4 + 5}{x - 1}$.

30. Differentiate $F(x) = \left(1 + \dfrac{2}{x}\right)\left(1 + \dfrac{2}{x^2}\right)$.

31. Find $f'(0)$ and $f'(1)$ for $f(x) = x^3(2x + 3)$.

32. Find $f'(0)$ and $f'(1)$ for $f(x) = \dfrac{4x + 1}{2x - 1}$.

33. Given that $h(0) = 4$ and $h'(0) = 3$, find $f'(0)$ for $f(x) = 2x^2 h(x) - 3x$.

34. Find an equation for the tangent to the graph of $f(x) = 2x^2 - \dfrac{5}{x}$ at the point $(-1, f(-1))$.

35. Find the points where the tangent to the curve is horizontal for $f(x) = (x + 1)(x^2 - 3x - 8)$.

36. Find the points where the tangent to the curve for $f(x) = -x^3 + 2x$ is parallel to the line $y = 2x + 5$.

37. Find the points where the tangent to the curve for $f(x) = 3x + x^2$ is perpendicular to the line $3x + 2y + 1 = 0$.

38. Find the area of the triangle formed by the x-axis and the lines tangent and normal to the curve $f(x) = 2x + 3x^2$ at the point $(-1, 3)$.

3.3 The d/dx Notation; Derivatives of Higher Order

39. Find $\dfrac{dy}{dx}$ for $y = \dfrac{3x^3 + 5y^2 + 2}{x^2}$.

40. Find $\dfrac{dy}{dx}$ for $y = 5x^3 + \dfrac{1}{\sqrt{x}}$.

41. Find $\dfrac{dy}{dx}$ for $y = \dfrac{x^2 + 3x}{7 - 2x}$.

42. Find $\dfrac{d}{dx}[-2(x^2 - 5x)(3 + x^7)]$.

43. Find $\dfrac{d}{dx}\left[\dfrac{x^2 - 5}{3x^2 - 1}\right]$.

44. Evaluate $\dfrac{dy}{dx}$ at $x = 2$ for $y = (x^2 + 1)(x^3 - x)$.

45. Find the second derivative for $f(x) = \dfrac{-8}{x^2} + \dfrac{1}{5}x^5$.

46. Find the second derivative for $f(x) = (x^2 - 2)(x^3 + 5x)$.

47. Find $\dfrac{d^3y}{dx^3}$ for $y = x^3 - \dfrac{1}{x}$.

48. Find $\dfrac{d^4y}{dx^4}$ for $y = \dfrac{-1}{x} - 5x^{-2}$.

49. Find $\dfrac{d}{dx}\left[x^2 \dfrac{d^2}{dx^2}(3x^2 - x^5)\right]$.

50. Determine the values of x for which (a) $f''(x) = 0$, (b) $f''(x) > 0$, and (c) $f''(x) < 0$ for $f(x) = 2x^4 + 2x^3 - x$.

3.4 The Derivative as a Rate of Change

51. Find the rate of change of the area of a circle with respect to the radius r when $r = 3$.

52. Find the rate of change of the volume of a cube with respect to the length s of a side when $s = 2$.

53. Find the rate of change of the area of a square with respect to the length z of a diagonal when $z = 5$.

54. Find the rate of change of the volume of a ball with respect to the radius r when $r = 4$.

55. Find the rate of change of $y = 6 - x - x^2$ with respect to x at $x = -1$.

56. Find the rate of change of the volume V of a cube with respect to the length w of a diagonal on one of the faces when $w = 2$.

57. The volume of a cylinder is given by the formula $V = \pi r^2 h$ where r is the base radius and h is the height.
    (a) Find the rate of change of V with respect to h if r remains constant.
    (b) Find the rate of change of V with respect to r if h remains constant.
    (c) Find the rate of change of h with respect to r if V remains constant.

58. An object moves along a coordinate line, its position at each time $t \geq 0$ given by $x(t) = 3t^2 - 7t + 4$. Find the position, velocity, acceleration, and speed at time $t_0 = 4$.

59. An object moves along a coordinate line, its position at each time $t \geq 0$ given by $x(t) = t^3 - 6t^2 - 15t$. Determine when, if ever, the object changes direction.

60. An object moves along the x-axis, its position at each time $t \geq 0$ given by $x(t) = t^4 - 12t^3 + 28t^2$. Determine the time interval(s), if any, during which the object moves left.

61. An object moves along the x-axis, its position at each time $t \geq 0$ given by $x(t) = 5t^4 - t^5$. Determine the time interval(s), if any, during which the object is speeding up to the right.

62. An object is dropped and hits the ground 5 seconds later. From what height was it dropped? Neglect air resistance.

63. A stone is thrown upward from ground level. The initial speed is 24 feet per second. (a) In how many seconds will the stone hit the ground? (b) How high will it go? (c) With what minimum speed should the stone be thrown to reach a height of 40 feet?

64. An object is projected vertically upward from ground level with a velocity of 32 feet per second. What is the height attained by the object? (Take g as 32 ft/sec^2.)

65. If $C(x) = 700 + 5x + \dfrac{100}{\sqrt{x}}$ is the cost function for a certain commodity, find the marginal cost at a production level of 400 units, and find the actual cost of producing the 401st unit.

66. If $C(x) = 25{,}000 + 30x + (0.003)x^2$ is the cost function for a certain commodity and $R(x) = 60x - (0.002)x^2$ is the revenue function, find:
    (a) the profit function
    (b) the marginal profit
    (c) the production level(s) at which the marginal profit is zero.

3.5 The Chain Rule

67. Differentiate $f(x) = (x^3 + 1)^3$: (a) by expanding before differentiation, (b) by using the chain rule. Then reconcile the results.

68. Differentiate $f(x) = (x - x^3)^3$.

69. Differentiate $f(x) = \left[\dfrac{x+1}{x-1}\right]^2$.

70. Differentiate $f(x) = \left[\dfrac{1}{x} + \dfrac{1}{x^2}\right]^4$.

71. Differentiate $f(x) = \left[\dfrac{x^2+7}{x^2-7}\right]^4$.

72. Differentiate $f(x) = (x+4)^4(3x+2)^3$.

73. Find $\dfrac{dy}{dx}$ at $x = 0$ for $y = \dfrac{1}{1+u}$ and $u = (3x+1)^3$.

74. Find $\dfrac{dy}{dt}$ at $t = 1$ for $y = u^3 - u^2$, $u = \dfrac{1-x}{1+x}$, and $x = 2t - 5$.

75. Find $\dfrac{dy}{dx}$ at $x = 1$ for $y = \dfrac{1+s}{1-s}$, $s = \sqrt{t+1}$, and $t = \dfrac{3x^2}{4}$.

76. Given that $f(1) = 2$, $g(1) = 1$, $f'(1) = 3$, $g'(1) = 2$, and $f'(2) = 0$, evaluate $(f \bullet g)'(1)$.

77. Find $\dfrac{d}{dx}[f(x^3 - 1)]$.

78. Given that $f(x) = (1 + 2x^2)^{-2}$, determine the values of x for which (a) $f'(x) = 0$, (b) $f'(x) > 0$, (c) $f'(x) < 0$.

79. An object moves along a coordinate line, its position at each time $t \geq 0$ given by $x(t) = (t^2 - 3)^3(t^2 + 1)^2$. Determine when the object changes direction.

80. Differentiate $f(x) = [(x^3 - x^{-3})^2 - x^2]^3$.

81. Find $f''(x)$ for $f(x) = (x^2 + 2x)^{17}$.

82. The edge of a cube is decreasing at the rate of 3 centimeters per second. How is the volume of the cube changing when the edge is 5 centimeters long?

83. The diameter of a sphere is increasing at the rate of 3 centimeters per second. How is the volume of the sphere changing when the diameter is 6 centimeters?

3.6 Differentiating the Trigonometric Functions

84. Differentiate $y = x \tan x$.

85. Differentiate $y = \sin x \tan x$.

86. Differentiate $y = \dfrac{\sin x}{1 - \cos x}$.

87. Differentiate $y = \dfrac{\sin x}{x^2}$.

88. Differentiate $y = \sec x \tan x$.

89. Find the second derivative for $y = x \sin x$.

90. Find the second derivative for $y = 5 \cos x + 7 \sin x + \dfrac{x^2}{2}$.

91. Find $\dfrac{d^3}{dx^3}(\sin x)$.

92. Find an equation for the tangent to the curve $y - \sin x$ at $x = \pi/6$.

93. Determine the numbers x between 0 and 2π on $y = \sin x$, where the tangent to the curve is parallel to the line $y = 0$.

94. An object moves along the y-axis, its position at each time t given by $x(t) = \sin 2t$. Determine those times from $t = 0$ to $t = \pi$ when the object is moving to the right with increasing speed.

95. Find $\dfrac{dy}{dt}$ for $y = \left[\dfrac{1}{2}(1 + u)\right]^3$, $u = \sin x$, and $x = 2\pi t$.

96. A rocket is launched 2 miles away from one observer on the ground. How fast is the rocket going when the angle of elevation of the observer's line of sight to the rocket is 50° (from the horizontal) and is increasing at 5°/sec?

97. An airplane at a height of 2000 meters is flying horizontally, directly toward an observer on the ground, with a speed of 300 meters per second. How fast is the angle of elevation of the plane changing when this angle is 45°?

3.7 Implicit Differentiation: Rational Powers

98. Use implicit differentiation to obtain $\dfrac{dy}{dx}$ in terms of x and y for $x^2 - 4xy + 2y^2 = 5$.

99. Use implicit differentiation to obtain $\dfrac{dy}{dx}$ in terms of x and y for $x^2y + y^2 = 6$.

100. Use implicit differentiation to obtain $\dfrac{dy}{dx}$ in terms of x and y for $xy^2 + \sqrt{xy} = 2$.

Calculus: One and Several Variables

101. Use implicit differentiation to obtain $\dfrac{dy}{dx}$ in terms of x and y for $y = \sin(x + y) + \cos x$.

102. Express $\dfrac{d^2y}{dx^2}$ in terms of x and y for $x^2 + 3y^2 = 10$.

103. Express $\dfrac{d^2y}{dx^2}$ in terms of x and y for $x^2 + 2xy - y^2 + 8 = 0$.

104. Express $\dfrac{dy}{dx}$ at the point $P(-1, -1)$ for $3x^2 + xy = y^2 + 3$.

105. Express $\dfrac{dy}{dx}$ and $\dfrac{d^2y}{dx^2}$ at the point $P(2, -1)$ for $x^2 - xy + y^2 = 7$.

106. Find the equations for the tangent and normal at the point $P(-1, -1)$ for $2x^2 - 3xy + 3y^2 = 2$.

107. Find $\dfrac{dy}{dx}$ for $y = (x^4 + x^3)^{3/2}$.

108. Find $\dfrac{dy}{dx}$ for $y = \sqrt[4]{3x^3 + 2}$.

109. Find $\dfrac{dy}{dx}$ for $y = (x^2 + 1)^{1/4}(x^2 + 2)^{1/2}$.

110. Compute $\dfrac{d}{dx}\left(\sqrt[4]{x} + \dfrac{1}{\sqrt[4]{x}}\right)$.

111. Compute $\dfrac{d}{dx}\sqrt{\dfrac{4x + 3}{2x - 5}}$.

112. Find the second derivative for $y = \sqrt{9 + x^3}$.

113. Find the second derivative for $y = 4\sqrt{4 - x^3}$.

114. Compute $\dfrac{d}{dx}\left[f(\sqrt{x} - 1)\right]$.

115. In economics, the elasticity of demand is given by the formula $\varepsilon = \dfrac{P}{Q}\left|\dfrac{dQ}{dP}\right|$ where P is price and Q quantity.

The demand is said to be $\begin{cases} \text{inelastic where } \varepsilon < 1 \\ \text{unitary where } \varepsilon = 1 \\ \text{elastic where } \varepsilon > 1 \end{cases}$. Describe the elasticity of $Q = (400 - P)^{3/5}$.

3.8 Rates of Change Per Unit Time

116. A shark, looking for dinner, is swimming parallel to a straight beach and is 90 feet offshore. The shark is swimming at a constant speed of 30 feet per second. At time $t = 0$, the shark is directly opposite a lifeguard station. How fast is the shark moving away from the lifeguard station when the distance between them is 150 feet?

117. A boat sails parallel to a straight beach at a constant speed of 12 miles per hour, staying 4 miles offshore. How fast is it approaching a lighthouse on the shoreline at the instant it is exactly 5 miles from the lighthouse?

118. A ladder 13 feet long is leaning against a wall. If the base of the ladder is moving away from the wall at the rate of ½ foot per second, at what rate will the top of the ladder be moving when the base of the ladder is 5 feet from the wall?

119. A spherical balloon is inflated so that its volume is increasing at the rate of 3 cubic feet per minute. How fast is the radius of the balloon increasing at the instant the radius is ½ foot? $V = \frac{4}{3}\pi r^3$

120. Sand is falling into a conical pile so that the radius of the base of the pile is always equal to one-half of its altitude. If the sand is falling at a rate of 10 cubic feet per minute, how fast is the altitude of the pile increasing when the pile is 5 feet deep? $V = \frac{1}{3}\pi r^2 h$

121. A spherical balloon is inflated so that its volume is increasing at the rate of 20 cubic feet per minute. How fast is the surface area of the balloon increasing at the instant the radius is 4 feet? $V = \frac{4}{3}\pi r^3$, $S = 4\pi r^2$

122. Two ships leave port at noon. One ship sails north at 6 miles per hour, and the other sails east at 8 miles per hour. At what rate are the two ships separating 2 hours later?

123. A conical funnel is 14 inches in diameter and 12 inches deep. A liquid is flowing out at the rate of 40 cubic inches per second. How fast is the depth of the liquid falling when the level is 6 inches deep? $V = \frac{1}{3}\pi r^2 h$

124. A baseball diamond is a square 90 feet on each side. A player is running from home to first base at the rate of 25 feet per second. At what rate is his distance from second base changing when he has run halfway to first base?

125. A ship, proceeding southward on a straight course at a rate of 12 miles/hr. is, at noon, 40 miles due north of a second ship, which is sailing west at 15 miles/hr.
    (a) How fast are the ships approaching each other 1 hour later?
    (b) Are the ships approaching each other or are they receding from each other at 2 o'clock and at what rate?

126. An angler has a fish at the end of his line, which is being reeled in at the rate of 2 feet per second from a bridge 30 feet above water. At what speed is the fish moving through the water toward the bridge when the amount of line out is 50 feet? (Assume the fish is at the surface of the water and that there is no sag in the line.)

127. A kite is 150 feet high and is moving horizontally away from a boy at the rate of 20 feet per second. How fast is the string being paid out when the kite is 250 feet from him?

128. An ice cube is melting so that its edge length x is decreasing at the rate of 0.1 meters per second. How fast is the volume decreasing when $x = 2$ meters?

129. Consider a rectangle where the sides are changing but the area is always 100 square inches. One side changes at the rate of 3 inches per second. When that side is 20 inches long, how fast is the other side changing?

130. The sides of an equilateral triangle are increasing at the rate of 5 centimeters per hour. At what rate is the area increasing when the side is 10 centimeters?

131. A circular cylinder has a radius r and a height h feet. If the height and radius both increase at the constant rate of 10 feet per minute, at what rate is the lateral surface area increasing? $S = 2\pi r h$

132. The edges of a cube of side x are contracting. At a certain instant, the rate of change of the surface area is equal to 6 times the rate of change of its edge. Find the length of the edge.

133. A particle is moving along the parabola $y = x^2$. If the x-coordinate of its position P is increasing at the rate of 10 m/sec, what is the rate of change of the angle of inclination of the line OP when $x = 3$ meters?

3.9 Differentials; Newton-Raphson Approximations

134. Estimate $\sqrt[4]{14}$ by differentials.

135. Estimate $\sqrt[3]{9}$ by differentials.

136. Estimate $\sqrt[5]{30}$ by differentials.

137. Estimate $\sqrt[3]{10}$ by differentials.

138. Use differentials to estimate $\cos 59°$.

139. Use differentials to estimate $\sin 31°$.

140. Use differentials to estimate $\tan 43°$.

141. Estimate $f(3.2)$, given that $f(3) = 2$ and $f'(x) = (x^3 + 5)^{1/5}$.

142. How accurately must we measure the edge of a cube to determine the volume within 1%?

143. Use differentials to estimate the volume of gold needed to cover a sphere of radius 10 cm with a layer of gold 0.05 cm thick.

Answers to Chapter 3, Part B Questions

1. 0
2. -4
3. $4x + 1$
4. $3x^2$
5. $\dfrac{1}{2\sqrt{x+2}}$
6. $\dfrac{-1}{(x+1)^2}$
7. $-1/x^3$
8. 28
9. 10
10. 9/4
11. $-6/25$
12. tangent: $y - 3 = 6(x - 1)$
    normal: $y - 3 = \dfrac{-1}{6}(x - 1)$
13. tangent: $y - 2 = 9(x - 2)$
    normal: $y - 2 = \dfrac{-1}{9}(x - 2)$
14. tangent: $y - 2 = \dfrac{1}{2}(x - 2)$
    normal: $y - 2 = -2(x - 2)$
15. tangent: $y - 2 = \dfrac{1}{2}\left(x - \dfrac{3}{2}\right)$
    normal: $y - 2 = -2\left(x - \dfrac{3}{2}\right)$
16. no derivative at $x = -1/2$

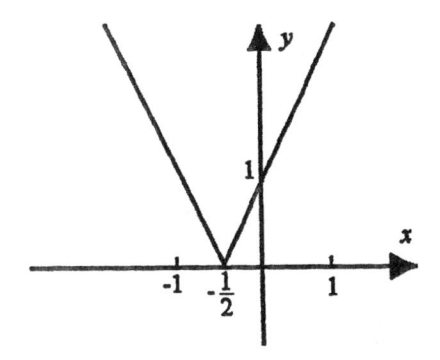

17. no derivative at $x = 1$

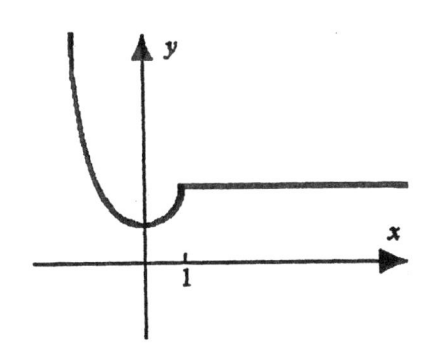

18. $f'(1) = 4$
19. $f'(-2) = 3$
20.
21. 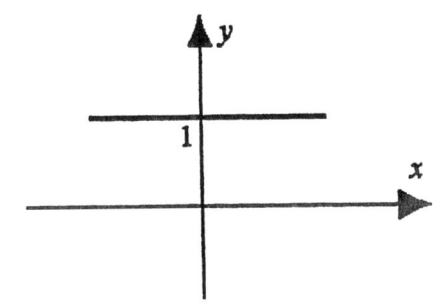
22. $f(x) = x^{2/3}$; $c = 8$ 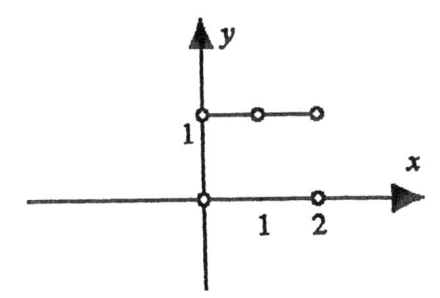
23. $f(x) = \sin x$; $c = 0$
24. -3
25. $20x^4 - 16x + 9$
26. $-8/x^5$
27. $18x^2 + 4x - 3$

28. $\dfrac{-4}{x^3} + \dfrac{20}{x^5}$

29. $\dfrac{9x^4 - 12x^3 - 5}{(x-1)^2}$

30. $\dfrac{-2}{x^2} - \dfrac{4}{x^3} - \dfrac{12}{x^4}$

31. $f'(x) = 8x^3 + 9x^2$
    $f'(0) = 0$
    $f'(1) = 17$

32. $f'(x) = \dfrac{-6}{(2x-1)^2}$
    $f'(0) = -6$
    $f'(1) = -6$

33. $f'(x) = 4xh(x) + 2x^2h'(x) - 3$
    $f'(0) = -3$

34. $y - 7 = (1)(x+1)$
    $y - 7 = x + 1$
    $y = x + 8$

35. $f'(x) = 0$ at $x = \dfrac{2 \pm \sqrt{37}}{3}$

36. at $x = 0$

37. at $x = -7/8$

38. $153/8$

39. $3 - 4/x^3$

40. $15x^2 - \dfrac{1}{2x\sqrt{x}}$

41. $\dfrac{21 + 14x - 2x^2}{(7 - 2x)^2}$

42. $-18x^8 + 80x^7 - 12x + 30$

43. $\dfrac{28x}{(3x^2 - 1)^2}$

44. 79

45. $\dfrac{-48}{x^4} + 4x^3$

46. $20x^3 + 18x$

47. $6 + 6/x^4$

48. $-24x^{-5} - 600x^{-6} = \dfrac{-24}{x^5} - \dfrac{600}{x^6}$

49. $12x - 100x^4$

50. (a) $x = 0$ or $x = -\tfrac{1}{2}$
    (b) $x < -\tfrac{1}{2}$ or $x > 0$
    (c) $-\tfrac{1}{2} < x < 0$

51. 6π

52. 12

53. 5

54. 64π

55. 1

56. $3\sqrt{2}$

57. (a) $\dfrac{dV}{dh} = \pi r^2$ (c) $\dfrac{dh}{dr} = \dfrac{-2V}{\pi r^3}$
    (b) $\dfrac{dV}{dr} = 2\pi rh$

58. $x(4) = 24$; $v(4) = 17$; $a(4) = 6$
    Speed $= |v(4)| = 17$

59. The object changes direction (from left to right) at $t = 5$.

60. $v(t) < 0$ when $2 < t < 7$.

61. $0 < t < 3$

62. 400 ft

63. (a) $1\tfrac{1}{2}$ sec   (c) $16\sqrt{10}$ ft/sec
    (b) 9 ft

64. 16 ft

65. $\$4.99$; $\$4.99$

66. (a) $P(x) = 30x - (0.005)x^2 - 25{,}000$
    (b) $P'(x) = 30 - (0.010)x$
    (c) $x = 3000$

67. $f(x) = x^9 + 3x^6 + 3x^3 + 1$
    $f'(x) = 9x^8 + 18x^5 + 9x^2$
    $f(x) = (x^3 + 1)^3$
    $f'(x) = 9x^2(x^3 + 1)^2 = 9x^8 + 18x^5 + 9x^2$

68. $3(x - x^3)^2(1 - 3x^2)$

69. $\dfrac{-4(x+1)}{(x-1)^3}$

70. $-4\left(\dfrac{1}{x}+\dfrac{1}{x^2}\right)^3\left(\dfrac{1}{x^2}+\dfrac{2}{x^3}\right)$

71. $\dfrac{-112x(x^2+7)^3}{(x^2-7)^5}$

72. $(x+4)^3(3x+2)^2(21x+44)$

73. $-9/4$

74. -16

75. $\dfrac{12}{\sqrt{7}\left(2-\sqrt{7}\right)^2}$ or $\dfrac{112+44\sqrt{7}}{21}$

76. 6

77. $3x^2 f'(x^3-1)$

78. (a) $x=0$ (c) $x>0$
    (b) $x<0$

79. The object changes direction (from left to right) at $t=\dfrac{\sqrt{15}}{5}$ and at $t=\sqrt{3}$ (from right to left).

80. $3[(x^3-x^{-3})^2 - x^2]^2\,[2(x^3-x^{-3})(3x^2+3x^{-4}) - 2x]$

81. $17(x^2+2x)^{15}(66x^2+132x+64)$

82. decreasing at a rate of 225 cm^3/sec

83. decreasing at a rate of 54π cm^3/sec

84. $\tan x + x\sec^2 x$

85. $\cos x \tan x + \sin x \sec^2 x$
    $= \sin x + \sin x \sec^2 x = \sin x(1+\sec^2 x)$

86. $\dfrac{1}{\cos x - 1}$

87. $\dfrac{x\cos x - 2\sin x}{x^3}$

88. $2\sec^3 x - \sec x$

89. $2\cos x - x\sin x$

90. $-5\cos x - 7\sin x + 1$

91. $-\cos x$

92. $y - \dfrac{1}{2} = \dfrac{\sqrt{3}}{2}\left(x - \dfrac{\pi}{6}\right)$

93. $x = \pi/2,\ 3\pi/2$

94. $3\pi/4 < t < \pi$

95. $\dfrac{3\pi}{4}(1+\sin 2\pi t)^2 \cos 2\pi t$

96. $\dfrac{\pi}{18}\sec^2 50° \cong 1521$ mph

97. $3/40$ radian/sec

98. $\dfrac{dy}{dx} = \dfrac{2y-x}{2y-2x}$

99. $\dfrac{dy}{dx} = \dfrac{-2xy}{x^2+2y}$

100. $\dfrac{dy}{dx} = \dfrac{-y-2y^2\sqrt{xy}}{4x\sqrt{xy}+x}$

101. $\dfrac{dy}{dx} = \dfrac{\cos(x+y)-\sin x}{1-\cos(x+y)}$

102. $\dfrac{d^2y}{dx^2} = \dfrac{-10}{9y^3}$

103. $\dfrac{16}{(y-x)^3}$

104. 7

105. at $(2,-1)$, $\dfrac{dy}{dx} = \dfrac{5}{4}$, $\dfrac{d^2y}{dx^2} = \dfrac{21}{32}$

106. tangent: $y+1 = \dfrac{-1}{3}(x+1)$
     normal: $y+1 = 3(x+1)$

107. $\dfrac{3x^2}{2}(4x+3(x^4+x^3)^{1/2}$

108. $\dfrac{9x^2}{4}(3x^3+2)^{-3/4}$

109. $\dfrac{3x^3+4x}{2(x^2+1)^{3/4}(x^2+2)^{1/2}}$

110. $\dfrac{1}{4}x^{-3/4} - \dfrac{1}{4}x^{-5/4} = \dfrac{1}{4\sqrt[4]{x^3}} - \dfrac{1}{4x\sqrt[4]{x}}$

111. $\dfrac{-13}{(2x-5)\sqrt{(4x+3)(2x-5)}}$

112. $\dfrac{3x^4 + 108x}{4(9+x^3)\sqrt{9+x^3}}$

113. $\dfrac{15x^5 - 96x^2}{4(4-x^3)\sqrt{4-x^3}}$

114. $\dfrac{1}{2\sqrt{x}} f'(\sqrt{x} - 1)$

115. with $0 < p < 400$, $\epsilon < 1$ for $P < 250$
$\epsilon = 1$ for $P = 250$
$\epsilon > 1$ for $P > 250$

116. 24 ft/sec

117. $\dfrac{36}{5}$ mi/hr

118. $\dfrac{-5}{24}$ ft/sec

119. $\dfrac{3}{\pi}$ ft/min

120. $\dfrac{8}{5\pi}$ ft/min

121. 10 ft²/min

122. 10 mi/hr

123. $\dfrac{-160}{49\pi}$ in/sec

124. $-5\sqrt{5}$ ft/sec

125. (a) $\dfrac{111}{\sqrt{1009}} \approx 3.498$ mi/hr

    (b) $\dfrac{129}{17}$ mi/hr (receding from each other)

126. 5/2 ft/sec

127. 16 ft/sec

128. -1.2 m³/sec

129. $\dfrac{-3}{4}$ in/sec

130. $25\sqrt{3}$ cm²/hr

131. $20\pi(r+h)$ ft²/min

132. $x = \tfrac{1}{2}$

133. 1 radian/sec (increasing)

134. 31/16

135. 25/12

136. 79/40

137. 13/6

138. $\approx \dfrac{1}{2} + \dfrac{\sqrt{3}\pi}{360} \approx 0.515$

139. $\approx \dfrac{1}{2} + \dfrac{\sqrt{3}}{2}\left(\dfrac{\pi}{180}\right) \approx 0.515$

140. $\approx 1 - \dfrac{\pi}{45} \approx 0.930$

141. 2.4

142. within $\dfrac{1}{3}$ %

143. approximately 20π cm³

Part C

1. The figure below is the graph of $f'(x)$. Sketch the graph of $f(x)$. (3.1)

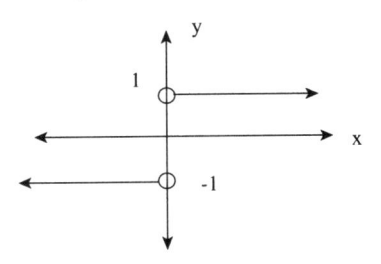

2. $f(x) = \begin{cases} 4 - 6x^2 & x \leq 0 \\ x^3 & x > 0 \end{cases}$ (3.1)
   (a) find $f'_-(0)$ if it exists
   (b) find $f'_+(0)$ if it exists
   (c) is $f(x)$ differentiable at 0?

3. Using (1) $f'(c) = \lim_{h \to 0} \dfrac{f(c+h) - f(c)}{h}$, let $x = c + h$ (3.1)
   we get (2) $f'(c) = \lim_{x \to c} \dfrac{f(x) - f(c)}{x - c}$
   Calculate the $f'(2)$ for $f(x) = x^3 + 1$ by both (1) and (2)

4. Using (1) and (2) of #3 calculate $f'(0)$ for $f(x) = 3x - 7$ (3.1)

5. Find A,B given that the derivative of $f(x)$ is continuous for all x (3.2)
   $f(x) = \begin{cases} Ax^2 + B & x < -1 \\ Bx^5 + Ax + 4 & x \geq -1 \end{cases}$

6. Show that, if f is differentiable, then
   $g(x) = [f(x)]^n$ has derivative $g'(x) = n\,[f(x)]^{n-1} f'(x)$ (3.2)

7. Given $h(x) = f(x) \cdot g(x)$ and f, g have an infinite number of derivatives (3.3)
   find (a) $h'(x) =$
   (b) $h''(x) =$
   (c) $h'''(x) =$
   (d) Speculate $h^5(x) =$

8. Given the polynomial function (3.3)
   $p(x) = a_n x^n + a_{n-1} x^{n-1} + \ldots + a_1 x + a_0$
   (a) find $\dfrac{d^n x}{dx^n}[p(x)]$
   (b) find $\dfrac{d^{n+1} x}{dx^{n+1}}[p(x)]$

9. For $f(x) = \dfrac{1}{x^2}$ (3.3)
   find (a) $f'(x)$
   (b) $f''(x)$
   (c) $f'''(x)$
   (d) write a formula for $f^n(x)$

10. An object moves along the x-axis, its position at each time $t \geq 0$ is given by $x(t)$. Determine the time intervals, if any, when it is moving right and slowing down $x(t) = t^3 - 6t^2 - 15t$.

11. The total revenue (in dollars) from the sale of x units of a certain commodity is given by
$$R(x) = 24x + 5x^2 - \frac{x^3}{3} \quad x \geq 0 \tag{3.4}$$
(a) find the marginal revenue function $R'(x)$ and determine the interval(s) on which $R'(x) > 0$
(b) for what value(s) of x is the marginal revenue a maximum?

12. Let f be a differentiable function, use the chain rule to show that (3.5)
(a) If f is even, then f' is odd
(b) If f is odd, then f' is even

13. Given f, g inverse functions, that is (3.5)
$f(g(x)) = g(f(x)) = x \quad y = g(x) \quad g'(x) \neq 0$
(a) Show that $f'(g) = \dfrac{1}{g'(x)}$
(b) for $f(x) = 2x - 2$ and $g(x) = \dfrac{1}{2}x + 1$ show that $f'(g) = \dfrac{1}{g'(f)}$

14. Given $\sin 2x = 2 \sin x \cos x$. Differentiate the formula to obtain the double angle formula for the cosine functions.

15. Let $f(x) = \begin{cases} \cos x & x \geq 0 \\ ax + b & x < 0 \end{cases}$ (3.6)
(a) determine a, b so that f is differentiable at 0
(b) sketch the graph (3.6)

16. A triangle has sides a, b constants and the angle between them is x radians. Using the law of cosines find the rate of change of the third side c.

17. Given the angle between two curves is the angle between their tangents at the point of intersection if m_1 is the slope of the first curve and m_2 is the slope of the second curve, then x the angle can be found using
$$\tan x = \left| \frac{m_2 - m_1}{1 + m_1 m_2} \right| \text{ find } x \text{ for } y = 2x \text{ and } x^2 - xy + 2y^2 = 28 \tag{3.7}$$

18. Prove that the sum of the x and y intercepts of any tangent line to the graph of (3.7)
$x^{1/2} + y^{1/2} = c^{1/2}$ is constant and equal to c.

19. In the special theory of relativity the mass of a particle moving at velocity v is $\dfrac{m}{\sqrt{1 - \dfrac{v^2}{c^2}}}$

m = mass at rest
c = speed of light
At what rate is the mass changing when the particle's velocity is $\frac{1}{2} c$ and the rate of the velocity is $0.01c/\text{sec}$?

21. Use differentials of estimate $f(5, 4)$ given $f(5) = 1$ and $f'(x) = \sqrt[3]{x^2 + 2}$. (3.9)

22. For what value of b does the graph of $y = x^2 + bx + 1$ have a horizontal tangent at $x = 3$?

23. Show that the following function is not differentiable at $x = 0$
$f(x) = \begin{cases} x & \text{if } x \text{ is rational} \\ 0 & \text{if } x \text{ is irrational} \end{cases}$

24. For $f(x) = \begin{cases} x^2 & \text{if } x \leq a \\ mx + b & \text{if } x > a \end{cases}$ a, b, m constants
Determine b, and m so that $f(x)$ is differentiable at all points (solve in terms of a).

25. Given $y = ax^2 + bx + c$ passes through the point $(-1, 0)$ and is tangent to the line $y = x$ at origin. Determine a, b, c.

26. Find $\dfrac{dy}{dx}$ for
   (a) $y = \cos[\sin(\cos x)]$
   (b) $y = \cos^5(1 - 3x^2)^3$
   (c) $y = \dfrac{\cos x}{1 - \sin x}$

27. Find the values of x for which the graph of $y = 2\sin x + \sin^2 x$ has a horizontal tangent.

28. Given $f(x) = \begin{cases} x \sin \dfrac{1}{x} & \text{if } x \neq 0 \\ 0 & \text{if } x = 0 \end{cases}$
   (a) find $f'(x)$ for $x \neq 0$
   (b) show that $f(x)$ is continuous at 0
   (c) show that $f'(0)$ does not exist
   (d) sketch the graph

29. Find $\dfrac{dy}{dx}$ by implicit differentiation and also by solving for y and the differentiating. Then show your answers are equivalent
   $$\dfrac{y}{x} - 2x = y$$

30. If a is a positive number and we calculate $\sqrt[3]{a}$ by using the Newton-Raphson method, show that
   $$x_2 = \dfrac{1}{3}\left(2x_1 + \dfrac{a}{x_1^2}\right)$$

31. For $ax^2 + bx + c = f(x)$, $a > 0$ show that $f(x) \geq 0$ iff $b^2 - 4ac \leq 0$.
   Hint: Calculate the minimum value of $f(x)$.

Answers to Chapter 3, Part C Questions

1.

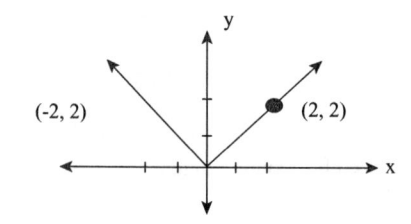

2. (a) 0
   (b) DNE
   (c) f is not differentiable or continuous at 0
   (d)

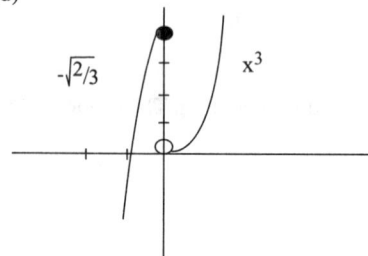

3. $f'(2) = \lim_{h \to 0} \dfrac{f(2+h) - f(2)}{h} = 12$

   $f'(2) = \lim_{x \to 2} \dfrac{f(x) - f(2)}{x - 2} = \lim_{x \to 2}(x^2 + 2x + 4)$
   $= 12$

4. $f'(0) = \lim_{h \to 0} \dfrac{f(0+h) - f(0)}{h} = 3$

   $f'(0) = \lim_{x \to 0} \dfrac{f(x) - f(0)}{x - 0}$
   $= \lim_{x \to 0} \left(\dfrac{3x - 7 + 7}{x}\right) = 3$

5. $\lim_{x \to -1^-} f(x) = \lim_{x \to -1^+} f(x),$
   $A + B = -B - A + 4 \iff A + B = 2$
   $\lim_{x \to -1^-} f'(x) = \lim_{x \to -1^+} f(x),$
   $-2A = 5B + A \iff 3A + 5B = 0$
   $A = 5, \; B = -3$

6. $n \in I'$ for $n = 1$
   $g(x) = [f(x)]^0 \quad g'(x) = 1[f(x)]^0 f'(x)$
   $\qquad\qquad\qquad\quad g'(x) = f'(x)$
   so 1 is in the set S
   for k in the set s, $k \in I'$ we have $g(x) = [f(x)]^{k+1}$
   so $g(x) = (f(x))(f(x))^k$ using product rule
   $g'(x) = f(x)[kf(x)]^{k-1}f'(x) + [f(x)]^k f'(x)$
   $\quad\;\; = (k+1)[f(x)]^k f'(x)$

   thus $k + 1 \in S$ valid for all $k \in I'$
   take $n \in I^-$
   $g(x) = [f(x)]^n = \dfrac{1}{[f(x)]^{-n}}$ use quotient rule
   $= \dfrac{[f(x)]^{-n} \cdot (0) - 1 \cdot [-n(f(x))]^{-n-1}}{[f(x)]^{-2n}}$
   $= n[f(x)]^{n-1}$ where $n \in I^-$

7. (a) $h'(x) = f(x) \cdot g'(x) + g(x) \cdot f'(x)$
   (b) $h''(x) = f(x)g''(x) + 2f'(x)g'(x)$
   $\qquad\qquad + g(x)f''(x)$
   (c) $h'''(x) = f(x)g'''(x) + 3f'(x)g''(x)$
   $\qquad\qquad + 3f''(x)g'(x) + g(x)f'''(x)$
   (d) $h^5(x) = f(x)g^5(x) + 5\,f'(x)g''(x)$
   $\qquad\qquad + 10f''(x)g'''(x) + f'''(x)\,g''(x)$
   $\qquad\qquad + 5f^4(x)g'(x) + f^5(x)g(x)$

8. (a) $a_n\, n!$
   (b) 0

9. (a) $-2/x^3$
   (b) $\dfrac{(-2)(3)}{x^4}$
   (c) $\dfrac{(-2)(-3)(-4)}{x^5}$
   (d) $\dfrac{(-1)^{2a-1}\, n!}{x^{n+2}}$

10. Moving to the right for $t > 5$ slowing down from $t = 2$ to $t = 5$

11. (a) $R'(x) = 24 + 10x - x^2$, $R'(x) > 0$ on $(0,12)$
    (b) $R''(x) = 10 - 2x \Rightarrow R'(x)$ is a maximum at $x = 5$

12. (a) For f an even function $f(x) = f(-x)$
    $[f(x)]' = [f(-x)]' = f'(-x)(-1) = -f'(-x)$
    hence $f'(x) = -f'(x)$
    (b) For f an odd function $f(-x) = -f(x)$
    $[f(x)]' = -[f(-x)]' = -f'(-x)(-1)$
    $\qquad\;\; = f'(-x)$
    hence $f'(-x) = f'(x)$

13. Let $y = g(x)$ and suppose $f(y) = f[g(x)] = x$
    $\dfrac{d}{dx}(f(g)) = f'(y)\, g'(x) = 1$
    $\Rightarrow f'(y) = \dfrac{1}{g'(x)}, \; g'(x) \neq 0$

14. $\sin 2x = 2 \sin x \cos x$
    Differentiating both sides
    $2 \cos 2x = 2\,[\sin x(-\sin x) + \cos x(\cos x)]$

$$= 2(\cos^2 x - \sin^{-2} x) \cdot \cos 2x$$
$$= (\cos^2 x - \sin^2 x)$$

15. (a) f must be continuous at 0:
$$\lim_{x \to 0^+} f(x) = \lim_{x \to 0^+} \cos x = 1,$$
$$\lim_{x \to 0^-} (ax + b) = b, \text{ so } b = 1$$

Differentiable at 0:
$$\lim_{x \to 0^+} \frac{f(h) - f(0))}{n} = \lim_{x \to 0^+} \frac{\cos h - 1}{n} = 0$$
and $\lim_{x \to 0^-} \frac{f(h) - f(0)}{n} = \lim_{x \to 0^-} \frac{ah + 1 - 1}{n} = a$
so $\Rightarrow a = 0$

16. $C = \sqrt{a^2 + b^2 - 2ab \cos x}$ then
$$\frac{dc}{dx} = \frac{1}{2}(a^2 + b^2 - 2ab \cos x)^{-1/2} (2ab \sin x)$$
$$= \frac{ab \sin x}{\sqrt{a^2 + b^2 - 2ab \cos x}}$$

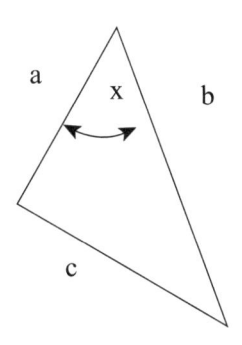

17. For $y = 2x$, The slope is $m_1 = 2$.
for $x^2 - xy + 2y^2 = 28$ we have
$$2x - y - y\frac{dx}{dy} + 4y\frac{dy}{dx} = 0 \Longrightarrow \frac{dy}{dx}$$
$$= \frac{y - 2x}{4y - x} = m_2 \text{ at the point of intersection of the}$$
line and the curve we have $m_2 = 0$ since
$y = 2x$, $\tan \alpha = |-m_1| = 2 \Longrightarrow \alpha \cong 1.107$
radians $\cong 63.4°$

18. $\frac{1}{2}x^{-1/2} + \frac{1}{2}y^{-1/2}\frac{dy}{dx} = 0 \Rightarrow \frac{dy}{dx} = -\left(\frac{y}{x}\right)^{1/2}$
equation of the tangent line at (x_0, y_0)
$$y - y_0 = -\left(\frac{y_0}{x_0}\right)^{1/2}(x - x_0)$$
x, y intercepts (a, b)
$$a = (x_0 y_0)^{1/2} + y_0 \quad b = (x_0 y_0)^{1/2} + y_0$$
$$a + b = 2(x_0 y_0)^{1/2} + x_0 + y_0 = \left(x_0^{1/2} + y_0^{1/2}\right)^2 = C$$

19. $M = \dfrac{m}{\sqrt{1 - \dfrac{v^2}{c^2}}}$;
$$\frac{dM}{dt} = m\,(-1/2)\left(1 - \frac{v^2}{c^2}\right)^{-3/2}\left(-\frac{2v}{c^2}\right)\frac{dv}{dt}$$
$$\frac{dM}{dt} = \frac{mv}{c^2\left(1 - \dfrac{v^2}{c^2}\right)^{3/2}}\frac{dv}{dt}$$
if $v = \dfrac{c}{2}$ and $\dfrac{dv}{dt} = \dfrac{c}{100}$ $\dfrac{dM}{dt} = \dfrac{\sqrt{3}}{225}m$

21. $f(5.4) \cong f(5) + (0.4)f'(5)$
$= 1 + (0.4)(3)$
$= 2.2$

22. $y' = 2x + b$, slope of the tangent line at $x = 3$ is
$2 \cdot 3 + b = b + 6$ true if $b = -6$

23. $\lim_{h \to 0} \dfrac{f(0+h) - f(0)}{h} = \lim_{h \to 0} \dfrac{f(h)}{h}$
$$= \begin{cases} 1 & \text{if } h \text{ is rational} \\ 0 & \text{if } h \text{ is irrational} \end{cases}$$
hence the limit does not exist since there exists both rationals and irrationals close to 0.

24. f must be continuous at a
$\lim_{x \to a^-} x^2 = \lim_{x \to a^+} (mx + b)$; $a^2 = ma + b$
and $\lim_{h \to 0^-} \dfrac{f(a + h) - f(a)}{h}$
$= \lim_{h \to 0^+} \dfrac{f(a + h) - fa}{h}$
$2a = m$
f is differentiable for $2a = m$ and $b = -a^2$

25. $y = ax^2 + bx + c$ contains $(-1, 0)$ $0 = a - b + c$
will pass through the origin if $c = 0$
tangent to $y = x$ at the origin $y'(0) = 1$
$y'(x) = 2a$
$x + b$ at $y(0) \Rightarrow b = 1$,
hence $a = b = 1, c = 0$

26. (a) $\sin x \cos(\cos x) \sin[\sin(\cos x)]$
(b) $90\, x(1 - 3x^2)^2 \sin(1 - 3x^2)^3 \cos^4(1 - 3x^2)^3$
(c) $\dfrac{1}{1 - \sin x}$

27. $\dfrac{dy}{dx} = 2\cos x + 2\sin x \cos x$
$= 2\cos x(1 + \sin x) = 0$
$\cos x = 0$ or $\sin x = -1$
$\Rightarrow \cos x = \pm\sqrt{1 - \sin^2 x} = 0$
in $[0, 2\pi]$ $x = \dfrac{\pi}{2}$, $3\pi/2 \Rightarrow$ general solution
$\pi/2 + n\pi, n \in I$

28. As $x \to 0$ $\sin \dfrac{1}{x}$ oscillates between $1, -1$

$f(x)$ oscillates between x and $-x$

for $x > \dfrac{1}{\pi} \Rightarrow 0 < \dfrac{1}{x} < \pi$

so $\sin \dfrac{1}{x} > 0$ and $f(x) > 0$, since $\dfrac{1}{x} \sin x$ is an even function $f(x) = f(-x)$

Now $\lim\limits_{\theta \to 0} \dfrac{\sin \theta}{\theta} = 1$

let $\theta = \dfrac{1}{x} \Rightarrow \lim\limits_{x \to \infty} x \sin \dfrac{1}{x} = 1$

so as $x \to \infty$ $f(x) \to 1$

(a) $\sin \dfrac{1}{x} - \dfrac{1}{x} \cos \dfrac{1}{x}$

(b) $x \neq 0$, $|f(x)| \leq |x| \Rightarrow \lim\limits_{x \to 0} f(x) = 0$
$= f(0)$

(c) $f'(\,\cdot\,) = \lim\limits_{x \to 0} \dfrac{f(0+h) - f(0)}{h} = \lim\limits_{x \to 0} \sin \dfrac{1}{h}$

since $\sin \dfrac{1}{h}$ oscidilation between $1, -1$ as

$h \to 0$ $f'(0) = $ DNE

29. Proof by verification

30. $f(x) = x^3 - a$
$f'(x) = 3x^2$

$x - \dfrac{f(x)}{f'(x)} = x - \dfrac{x^3 - a}{3x^2} = \dfrac{2x^3 + a}{3x^2}$

$= \dfrac{1}{3}\left(2x + \dfrac{a}{x^2}\right)$

so $x_2 = x_1 - \dfrac{f(x_1)}{f'(x_1)} = \dfrac{1}{3}\left(2x_1 + \dfrac{a}{x_1^2}\right)$

31. For $ax^2 + bx + c = f(x)$ $a > 0$
show that $f(x) \geq 0$ if $fb^2 - 4ac \geq 0$
Hint: Calculate the maximum value
solution
$f'(x) = 2ax + b = 0$ at $x = \dfrac{-b}{2a}$

$f\left(\dfrac{-b}{2a}\right) = a\left(\dfrac{-b}{2a}\right)^2 + b\left(\dfrac{-b}{2a}\right) + C$

$= \dfrac{b^2}{4a} - \dfrac{b^2}{2a} + C$

$= \dfrac{-b^2 - 4ac}{4a} = c - \dfrac{b^2}{4a}$

$c - \dfrac{b^2}{4a} \geq 0$ iff $b^2 - 4ac \leq 0$

CHAPTER 4 The Mean-Value Theorem and Applications
Part A

1. Identify the following as results of (a) Rolle's theorem, (b) mean value theorem, (c) both (a) and (b) (4.1)

   (i) $f(x) = x^2 - 4$ $f(2) = 0,$ $f(-2) = 0,$ $\therefore$ $\exists\, c$ such that
   $f'(c) = 0$ _____

   (ii) $f(x) = x^2 - 4$ $f(1) = -3,$ $f(0) = -4$ $\therefore$ $\exists\, c$ such that
   $f'(c) = \dfrac{f(0) - f(1)}{1 - 0} = \dfrac{-7}{1} = -7$ _____

   (iii) $f(x) = x^2 - 4$ $f(1) = -3$ $f(-1) = -3$ $\therefore$ $\exists\, c$ such that
   $f'(c) = 0$ _____.

2. For #1 above find an appropriate value of c (4.1)
   (i) $c =$ _____
   (ii) $c =$ _____
   (iii) $c =$ _____

3. For $f(x) = x^2 - 4$ and domain of f is $(0, \infty)$ (4.2)
   (a) then for $h > 0$ $f(x)$ _____ $f(x + h)$
   (b) for $f(x) = x^2 - 4$ and domain of f is $(-\infty, 0)$
   then for $h > 0$ $f(x)$ _____ $f(x + h)$

4. For any function $f(x)$ differentiable on I (4.2)
   (a) If $x_1 < x_2$ implies that $f(x_1) < f(x_2)$ we say
   f is _____ on I
   (b) If $x_1 < x_2$ implies that $f(x_1) > f(x_2)$ we say
   f is _____ on I
   (c) for part (a) what can we conclude about $f'(x)$?
   (d) for part (b) what can we conclude about $f'(x)$?

5. If $f'(x) = 0$ for all $x \in I$ what can we conclude about $f(x)$ on I? (4.2)

6. Given $f'(x) = 2x$ and $g'(x) = 2x$ for all $x \in I$ (4.2)
   (a) then $f(x) = g(x) +$ ☐ on I
   (b) if $f(0) = 7$ then $g(x) =$ _____

7. (a) Define f has a local maximum at c iff _____.
   (b) Define f has a local minimum at c iff _____.

8. If c is a critical number in the domain of f, then either $f'(c) =$ _____ or $f'(c)$ _____. (4.3)

9. Given $f(x) = x^3 - 3x^2$, find the critical number(s) and (4.3)
   apply: (a) the first derivative test to classify them
           (b) the second derivative test to confirm part (a)

10. When will the second derivative test not be useful or practical? Give an example. (4.3)

11. (a) State the extreme value theorem. (4.4)
    (b) Apply (a) to the function $f(x) = 2x - 1$ on $[0, 4]$
    (c) Apply (a) to the function $f(x) = x^2 - 4$ on $[-2, 1]$

Calculus: One and Several Variables

12. If f is continuous on I and has only one critical value, c, on I, then if $f(c)$ is a local maximum $f(c)$ will also be the _____ . (4.4)

13. Given f is differentiable on (a, b). The graph of f is concave up iff f' is _____ on (a, b) and is concave down iff f' is _____ on (a, b) (4.6)

14. f is continuous at c, $(c, f(c))$ is a point of inflection if $f''(c) =$ _____ or $f''(c)$ _____ . (4.6)

15. If f is at least twice differentiable on (a, b), $f''(x) > 0$ for all x in (a, b), then f is _____ . (4.6)

16. Give an example of a function that
    (a) has a vertical asymptote a $x = 1$
    (b) has vertical asymptotes at $x = 1, x = -2$.

17. If $x \to c^-$, $f'(x) \to -\infty$ and $x \to c^+$, $f'(x) \to \infty$ then f has _____ at $(c, f(c))$. (4.7)

18. If $x \to c^-$, $f'(x) \to -\infty$ and $x \to c^+$, $f'(x) \to -\infty$ and $f(c)$ exists then, f has _____ at $(c, f(c))$. (4.7)

19. If $f(x) = L$ as $x \to \infty$, then L is _____ for $f(x)$. (4.8)

20. Given: $f(x) = x^{2/3}$, $g(x) = \dfrac{1}{x-1}$, $h(x) = x^{1/3}$, $r(x) = \dfrac{x}{x^2 - 1}$ match f, g, h, r with the following descriptions (more than one answer is possible for each.) (4.8)
    (a) has vertical asymptote at $x = 1$ _____
    (b) has vertical tangent at $x = 0$ _____ and $f'(0^-) \neq f'(0^+)$
    (c) has vertical tangent at $x = 0$ _____ and $f'(0^-) = f'(0^+)$
    (d) has a horizontal asymptote _____ at $f(x) = 0$
    (e) has a root at $(0, 0)$ _____

Answers to Chapter 4, Part A Questions

1. (i) a)
   (ii) b)
   (iii) c)

2. (i) $c = 0$
   (ii) $c = -\frac{7}{2}$
   (iii) $c = 0$

3. (a) $\leq$ (b) $\geq$

4. (a) increase   (c) $f'(x) > 0$
   (b) decrease   (d) $f'(x) < 0$

5. $f(x) = c$ where $c \in \mathbb{R}$

6. (a) $f(x) = g(x) + c$ where $c \in \mathbb{R}$
   (b) $f(x) = x^2 + 7$
   $\therefore g(x) = x^2 + 7 + c$ where $c \in \mathbb{R}$

7. (a) $f(c) \geq f(x)$ for all x sufficiently does to c
   (b) $f(c) \leq f(x)$ for all x sufficiently does to c

8. $f'(c) = 0$ or $f'(c)$ doesn't exist.

9. $f'(x) = 2x^2 - 6x = 0$ $\therefore x = 0$ or $x = 2$

   f' $+++$ 0 $---$ 0 $++++$
                 0 2

   $f''(x) = 6x - 6$

   $f''(0) = -6 < 0$ Local maximum
   $f''(2) = 6 > 0$ Local minimum

10. Because the function must be twice differentiable; for example
    $f(x) = x^{4/3}$ $f'(x) = \frac{4}{3} x^{1/3}$

    but the second derivative
    $$f''(x) = \frac{4}{9} x^{-2/3}$$
    isn't defined at $x = 0$.

11. (a) If f is continuous on [a, b] then (i) f is bounded on [a, b] and (ii) f attains its absolute maximum M, and its absolute minimum m on [a, b]
    (b) min $f(0) = -1$ max $f(4) = 7$
    (c) min $f(0) = -4$ max $f(-2) = 0$

12. absolute maximum

13. increases
    decreases

14. $f''(c) = 0$ $f''(c)$ doesn't exist

15. concave up in (a, b)

16. (a) $f(x) = -\frac{1}{|x-1|}$ answers vary
    (b) $f(x) = \frac{x^2}{x^2 + x - 2}$ other answers possible

17. vertical cusp

18. vertical tangent

19. horizontal asymptote

20. (a) $g(x) = \frac{1}{x-1}$ (c) $h(x) = x^{1/3}$
    $r(x) = \frac{x}{x^2 - 1}$ (d) $r(x) = \frac{x}{x^2 - 1}$
    (b) $f(x) = x^{2/3}$ (e) $f(x), h(x), r(x)$

Part B

4.1 The Mean-Value Theorem

1. Determine whether the function $f(x) = x^3 - 3x + 2$ satisfies the conditions of the mean-value theorem on the interval $[-2, 3]$. If so, find the admissible values of c.

2. Determine whether the function $f(x) = x^2 + 2x - 1$ satisfies the conditions of the mean-value theorem on the interval $[0, 1]$. If so, find the admissible values of c.

3. Determine whether the function $f(x) = 1/x^2$ satisfies the conditions of the mean-value theorem on the interval $[-1, 1]$. If so, find the admissible values of c.

4. Determine whether the function $f(x) = x^2 + 4$ satisfies the conditions of the mean-value theorem on the interval $[0, 2]$. If so, find the admissible values of c.

5. Determine whether the function $f(x) = x^3 - 3x + 1$ satisfies the conditions of the mean-value theorem on the interval $[-2, 2]$. If so, find the admissible values of c.

6. Determine whether the function $f(x) = x^3 - 2x + 4$ satisfies the conditions of the mean-value theorem on the interval $[1, 2]$. If so, find the admissible values of c.

7. Determine whether the function $f(x) = x^3 - 3x^2 - 3x + 1$ satisfies the conditions of the mean-value theorem on the interval $[0, 2]$. If so, find the admissible values of c.

8. Determine whether the function $f(x) = \sqrt{x}$ satisfies the conditions of the mean-value theorem on the interval $[0, 4]$. If so, find the admissible values of c.

9. Determine whether the function $f(x) = \sqrt[3]{x}$ satisfies the conditions of the mean-value theorem on the interval $[-1, 1]$. If so, find the admissible values of c.

10. Determine whether the function $f(x) = x^3 - x$ satisfies the conditions of the mean-value theorem on the interval $[-1, 1]$. If so, find the admissible values of c.

11. Determine whether the function $f(x) = x^3 - 4x$ satisfies the conditions of Rolle's theorem on the interval $[-2, 2]$. If so, find the admissible values of c.

12. Determine whether the function $f(x) = \sqrt[3]{x}$ satisfies the conditions of the mean-value theorem on the interval $[0, 1]$. If so, find the admissible values of c.

4.2 Increasing and Decreasing Functions

13. Find the intervals on which $f(x) = x^4 - 24x^2$ increases and the intervals on which f decreases.

14. Find the intervals on which $f(x) = x^4 - 4x^3$ increases and the intervals on which f decreases.

15. Find the intervals on which $f(x) = x^4 - 6x^2 + 2$ increases and the intervals on which f decreases.

16. Find the intervals on which $f(x) = 5x^4 - x^5$ increases and the intervals on which f decreases.

17. Find the intervals on which $f(x) = 4x^3 - 15x^2 - 18x + 10$ increases and the intervals on which f decreases.

18. Find the intervals on which $f(x) = x(x - 6)^2$ increases and the intervals on which f decreases.

19. Find the intervals on which $f(x) = x^2 + \dfrac{2}{x}$ increases and the intervals on which f decreases.

20. Find the intervals on which $f(x) = x(x-4)^4 + 4$ increases and the intervals on which f decreases.

21. Find the intervals on which $f(x) = \sin 2x$, $0 \leq x \leq \pi$, increases and the intervals on which f decreases.

22. Find f given that $f'(x) = 3x^2 - 10x + 3$ for all real x and $f(0) = 1$.

23. Find f given that $f'(x) = 12x^3 - 12x^2$ for all real x and $f(0) = 1$.

24. Find f given that $f'(x) = 3(x-2)^2$ for all real x and $f(0) = 1$.

25. Find the intervals on which f increases and the intervals on which f decreases given that
$$f(x) = \begin{cases} x+2, & x < 0 \\ 7-2x, & 0 \leq x < 3. \\ (x-1)^2, & 3 \leq x \end{cases}$$

26. Given the graph of $f'(x)$ below, and given that $f(0) = 0$, sketch the graph of f.

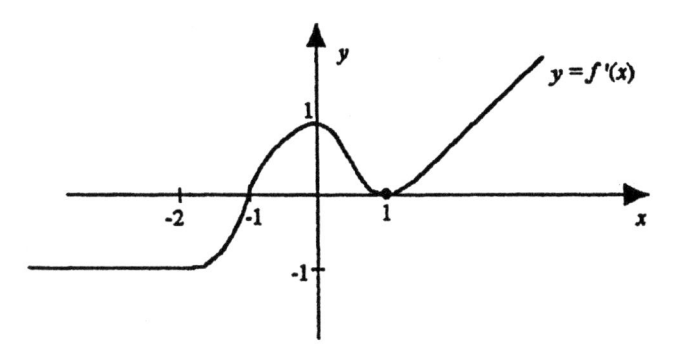

27. Sketch the graph of a differentiable function f that satisfies $f(2) = 1$, $f(-1) = 0$, and $f'(x) > 0$, for all x, if possible.

28. Sketch the graph of a differentiable function f that satisfies $f(0) = 0$, $f'(x) < 0$ for $x < 0$, and $f'(x) > 0$ for $x > 0$, if possible.

4.3 Local Extreme Values

29. Find the critical numbers and the local extreme values of $f(x) = 3x^5 - 5x^4$.

30. Find the critical numbers and the local extreme values of $f(x) = 12x^{2/3} - 16x$.

31. Find the critical numbers and the local extreme values of $f(x) = x^{2/3}(5-x)$.

32. Find the critical numbers and the local extreme values of $f(x) = \frac{1}{3}x^{4/3} - \frac{4}{3}x^{1/3}$.

33. Find the critical numbers and the local extreme values of $f(x) = \frac{x^4}{4} - 2x^2 + 1$.

34. Find the critical numbers and the local extreme values of $f(x) = (x+1)(x-1)^3$.

35. Find the critical numbers and the local extreme values of $f(x) = 2x + 2x^{2/3}$.

36. Find the critical numbers and the local extreme values of $f(x) = \frac{1}{x} - \frac{1}{3x^3}$.

70 Calculus: One and Several Variables

37. Find the critical numbers and the local extreme values of $f(x) = x^{4/3} - 4x^{-1/3}$.

38. Find the critical numbers and the local extreme values of $f(x) = 6x^2 - 9x + 5$.

39. Find the critical numbers and the local extreme values of $f(x) = x^4 - 6x^2 + 17$.

40. Find the critical numbers and the local extreme values of $f(x) = x - \dfrac{1}{x}$.

41. Find the critical numbers and the local extreme values of $f(x) = (x+1)^{2/3}$.

42. Find the critical numbers and the local extreme values of $f(x) = x - \sin 2x,\ 0 < x < \pi$.

43. Show that $f(x) = x^3 - 4x^2 + 2x - 5$ has exactly one critical number in $(0, 1)$.

4.4 Endpoint and Absolute Extreme Values

44. Find the critical numbers and classify the extreme values for $f(x) = \dfrac{x}{2} + 2,\ x \in [0, 100]$.

45. Find the critical numbers and classify the extreme values for $f(x) = 2x^3 - 3x^2 - 12x + 8,\ x \in [-2, 2]$.

46. Find the critical numbers and classify the extreme values for $f(x) = \dfrac{x^3}{3} - x^2 - 3x + 1,\ x \in [-1, 2]$.

47. Find the critical numbers and classify the extreme values for $f(x) = x^3 - 6x^2 + 5,\ x \in [-1, 5]$.

48. Find the critical numbers and classify the extreme values for $f(x) = 2x^3 - 3x^2 - 12x + 5,\ x \in [0, 4]$.

49. Find the critical numbers and classify the extreme values for $f(x) = 4x^3 - 6x^2 - 9x,\ x \in [-1, 2]$.

50. Find the critical numbers and classify the extreme values for $f(x) = x^3 - 3x + 6,\ x \in [0, 3/2]$.

51. Find the critical numbers and classify the extreme values for $f(x) = 1 - x^{2/3},\ x \in [-1, 1]$.

52. Find the critical numbers and classify the extreme values for $f(x) = x^3 - 12x + 8,\ x \in [-4, 3]$.

53. Find the critical numbers and classify the extreme values for $f(x) = x^{4/3} - 3x^{1/3},\ x \in [-1, 8]$.

54. Find the critical numbers and classify the extreme values for $f(x) = \dfrac{\sqrt{x}}{x^2 + 3},\ x \in [0, \infty]$.

55. Find the critical numbers and classify the extreme values for $f(x) = \dfrac{x}{x^2 + 1},\ x \in [0, 2]$.

56. Find the critical numbers and classify the extreme values for $f(x) = \dfrac{1}{x - x^2},\ x \in [0, 1]$.

57. Find the critical numbers and classify the extreme values for $f(x) = \begin{cases} x^2, & x < 0 \\ x^3, & x \geq 0 \end{cases}$.

58. Find the critical numbers and classify the extreme values for $f(x) = \begin{cases} -x - 1, & x < -1 \\ 1 - x^2, & -1 \leq x \leq 1,\ x \in [-2, 2]. \\ x - 1, & x > 1 \end{cases}$

59. Find the critical numbers and classify the extreme values for $f(x) = \begin{cases} -1 - x^2, & x < 0 \\ x^3 - 1, & x \geq 1 \end{cases}$, $x \in [-2, 1]$.

4.5 Some Max-Min Problems

60. Find the dimensions of the rectangle of greatest area that can be inscribed in a circle of radius a.

61. Find the dimensions of the rectangle of greatest area that can be inscribed in a semicircle of radius 1.

62. An open field is to be surrounded with a fence that also divides the enclosure into three equal areas as shown in the figure below. The fence is 4,000 feet long. For what value of x will the total area be a maximum?

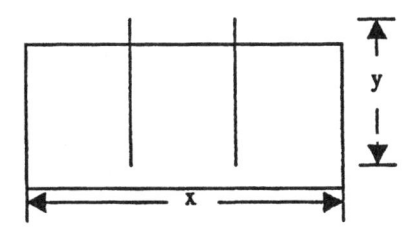

63. Find the dimension of the rectangle of maximum area that may be embedded in a right triangle with sides of length 12, 16, and 20 feet as shown in the figure below.

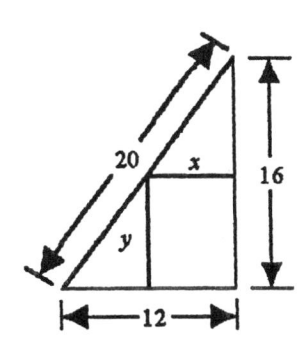

64. The infield of a 440-yard track consists of a rectangle and two semicircles as shown below. To what dimensions should the track be built in order to maximize the area of the rectangle?

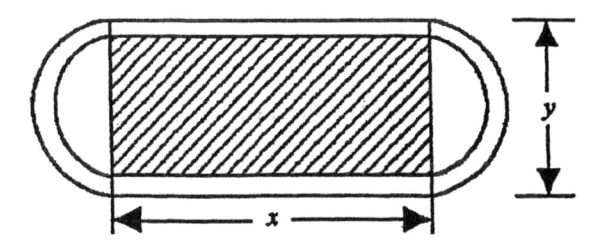

65. Find the dimensions of the largest circular cylinder that can be inscribed in a hemisphere of radius 1.

66. A long strip of copper 8 inches wide is to be made into a rain gutter by turning up the sides to form a trough with a rectangular cross section. Find the dimensions of the cross section if the carrying capacity of the trough is to be a maximum.

67. An isosceles triangle is drawn with its vertex at the origin and its base parallel to the x-axis. The vertices of the base are on the curve $5y = 25 - x^2$. Find the area of the largest such triangle.

72 **Calculus: One and Several Variables**

68. The strength of a beam with a rectangular cross section varies directly as x and as the square of y. What are the dimensions of the strongest beam that can be sawed out of a round log whose diameter is d? See the figure below.

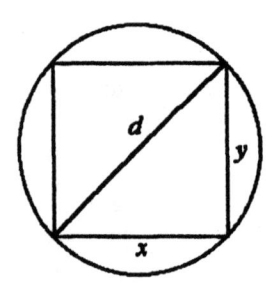

69. Find the area of the largest possible isosceles triangle with 2 sides equal to 6.

70. A lighthouse is 8 miles off a straight coast and a town is located 18 miles down the seacoast. Supplies are to be moved from the town to the lighthouse on a regular basis and at a minimum time. If the supplies can be moved at the rate of 7 miles/hour on water and 25 miles/hour over land, how far from the town should a dock be constructed for shipment of supplies?

71. Find the circular cylinder of largest lateral area that can be inscribed in a sphere of radius 4 feet. [Surface area of a cylinder, $S = 2\pi rh$, where r = radius, h = height.]

72. If three sides of a trapezoid are 10 inches long, how long should the fourth side be if the area is to be a maximum? [Area of a trapezoid $= (a+b)h/2$ where a and b are the lengths of the parallel sides and h = height.]

73. The stiffness of a beam of rectangular cross section is proportional to the product xy^3. Find the stiffest beam that can be cut from a round log 2 feet in diameter. See the figure below.

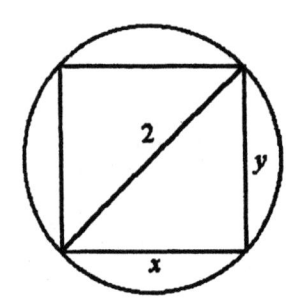

74. Find the dimensions of the maximum rectangular area that can be laid out within a triangle of base 12 and altitude 4 if one side of the rectangle lies on the base of the triangle.

75. Find the dimensions of the rectangle of greatest area with its base on the x-axis and its other two vertices above the x-axis and on $4y = 16 - x^2$.

76. Find the dimensions of the trapezoid of greatest area with its longer base on the x-axis and its other two vertices above the x-axis on $4y = 16 - x^2$. [Area of a trapezoid $= (a+b)h/2$ where a and b are the lengths of the parallel sides and h = height.]

77. A poster is to contain 50 in.^{2} of printed matter with margins of 4 inches each at top and bottom and 2 inches at each side. Find the overall dimensions if the total area of the poster is to be a minimum.

The Mean-Value Theorem and Applications

78. A rancher is going to build a three-sided cattle enclosure with a divider down the middle as shown below.

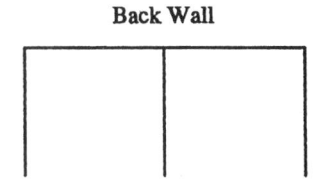

Back Wall

The cost per foot of the three side walls will be $6/foot, while the back wall, being taller, will be $10/foot. If the rancher wishes to enclose an area of 180 ft², what dimensions of the enclosure will minimize his cost?

79. A can containing 16 in.³ of tuna and water is to be made in the form of a circular cylinder. What dimensions of the can will require the least amount of material? ($V = \pi r^2 h$, $S = 2\pi rh$, $A = \pi r^2$)

80. Find the maximum sum of two numbers given that the first plus the square of the second is equal to 30.

81. An open-top shipping crate with a square bottom and rectangular sides is to hold 32 in.³ and requires a minimum amount of cardboard. Find the most economical dimensions.

82. Find the minimum distance from the point (3, 0) to $y = \sqrt{x}$.

83. The product of two positive numbers is 48. Find the numbers if the sum of one number and the cube of the other is to be minimized.

84. Find the values for x and y such that their product is a minimum, if $y = 2x - 10$.

85. A container with a square base, vertical sides, and an open top is to be made from 192 ft² of material. Find the dimensions of the container with greatest volume.

86. The cost of fuel used in propelling a dirigible varies as the square of its speed and costs $200/hour when the speed is 100 miles/hour. Other expenses amount to $300/hour. Find the most economical speed for a voyage of 1,000 miles.

87. A rectangular garden is to be laid out with one side adjoining a neighbor's lot and is to contain 675 ft². If the neighbor agrees to pay for half of the dividing fence, what should the dimensions of the garden be to ensure a minimum cost of enclosure?.

88. A rectangle is to have an area of 32 in². What should be its dimensions if the distance from one corner to the midpoint of the nonadjacent side is to be a minimum?

89. A slice of pizza, in the form of a sector of a circle, is to have a perimeter of 24 inches. What should be the radius of the pan to make the slice of pizza the largest? (*Hint:* The area of a sector of circle is $A = r^2\theta/2$ where θ is the central angle in radians and the arc length along a circle is $C = r\theta$ with θ in radians.)

90. Find the minimum value for the slope of the tangent to the curve of $f(x) = x^5 + x^3 - 2x$.

91. A line is drawn through the point $P(3, 4)$ so that it intersects the y-axis at $A(0, y)$ and the x-axis at $B(x,0)$. Find the triangle formed if x and y are positive.

92. An open cylindrical trash can is to hold 6 ft³ of material. What should be its dimension if the cost of material used is to be a minimum? [Surface area, $S = 2\pi rh$ where r = radius and h = height.]

93. Two fences, 16 feet apart, are to be constructed so that the first fence is 2 feet high and the second fence is higher than the first. What is the length of the shortest pole that has one end on the ground, passes over the first fence and reaches the second fence? See the figure below.

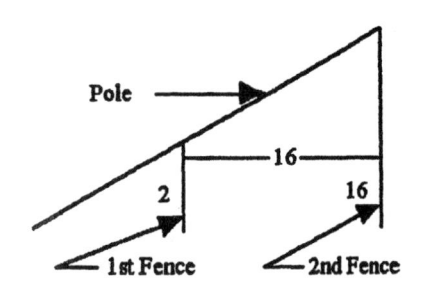

94. A line is drawn through the point (3, 4) so that it intersects the y-axis at $A(0, y)$ and the x-axis at $B(x, 0)$. Find the equation of the line through AB if the triangle is to have a minimum area and both x and y are positive.

4.6 Concavity and Points of Inflection

95. Describe the concavity of the graph of $f(x) = x^4 - 24x^2$ and find the points of inflection, if any.

96. Describe the concavity of the graph of $f(x) = x^4 - 4x^3$ and find the points of inflection, if any.

97. Describe the concavity of the graph of $f(x) = x^4 - 6x^2 + 2$ and find the points of inflection, if any.

98. Describe the concavity of the graph of $f(x) = 5x^4 - x^5$ and find the points of inflection, if any.

99. Describe the concavity of the graph of $f(x) = 4x^3 - 15x^2 - 18x + 10$ and find the points of inflection, if any.

100. Describe the concavity of the graph of $f(x) = x(x - 6)^2$ and find the points of inflection, if any.

101. Describe the concavity of the graph of $f(x) = x^3 - 5x^2 + 3x + 1$ and find the points of inflection, if any.

102. Describe the concavity of the graph of $f(x) = 3x^4 - 4x^3 + 1$ and find the points of inflection, if any.

103. Describe the concavity of the graph of $f(x) = x^2 + 2/x$ and find the points of inflection, if any.

104. Describe the concavity of the graph of $f(x) = (x - 2)^3 + 1$ and find the points of inflection, if any.

105. Describe the concavity of the graph of $f(x) = (x - 4)^4 + 4$ and find the points of inflection, if any.

106. Describe the concavity of the graph of $f(x) = \sin 2x$, $x \in (0, \pi)$ and find the points of inflection, if any.

4.7 Vertical and Horizontal Asymptotes; Vertical Tangents and Cusps

107. Find the vertical and horizontal asymptotes for $f(x) = \left(\dfrac{x-3}{x-1}\right)^2$.

108. Find the vertical and horizontal asymptotes for $f(x) = \dfrac{x^2}{x^2+1}$.

109. Find the vertical and horizontal asymptotes for $f(x) = \dfrac{x^2 - x}{(x+1)^2}$.

110. Find the vertical and horizontal asymptotes for $f(x) = \dfrac{3x^2}{x^2 - 4}$.

111. Find the vertical and horizontal asymptotes for $f(x) = \dfrac{8}{4 - x^2}$.

112. Find the vertical and horizontal asymptotes for $f(x) = \dfrac{x^2}{x^2 - 9}$.

113. Find the vertical and horizontal asymptotes for $f(x) = \dfrac{x - 1}{x - 2}$.

114. Determine whether the graph of $f(x) = 1 + (x - 2)^{1/3}$ has a vertical tangent or a vertical cusp at $c = 2$.

115. Determine whether the graph of $f(x) = (x + 1)^{1/3}(x - 4)$ has a vertical tangent or a vertical cusp at $c = -1$.

116. Determine whether the graph of $f(x) = (x + 1)^{2/3}$ has a vertical tangent or a vertical cusp at $c = -1$.

117. Determine whether the graph of $f(x) = (x - 2)^{2/3} - 1$ has a vertical tangent or a vertical cusp at $c = 2$.

4.8 Some Curve Sketching

When graphing the following functions, you need not indicate the extrema or inflection points, but show all asymptotes (vertical, horizontal, or oblique).

118. Sketch the graph of $f(x) = 5 - 2x - x^2$.

119. Sketch the graph of $f(x) = x^3 - 9x^2 + 24x - 7$.

120. Sketch the graph of $f(x) = x^3 + 6x^2$.

121. Sketch the graph of $f(x) = x^3 - 5x^2 + 8x - 4$.

122. Sketch the graph of $f(x) = x^3 - 12x^2 + 6$.

123. Sketch the graph of $f(x) = x^3 - 6x^2 + 9x + 6$.

124. Sketch the graph of $f(x) = 3x^4 - 4x^3 + 1$.

125. Sketch the graph of $f(x) = x^2(9 - x^2)$.

126. Sketch the graph of $f(x) = x^4 - 2x^2 + 7$.

127. Sketch the graph of $f(x) = x^3 + \dfrac{3}{2}x^2 - 6x + 12$.

128. Sketch the graph of $f(x) = x^{1/3}(x + 4)$.

129. Sketch the graph of $f(x) = x^{2/3}(x + 5)$.

130. Sketch the graph of $f(x) = x(x - 3)^{2/3}$.

131. Sketch the graph of $f(x) = \sqrt{1 - x}$.

132. Sketch the graph of $f(x) = \sqrt{1 - x^2}$.

133. Sketch the graph of $f(x) = \sqrt{4 - x^2}$.

134. Sketch the graph of $f(x) = \sqrt{\dfrac{x}{4-x}}$.

135. Sketch the graph of $f(x) = \dfrac{x-3}{x+2}$.

136. Sketch the graph of $f(x) = \dfrac{x^3 - 1}{3x^2 - 3x - 6}$.

Answers to Chapter 4, Part B Questions

1. $c = \pm\sqrt{\dfrac{7}{3}} = \pm\sqrt{\dfrac{21}{3}}$

2. $c = 1/2$

3. Since f is not differentiable at $x = 0$, which is in $(-1, 1)$, the function $f(x)$ does not satisfy the conditions of the mean-value theorem.

4. $c = 1$

5. $c = \pm\dfrac{2\sqrt{3}}{3}$

6. $c = \sqrt{\dfrac{7}{3}} = \dfrac{\sqrt{21}}{3}$

7. $c = \dfrac{3 \pm \sqrt{3}}{3}$

8. $c = 1$

9. Since f is not differentiable at $x = 0$, which is in $(-1, 1)$, the function $f(x)$ does not satisfy the conditions of the mean-value theorem.

10. $c = \pm\dfrac{\sqrt{3}}{3}$

11. $c = \pm\dfrac{2\sqrt{3}}{3}$

12. $c = \dfrac{\sqrt{3}}{9}$

13. f increases on $[-2\sqrt{3},\ 0]$ and $[2\sqrt{3}, \infty]$
    f decreases on $[-\infty,\ -2\sqrt{3}]$ and $[0, 2\sqrt{3}]$

14. f increases on $[3, \infty)$
    f decreases on $(-\infty,\ 0)$ and $[0, 3]$

15. f increases on $[-\sqrt{3},\ 0]$ and $[\sqrt{3}, \infty]$
    f decreases on $[-\infty,\ -\sqrt{3}]$ and $[0,\ \sqrt{3}]$

16. f increases on $[0, 4]$
    f decreases on $(-\infty,\ 0]$ and $[4,\ \infty)$

17. f increases on $(-\infty,\ -1/2]$ and $[3,\ \infty)$
    f decreases on $[-1/2,\ 3]$

18. f increases on $(-\infty,\ 2]$ and $[6,\ \infty)$
    f decreases on $[2, 6]$

19. f increases on $[1, \infty)$
    f decreases on $(-\infty, 0)$ and $(0, 1]$

20. f increases on $[4, \infty)$
    f decreases on $(-\infty, 4]$

21. f increases on $[0, \pi/4]$ and $[3\pi/4, \pi]$
    f decreases on $[\pi/4, 3\pi/4]$

22. $f(x) = x^3 - 5x^2 + 3x + 1$

23. $f(x) = 3x^4 - 4x^3 + 1$

24. $f(x) = (x - 2)^3 + 9$

25. f increases on $(-\infty, 0)$ and $[3, \infty)$
    f decreases on $[0, 3)$

26.

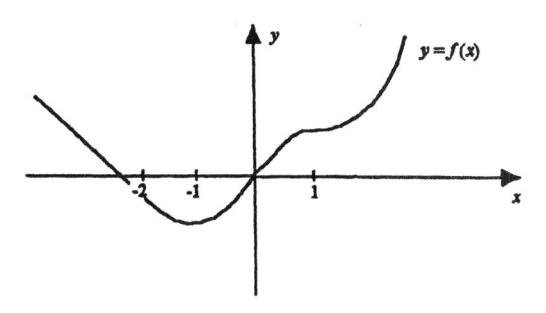

27. impossible

28.

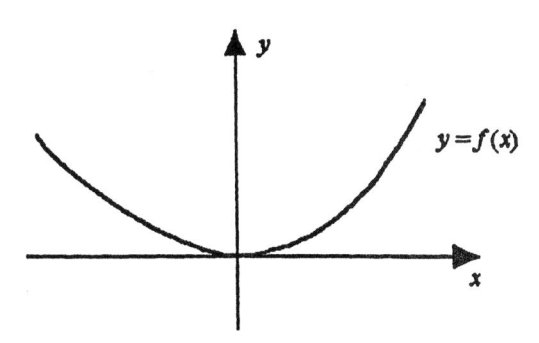

29. critical numbers $x = 0,\ 4/3$;
    local maximum $f(0) = 0$;
    local minimum $f(4/3) = -256/81$

30. critical numbers $x = 0,\ 1/8$;
    local maximum $f(1/8) = 1$;
    local minimum $f(0) = 0$

31. critical numbers $x = 0,\ 2$;
    local maximum $f(2) = 3(2)^{2/3}$;
    local minimum $f(0) = 0$

32. critical numbers $x = 0$, 1;
    local minimum $f(1) = -1$;
    no local extreme at $x = 0$

33. critical numbers $x = -2$, 0, 2;
    local minimum $f(-2) = -3$;
    local maximum $f(0) = 1$;
    local minimum $f(2) = -3$

34. critical numbers $x = -1/2$, 1;
    local minimum $f(-1/2) = -27/16$;
    no local extreme at $x = 1$

35. critical numbers $x = -8/27$, 1;
    local maximum $f(-8/27) = 8/27$;
    local minimum $f(0) = 0$

36. critical numbers $x = -1$, 1;
    local minimum $f(-1) = -2/3$;
    local maximum $f(1) = 2/3$

37. critical number $x = -1$;
    local minimum $f(-1) = 5$

38. critical number $x = 3/4$;
    local minimum $f(3/4) = 13/8$

39. critical numbers $x = -\sqrt{3}$, 0, $\sqrt{3}$;
    local minimum $f(-\sqrt{3}) = 8$;
    local maximum $f(0) = 17$;
    local minimum $f(\sqrt{3}) = 8$

40. no critical numbers; no local extreme values

41. critical number $x = -1$;
    local maximum $f(-1) = 0$

42. critical numbers $x = \pi/6$, $5\pi/6$;
    local minimum $f(\pi/6) = \pi/6 - \sqrt{3}/2$;
    local maximum $f(5\pi/6) = 5\pi/6 + \sqrt{3}/2$

43. $f'(x) = 3x^2 - 8x + 2$, $f'(0) = 2$, $f'(1) = -3$,
    so f' has at least one zero in $(0,1)$. $f''(x) = 6x - 8$
    < 0 on $(0, 1)$ so f' is decreasing on $(0, 1)$, and
    therefore, it has exactly one zero in $(0, 1)$. Hence
    f has exactly one critical number in $(0, 1)$.

44. no critical numbers; $f(0) = 2$ endpoint minimum
    and absolute minimum; $f(100) = 52$ endpoint
    maximum and absolute maximum

45. critical numbers $x = -1$, 2; $f(-2) = 4$ endpoint
    minimum; $f(-1) = 15$ local and absolute
    maximum; $f(2) = -12$ endpoint and absolute
    minimum

46. critical numbers $x = -1$, 3 but $x = 3$ is outside
    the interval; $f(-1) = 8/3$ endpoint and absolute
    maximum; $f(2) = -19/3$ endpoint and absolute
    minimum

47. critical numbers $x = 0$, 4; $f(-1) = -2$ endpoint
    minimum; $f(0) = 5$ local and absolute maximum;
    $f(4) = -27$ endpoint and absolute minimum;
    $f(5) = -20$ endpoint maximum

48. critical numbers $x = -1$, 2 but $x = -1$ is outside
    the interval; $f(0) = 5$ endpoint maximum;
    $f(2) = -15$ local and absolute minimum;
    $f(4) = 37$ local and absolute maximum

49. critical numbers $x = -1/2$, $3/2$; $f(-1) = -1$
    endpoint minimum; $f(-1/2) = 5/2$ local and
    absolute maximum; $f(3/2) = -27/2$ local and
    absolute minimum; $f(2) = -10$ endpoint
    maximum

50. critical numbers $x = -1$, 1 but $x = -1$ is outside
    the interval; $f(0) = 6$ endpoint maximum;
    $f(1) = 4$ local and absolute minimum;
    $f(3/2) = 39/8$ endpoint minimum

51. critical number $x = 0$; $f(-1) = 0$ endpoint
    minimum; $f(0) = 1$ absolute maximum; $f(1) = 0$
    endpoint minimum

52. critical numbers $x = -2$, 2; $f(-4) = -8$
    endpoint and absolute minimum; $f(-2) = 24$
    local and absolute maximum; $f(2) = -8$ local and
    absolute minimum; $f(3) = -1$ endpoint maximum

53. critical numbers $x = 3/4$, 0; $f(-1) = 4$ endpoint
    maximum; $f(3/4) = 9/4(3/4)^{1/3} \approx -2.04$ local
    and absolute minimum; $f(8) = 10$ endpoint and
    absolute maximum

54. critical numbers $x = -1$, 0, 1 but $x = -1$, 0 are
    outside the interval; $f(1) = 1/4$ local and absolute
    maximum; no minimum

55. critical numbers $x = -1$, 1 but $x = -1$ is outside
    the interval; $f(0) =$ endpoint and absolute
    minimum; $f(1) = \frac{1}{2}$ local and absolute
    maximum; $f(2) = 2/5$ endpoint minimum

56. critical number $x = \frac{1}{2}$; $f(\frac{1}{2}) = 4$ local and
    absolute minimum; no maximum

57. critical number $x = 0$; $f(x) = 0$ local and absolute
    minimum; no maximum

58. critical numbers $x = -1$, 0, 1; $f(-2) = 1$
    endpoint maximum; $f(-1) = 0$ local and absolute

minimum; $f(0) = 1$ local and absolute maximum; $f(1) = 0$ local and absolute minimum; $f(2) = 1$ endpoint maximum

59. critical number $x = 0$; $f(-2) = -5$ endpoint minimum; $f(1) = 0$ endpoint maximum

60. $a\sqrt{2}$ by $a\sqrt{2}$

61. $\sqrt{2}$ by $\sqrt{2}/2$

62. 1000 ft

63. $x = 6$, $y = 8$

64. $x = 110$, $y = 220/\pi$

65. $\sqrt{3}/3$ by $\sqrt{6}/3$

66. 2 by 4

67. $50\sqrt{3}/9$

68. $x = \dfrac{\sqrt{3}}{3}d$, $y = \dfrac{\sqrt{6}}{3}d$

69. 18

70. 15 2/3 mi

71. largest lateral area = 32π $h = 4\sqrt{2}$, $r = 2\sqrt{2}$

72. 20

73. $x = 1$, $y = \sqrt{3}$

74. 6 by 2

75. $x = \dfrac{4\sqrt{3}}{3}$, $y = \dfrac{8}{3}$; $\dfrac{8\sqrt{3}}{3}$ by $\dfrac{8}{3}$

76. $x = 4/3$, $y = 32/9$; vertices of the trapezoid are at $(-4, 0)$, $(-4/3, 32/9)$, $(4/3, 32/9)$, $(4, 0)$; Lengths of the bases are 8 8/3. Lengths of the sides are 40/9 and 40/9.

77. 5 in. by 10 in.

78. 18 ft by 10 ft

79. $r = \dfrac{2}{\sqrt[3]{\pi}}$ in., $h = \dfrac{4}{\sqrt[3]{\pi}}$ in.

80. Numbers are $1/2$ and $119/4$.

81. 4 in. by 4 in. by 2 in.

82. Minimum distance is $\dfrac{26}{\sqrt{2}}$, when $x = \dfrac{5}{2}$ and $y = \sqrt{\dfrac{5}{2}}$

83. Numbers are 24 and 2.

84. $x = 5/2$ and $y = -5$

85. 8 ft by 8 ft by 4 ft

86. $50\sqrt{6}$ miles/hr

87. 30 ft by 45/2 ft

88. 4 in. by 8 in.

89. $r = 6$ in.

90. Minimum slope of the tangent is -2 when $x = 0$.

91. $x = 6$ and $y = 8$

92. $r = \sqrt[3]{\dfrac{6}{\pi}}$ ft and $h = \sqrt[3]{\dfrac{6}{\pi}}$ ft

93. $10\sqrt{5}$ ft

94. $3y - 4x - 48 = 0$

95. concave up on $(-\infty, -2)$; concave down on $(-2, 2)$; concave up on $(2, \infty)$; points of inflection $(-2, -80)$ and $(2, 80)$

96. concave up on $(-\infty, -0)$; concave down on $(0, 2)$; concave up on $(2, \infty)$; points of inflection $(0,0)$ and $(2, -16)$

97. concave up on $(-\infty, -1)$; concave down on $(-1, 1)$; concave up on $(1, \infty)$; points of inflection $(-1, -3)$ and $(1, -3)$

98. concave up on $(-\infty, 0)$; concave up on $(0, 3)$; concave down on $(3, \infty)$; point of inflection $(3, 162)$

99. concave down on $(-\infty, 5/4)$; concave up on $(5/4, \infty)$; point of inflection $(5/4, -28\ 1/4)$

100. concave down on $(-\infty, 4)$; concave up on $(4, \infty)$; point of inflection $(4, 16)$

101. concave down on $(-\infty, 5/3)$; concave up on $(5/3, \infty)$; point of inflection $(5/3, -88/27)$

80   Calculus: One and Several Variables

102. concave up on $(-\infty, 0)$; concave down on $(0, 2/3)$; concave up on $(2/3, \infty)$; points of inflection $(0,1)$ and $(2/3, 11/27)$

103. concave up on $(-\infty, -\sqrt[3]{2})$; concave down on $(-\sqrt[3]{2}, 0)$; concave up on $(0, \infty)$; point of inflection $(-\sqrt[3]{2}, 0)$

104. concave down on $(-\infty, 2)$; concave up on $(2, \infty)$; point of inflection $(2, 1)$

105. concave down on $(-\infty, 4)$; concave up on $(4, \infty)$; no points of inflection

106. concave down on $(0, \pi/2)$; concave up on $(\pi/2, \pi)$; point of inflection $(\pi/2, 0)$

107. vertical: $x = 1$; horizontal: $y = 1$

108. vertical: none; horizontal: $y = 1$

109. vertical: $x = -1$; horizontal: $y = 1$

110. vertical: $x = \pm 2$; horizontal: $y = 3$

111. vertical: $x = \pm 2$; horizontal: $y = 0$

112. vertical: $x = \pm 3$; horizontal: $y = 1$

113. vertical: $x = 2$; horizontal: $y = 1$

114. vertical tangent at $(2, 1)$

115. vertical tangent at $(-1, 0)$

116. cusp at $(-1, 0)$

117. cusp at $(2, -1)$

118. $f(x) = 5 - 2x - x^2$

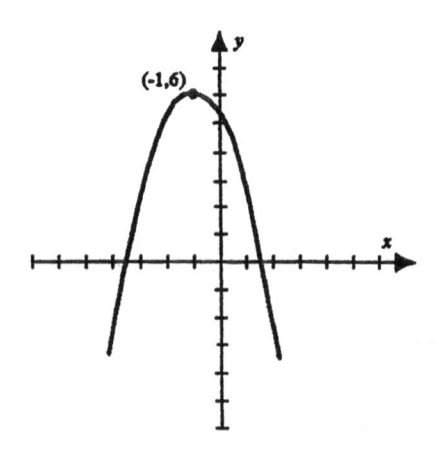

119. $f(x) = x^3 - 9x^2 + 24x - 7$

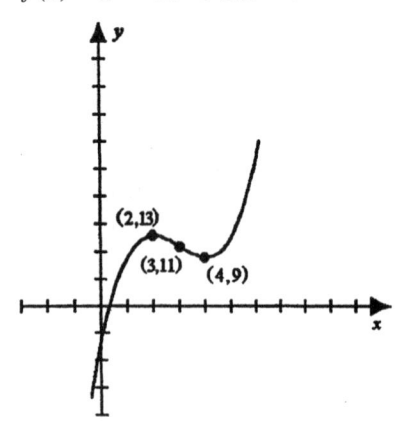

120. $f(x) = x^3 + 6x^2$

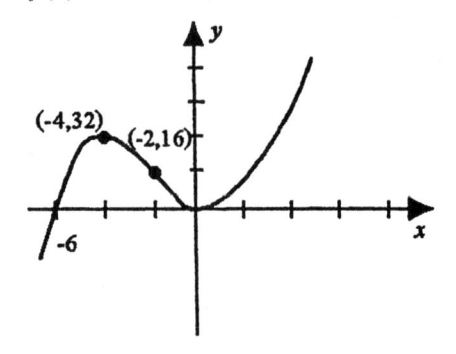

121. $f(x) = x^3 - 5x^2 + 8x - 4$

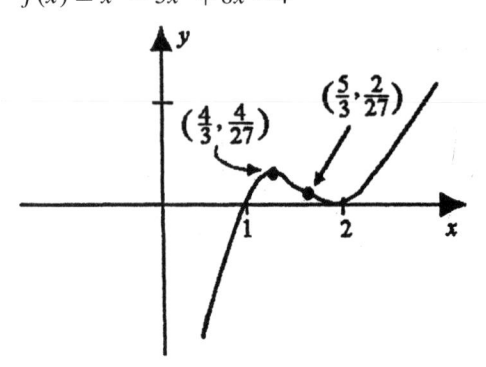

122. $f(x) = x^3 - 12x + 6$

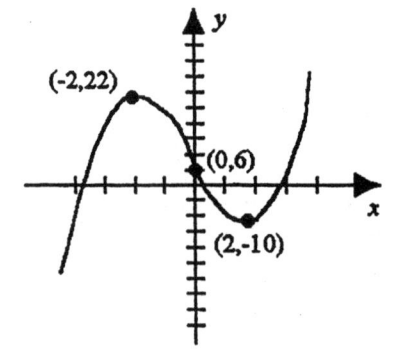

Answers   81

123. $f(x) = x^3 - 6x^2 + 9x + 6$

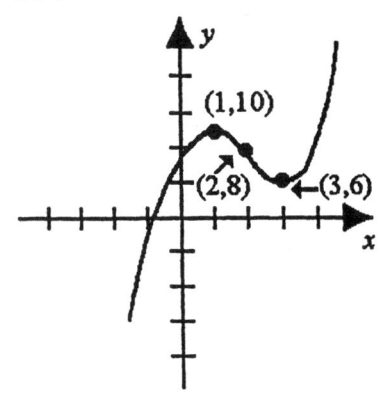

(1,10)
(2,8)
(3,6)

124. $f(x) = 3x^4 - 4x^3 + 1$

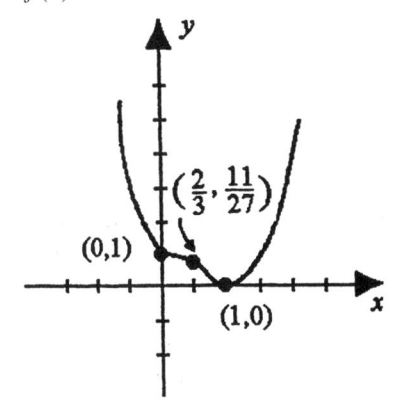

$(\frac{2}{3}, \frac{11}{27})$
(0,1)
(1,0)

125. $f(x) = x^2(9 - x^2)$

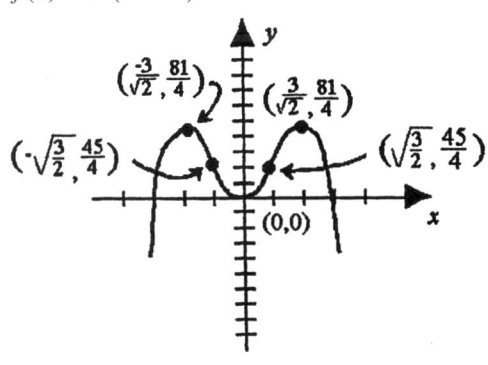

$(\frac{3}{\sqrt{2}}, \frac{81}{4})$ $(\frac{3}{\sqrt{2}}, \frac{81}{4})$
$(-\sqrt{\frac{3}{2}}, \frac{45}{4})$ $(\sqrt{\frac{3}{2}}, \frac{45}{4})$
(0,0)

126. $f(x) = x^4 - 2x^2 + 7$

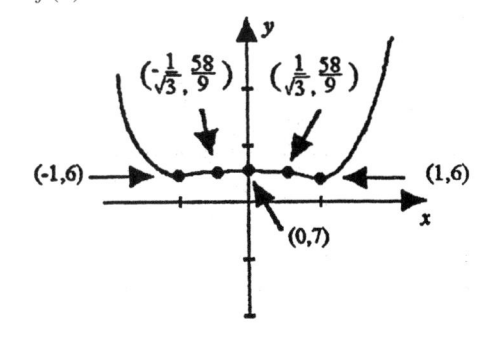

$(-\frac{1}{\sqrt{3}}, \frac{58}{9})$ $(\frac{1}{\sqrt{3}}, \frac{58}{9})$
(-1,6)   (1,6)
(0,7)

127. $f(x) = x^3 + \frac{3}{2}x^2 - 6x + 12$

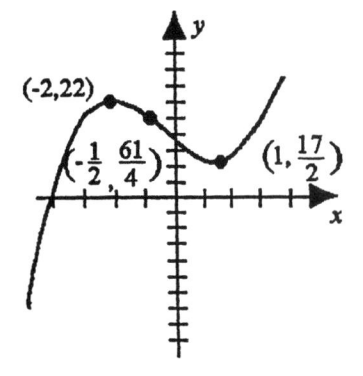

(-2,22)
$(-\frac{1}{2}, \frac{61}{4})$
$(1, \frac{17}{2})$

128. $f(x) = x^{1/3}(x + 4)$

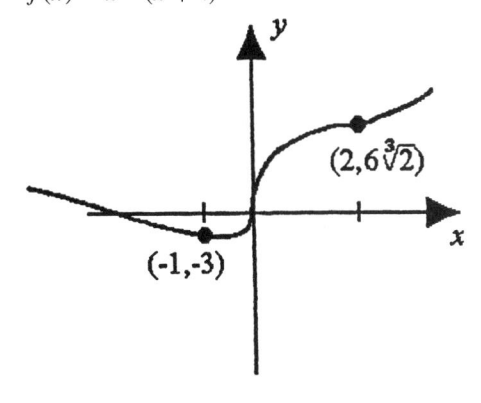

$(2, 6\sqrt[3]{2})$
(-1,-3)

129. $f(x) = x^{2/3}(x + 5)$

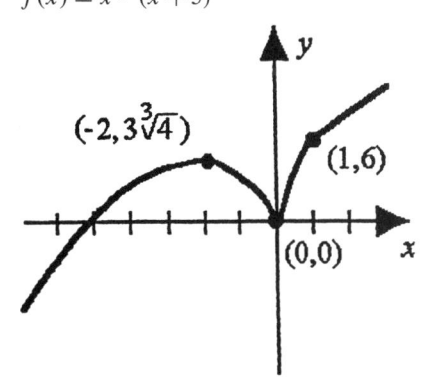

$(-2, 3\sqrt[3]{4})$
(1,6)
(0,0)

130. $f(x) = x(x - 3)^{2/3}$

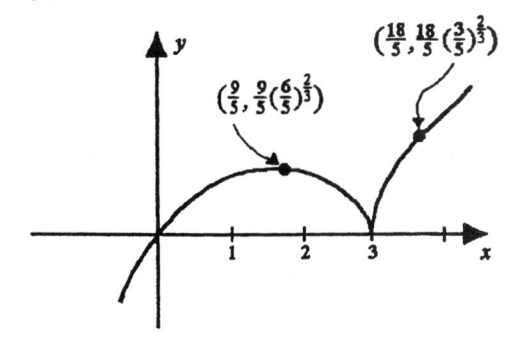

$(\frac{18}{5}, \frac{18}{5}(\frac{3}{5})^{\frac{2}{3}})$
$(\frac{9}{5}, \frac{9}{5}(\frac{6}{5})^{\frac{2}{3}})$

131. $f(x) = \sqrt{1+x}$

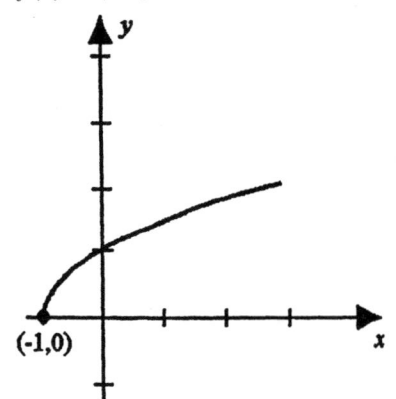

132. $f(x) = \sqrt{1-x^2}$

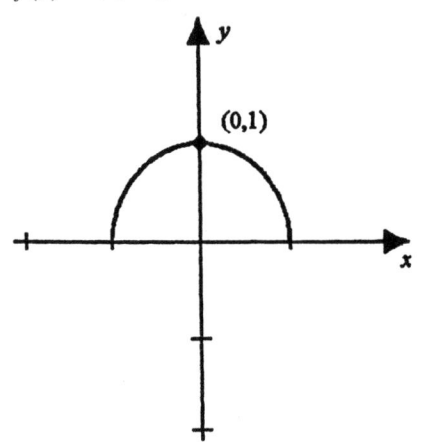

133. $f(x) = \sqrt{4-x^2}$

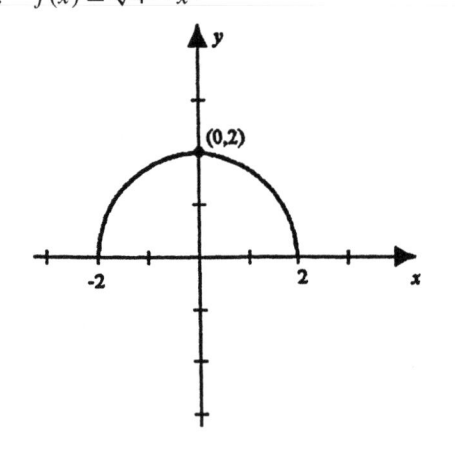

134. $f(x) = \sqrt{\dfrac{x}{4-x}}$

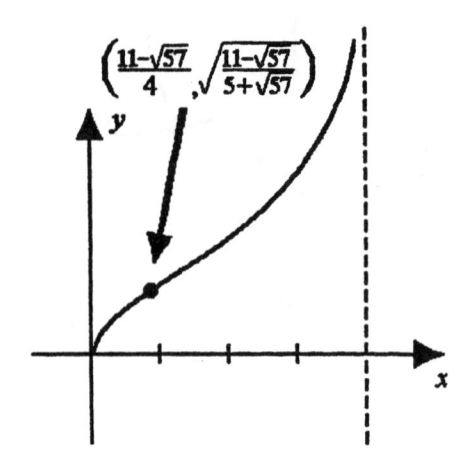

135. $y = \dfrac{x-3}{x+2}$

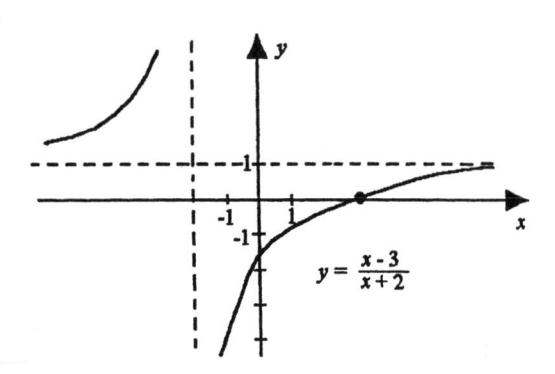

136. $y = \dfrac{x}{3}$

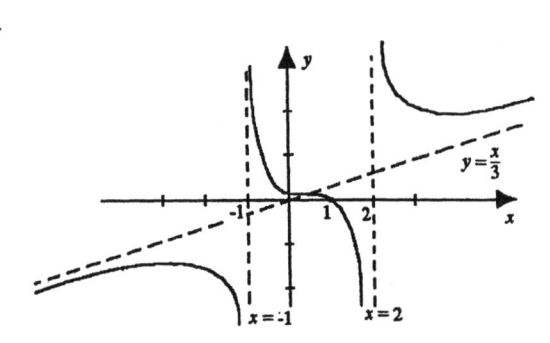

Part C

1. Use the mean-value theorem to show that, if f is continuous at x and $x + h$ and differentiable between them, then $f(x + h) - f(x) = f'(x + \theta h)h$ for some number θ between 0 and 1. (4.1)

2. Let $f(x) = x^3 - 3a^2 x + b$, $a > 0$. Show that $f(x) = 0$ for at most one x in $[-a, a]$. (4.1)

3. Prove that for all real x and y space, $|\sin x - \sin y| \leq |x - y|$. (4.2)

4. Sketch the graph of f, a differentiable function, that satisfies the given conditions, if possible. If not possible, explain why. f has x intercepts only at $x = -1$ and $x = 2$, $f(3) = 4$, and $f(5) = -1$ (4.3)

5. Prove that a polynomial of degree n has at most $n - 1$ local extreme values. (4.3)

6. Let P be a nonconstant polynomical with $a_n > 0$ $p(x) = a_n x^n + a_{n-1} x^{n-1} + \cdots + a_1 x + a_0$ $n \geq 1$. Show that as $x \to \infty$ $p(x) \to \infty$ by showing that, given any $M > 0$ there exist $K > 0$ such that, if $x \geq k$, then $f(x) \geq M$. (4.4)

7. The sum of 2 numbers is 16. Find the numbers if the sum of their cubes is an absolute minimum. (4.5)

8. $c_1 < c_2$ and $f(c_1)$, $f(c_2)$ are local maximum. Prove that, if f is continuous on $[c_1, c_2]$ then there exist $c \in (c_1, c_2)$ such that $f(c)$ is a local minimum. (4.4)

9. Find the point(s) on the parabola $y = \dfrac{1}{8} x^2$ closest to the point $(0, 6)$. (4.5)

10. Find A and B given the function $y = Ax^{-1/2} + Bx^{1/2}$ that will have a minimum of 6 at $x = 9$. (4.5)

11. What is the maximum volume for a rectangular box (square base, no top) made from 12 square feet of cardboard? (4.6)

12. What is the maximum possible area for a triangle inscribed in a circle of radius r? (4.6)

13. Sketch the graph of $f(x) = \dfrac{2x^3 + 3x - 2}{x + 1}$ (4.7)

14. Sketch the graph of $f(x) = \dfrac{1}{1 - \cos x}$ $x \in (-\pi, \pi)$ (4.8)

15. Sketch the graph of $f(x) = x - x^{1/3}$

16. Find a so that $y = x^3 - ax^2 + 1$ will have a point of inflection at $x = 1$. (4.6)

17. (a) When is the graph $f(x) = \sin x$ concave up?
    (b) When is the graph $f(x) = \sin x$ concave down? (4.6)

18. The velocity of a wave on the surface of a calm body of liquid is a function on its wavelength λ

$$V = \sqrt{\dfrac{g}{2\pi}\lambda + \dfrac{2\pi\sigma}{\delta\lambda}}$$ (4.5)

Constants g = acceleration due to gravity
σ = surface tension
δ = density of the liquid

Find the minimum speed of a wave and the corresponding wavelength. (4.5)

19. Prove if the perimeter of an isosceles triangle is fixed, the triangle of greatest area is an equilateral triangle. (4.5)

20. If the sum of the surface area of a sphere and cube is fixed, what is the ratio of the edge of the cube to the radius of the sphere in order to maximize the volumes? (4.5)

21. Find $a > 0$ so that $a - a^3$ is a maximum. (4.5)

22. Find two positive numbers a, b so that $a + b = 30$ and ab^4 is a maximum. (4.5)

23. Given a point (x_0, y_0) and a line $ax + by + c = 0$ use calculus to prove the shortest distance from (x, y) to $ax + by + c = 0$ is $\dfrac{|ax_0 + by_0 + c|}{\sqrt{a^2 + b^2}}$ (4.5)

Answers to Chapter 4, Part C Questions

1. Pf now if $h > 0$, if f is differentiable on $(x, x+h)$ and continuous by mean-value theorem
$$\exists c \in (x, x+h) \frac{f(x+h) - f(x)}{x+h-x} = f'(c)$$
$$\Rightarrow f(x+h) - f(x) = f'(c)h$$
Since $c \in (x, x+h)$, so c can be expressed as
$$c = x + \theta h \text{ with } 0 < \theta < 1$$
hence
$$f(x+h) - f(x) = f'(x + \theta h)h$$
the situation when $h < 0$ is similar.

2. $f'(x) = 3x^2 - 3a^2 = 3(x^2 - a^2)$ so when $x \in (-a, a)$ $x^2 - a^2 < 0$ so $f'(x) < 0$ also $f(x)$ is continuous and differentiable so by Rolles theorem if
$$\exists b, c, f(b) = f(c) = 0 \ b \in [-a, a] \quad c \in [-a, a]$$
then these must be some $d \in (b, c)$, such that $f'(d) = 0$ but this contradicts
$$f'(x) < 0 \quad \text{for} \quad x \in (-a, a).$$

3. Let $f(x) = \sin x$ if $\forall x, y x < y$, by mean-value theorem
$$\frac{|f(y) - f(x)|}{|y - x|} = |f'(c)| \le |\cos c| \le 1$$
$$\Rightarrow |\sin y - \sin x| \le |y - x|$$
same as $|\sin x - \sin y| \le |x - y|$

4. Impossible; f must have intercepts x on $(3, 5)$ by intermediate value theorem.

5. Pf. if p is a polynomial of degree n, the p' must has degree $n - 1$, so p' has at most $n - 1$ different zeros, hence p has at most $n - 1$ local extreme values.

6. Given any $M > 0$ then
$$P(x) - M \ge a_n x^n - (|a_{n-1}|x^{n-1} + |a_{n-2}|x^{n-2} + \cdots + |a_1|x + |a_0| + M) \quad \text{for } x > 0$$
for $x > 1 \Rightarrow P(x) - M \ge a_n x^n - (|a_{n-1}| + |a_{n-2}| + \cdots + |a_1| + |a_0| + M)$
It now follows that
$$P(x) - M \ge 0 \quad \text{for } x \ge k,$$
$$k = \left(\frac{|a_{n-1}| + |a_{n-2}| + \cdots |a_1| + |a_0| + M}{a_n}\right) + 1$$
Hence if $x \ge k$ then $f(x) \ge M$

7. $x + y = 16 \Rightarrow y = 16 - x$
Let $c = x^3 + y^3$
$\Rightarrow c(x) = x^3 + (16 - x)^3$

$c'(x) = 3x^{2+3}(16 - x)^2 = 96(x - 8)$
$c'(x) = 0 \Rightarrow x = 8 \Rightarrow y = 8$
$c''(x) = 6x + 6(16 - x) \quad c''(8) = 96$
$\Rightarrow c$ has local minimum at $x = 8$ it turns out $c(8)$ is the absolute minimum

8. Suppose no such c in (c_1, c_2) is a local minimum f continuous hence f has minimum on $[c_1, c_2]$ thus either c_1 or c_2 admits the minimum. Suppose $f(c_1)$ is the endpoint minimum, then for some $\delta_1 > 0$
$$f(x) \ge f(c_1), x \in [c_1, c_1 + \delta_1]$$
and since $f(c_1)$ is a local maximum $\exists \delta_2 > 0$ such that
$$f(x) \le f(c_1), \text{ for } x \in [c_1 - \delta_1, c_1 + \delta_2]$$
let $\delta = \min(\delta_1, \delta_2)$ so
$$f(c_1) \le f(x) \le f(c_1) \to f(x) = f(c_1)$$
where $x \in (c_1, c_1 + \delta)$
This means f has a local minimum on (c_1, c_2); the argument at c_2 is similar.

9. Let d be the distance, it is sufficient the square of d admits the minimum.
$$s = d^2 = (x - 0)^2 + (y - 6)^2 = 8y + (y - 6)^2$$
since $y = \frac{1}{8} x^2$
$$s'(y) = 2y - 4 = 0 \quad \Rightarrow y = 2.$$
so the closest points on the parabola are $(4, 2)(-4, 2)$

10. $\dfrac{dy}{dx} = -\dfrac{A}{2x^{3/2}} + \dfrac{B}{2x^{1/2}}$ when $x = 9$, $\dfrac{dy}{dx} = 0$
$$\to -\frac{A}{54} + \frac{B}{6} = 0 \tag{1}$$
also $y = Ax^{-1/2} + Bx^{1/2}$ when $x = 9$ $y = 6$
$$\to \frac{1}{3} A + 3B = 6 \tag{2}$$
solving (1), (2) gives: $A = 9 \ B = 1$

11. $V = x^2 h$ maximize at $x^2 + 4xh = 12$
$$\Rightarrow h = \frac{12 - x^2}{4x}$$
$$V(x) = x^2 \left(\frac{12 - x^2}{4x}\right) = 3x - \frac{1}{4} x^3$$
$0 < x < \sqrt{12}$
$V'(x) = 3 - \dfrac{3}{4} x^2, \quad V'(x) = 0 \to x = 2$
so V has an absolute max at $x = 2$, the maximum volume is $V(2) = 4 \text{ ft}^3$

12. $A(x) = \frac{1}{2}(r+x) \cdot 2\sqrt{r^2 - x^2}$
    $= (r+x)\sqrt{r^2 - x^2} \quad 0 \le x \le r$
    $A'(x) = \frac{r^2 - rx - 2x^2}{\sqrt{r^2 - x^2}}$
    $A'(x) = 0 \Rightarrow x = \frac{r}{2}$

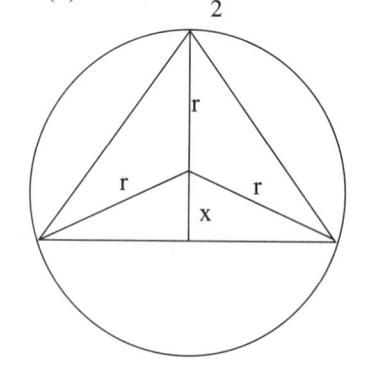

13. vertical asymptote $x = -1$
    oblique asymptote: $y = 2x + 3$

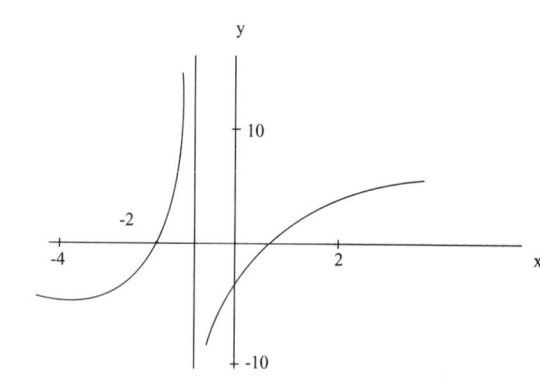

14. $f(x) = \frac{1}{1 - \cos x} \quad xt(-\pi, \pi)$
    $f'(x) = \frac{-\sin x}{(1 - \cos x)^2}$
    $f''(x) = \frac{1 - \cos x - \cos^2 x}{(1 - \cos x)^3}$
    asymptote $x = 0$

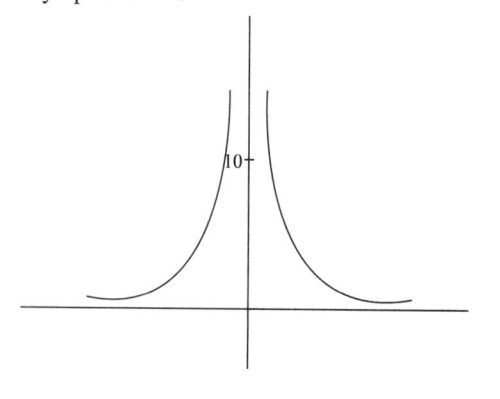

15. $f(x) = x - x^{1/3}$
    $f'(x) = 1 - \frac{1}{3}x^{-2/3} \Rightarrow x = \pm\frac{1}{9}\sqrt{3}$
    $f''(x) = \frac{2}{9}x^{-5/3}$
    vertical tangent at (0, 0)

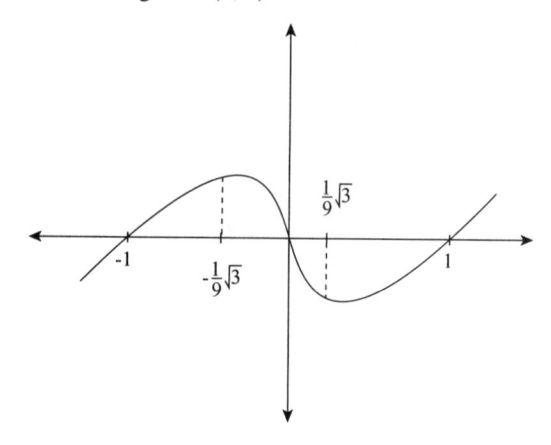

16. Let $f(x) = x^3 - ax^2 + 1$
    $f'(x) = 3x^2 - 2ax \quad f''(x) = 6x - 2a$
    $f''(1) = 6 - 2a = 0 \quad \therefore \quad a = 3$

17. $f'(x) = \cos x \quad f''(x) = -\sin x = -f(x)$
    thus f is concave down when
    $f''(x) < 0 \rightarrow f(x) > 0$
    same, f concave up when $f(x) < 0$

18. Let
    $$z = V^2 = \frac{g}{2\pi}\lambda + 2\pi\frac{\sigma}{\delta\lambda}$$
    $$\frac{dz}{d\lambda} = \frac{g}{2\pi} - \frac{2\pi\sigma}{\delta\lambda^2} = \frac{g\delta\lambda^2 - 4\pi^2\sigma}{2\pi\delta\lambda^2}$$
    Set $\frac{dz}{d\lambda} = 0 \Rightarrow \lambda = 2\pi\left(\frac{\sigma}{g\delta}\right)^{1/2}$
    thus minimum velocity v
    $= \left[\frac{g}{2\pi} \cdot 2\pi\left(\frac{\sigma}{g\delta}\right)^{1/2} + \frac{2\pi\sigma}{\delta} \cdot \frac{1}{2\pi}\left(\frac{g\delta}{\sigma}\right)^{1/2}\right]^{1/2}$
    $= \left(\frac{4g\sigma}{\delta}\right)^{1/4}$

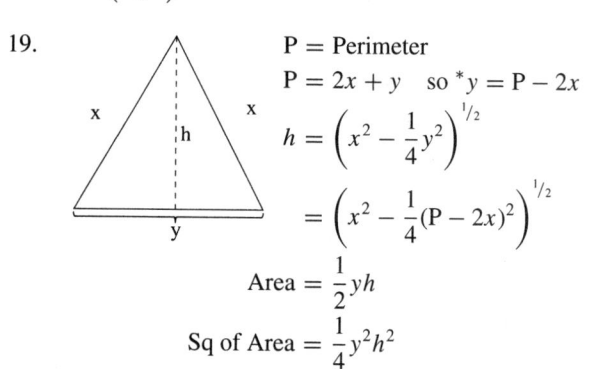

19. $P = $ Perimeter
    $P = 2x + y \quad$ so $^*y = P - 2x$
    $h = \left(x^2 - \frac{1}{4}y^2\right)^{1/2}$
    $= \left(x^2 - \frac{1}{4}(P - 2x)^2\right)^{1/2}$
    Area $= \frac{1}{2}yh$
    Sq of Area $= \frac{1}{4}y^2h^2$

Answers

$$= \frac{1}{4}(P-2x)^2 \left[x^2 - \frac{1}{4}(P-2x)^2 \right]$$

$$= \frac{1}{16}(P-2x)^2(4Px - P^2)$$

$$\frac{d(\text{Sq of Area})}{dx} = -\frac{1}{2}P(P-2x)(3x - P)$$

$= 0$ when $x = P/3$

Now $x < P/3$ is positive
$x > P/3$ is negative
So at $P/3$ we have the maximum

and $x = P/3$
$3x = P$
*$y = P - 2x$
$y = 3x - 2x$
$y = x$ and the triangle is equilateral

20. Fixed
$S = 4\pi R^2 + 6x^2$ max $x = 0$
$V = x^3 + \frac{4}{3}\pi R^3$ min $\frac{x}{2r} = 1$

$$X = \sqrt{\frac{S - 4\pi R^2}{6}}$$

$$V = \left(\frac{S - 4\pi R^2}{6} \right)^{3/2} + \frac{4}{3}\pi R^3$$

$$V'(R) = \frac{3}{2}\left(\frac{S - 4\pi R^2}{6} \right)^{1/2} \cdot \left(-\frac{4}{3}\pi \right) R$$

$$+ 4\pi R^2 = 0$$

$$\therefore R = \sqrt{\frac{S}{24 + 4\pi}}$$

$$X = \sqrt{\frac{S - 4\pi \left(\frac{S}{24 + 4\pi} \right)}{6}}$$

$$\frac{X}{R} = \sqrt{\frac{S \cdot \frac{24 + 4\pi}{S} - 4\pi}{6}} = 2$$

21. $f(a) = a - a^3$ $f'(a) = 1 - 3a^2 = 0$

$a > 0 \therefore a = \sqrt{\frac{1}{3}}.$

22. Let $S = ab^4$

$a + b = 30 \to a = 30 - b$

so
$S(b) = (30 - b)b^4 = 30b^4 - b^5$
$S'(b) = 120 b^3 - 5b^4 = b^3(120 - 5b) = 0$

so $b = 0$ or $b = 24$

when $a = 6$ $b = 24$ S admits maximum.

23. (x_0, y_0) the distance between (x_0, y_0) and a point (x, y) on $ax + by + c = 0$ is

$$f(x) = s^2 = (x_0 - x)^2 + (y_0 - y)^2$$

where $y = -\frac{a}{b}x - \frac{c}{b}$

$$f'(x) = -2(x_0 - x) + 2 \cdot \frac{a}{b}\left(y_0 + \frac{a}{b}x + \frac{c}{b} \right) = 0$$

$$\frac{f'(y)}{2} = -x_0 + x + \frac{a}{b}y_0 + \frac{a^2}{b^2}x + \frac{ac}{b^2} = 0$$

$$x = \frac{b^2 x_0 - aby_0 - ac}{a^2 + b^2}$$

also $y = \frac{a^2 y_0 - abx_0 - bc}{a^2 + b^2}$

$\therefore S^2 = (x_0 - x)^2 + (y_0 - y)^2$

$$= \left(x_0 - \frac{b^2 x_0 - aby_0 - ac}{a^2 + b^2} \right)^2$$

$$+ \left(y_0 - \frac{a^2 y_0 - abx_0 - bc}{a^2 + b^2} \right)^2$$

$$= \frac{a^2(ax_0 + by_0 + c)^2 + b^2(ax_0 + by_0 + c)^2}{(a^2 + b^2)^2}$$

$$= \frac{(a^2 + b^2)(ax_0 + by_0 + c)^2}{(a^2 + b^2)^2}$$

$$= \frac{(ax_0 + by_0 + c)^2}{a^2 + b^2}$$

$$\therefore S = \sqrt{\frac{(ax_0 + by_0 + c)^2}{a^2 + b^2}} = \frac{|ax_0 + by_0 + c|}{\sqrt{a^2 + b^2}}$$

CHAPTER 5 Integration

Part A

1. Given [2, 4] which of the following is a partition of [2, 4]? (5.1)
   (a) {2, 2.1, 2.2, ... 3.9}
   (b) {2, 2.01, 2.02, ... 4.00}
   (c) {1.999, 2.000, 2.001, ..., 4.000}

2. True or false
   (a) If f is continuous on $[a, b]$ then f will attain m a minimum value and M a maximum value.
   (b) M can never equal m.

3. Define $\int_a^b f(x)\,dx$ using $H_f(P)$ and $U_f(P)$ (5.2)

4. True or false
   If $f(x) = c$, c a constant, for all $x \in [a, b]$
   then $\int_a^b f(x)\,dx = -c(a - b)$

5. Explain why both statements are true. (5.2)
   Given $f(x) = x$ $x \in [0, 4]$ then the area bounded by $f(x)$ and $x = 0$, $x = 4$ is
   (1) $\frac{1}{2}(4)(4)$ or (2) $\int_0^4 f(x)\,dx = \frac{1}{2}(4^2 - 0^2)$

6. Given the interval Δx_i and $f(x)$ continuous and increasing on Δx_i (5.2)

   relate $f(x_{i-1})\Delta x_i$, $f(x_i^*)\Delta x_i$, $f(x_i)\Delta x_i$

7. Given $f(x) = 2x + 1$ on $[1, 3]$ and partition (5.2)
   $P = \{1, 2, 3\}$ (5.3)
   (a) find $L_f(P)$ (d) Does $L_f(P) = U_f D$?
   (b) find $U_f(P)$ (e) Can $L_f(P) = S^*(P)$?
   (c) find an $S^*(P)$ (f) relate $L_f(P)$, $U_f(P)$ and $S^*(P)$

8. Given $f(x) = x$ on $[0, 4]$ (5.3)
   Partition $P = \{1, 2, 3, 4\}$
   Partition $Q = \{1, 1.5, 2, 2.5, 3, 3.5, 4\}$
   Find:
   (a) $L_f(P)$ (d) $U_f(Q)$
   (b) $L_f(Q)$ (e) $U_f(P)$
   (c) relate $L_f(P)$ and $L_f(Q)$ (f) relate $U_f(Q)$ and $U_f(P)$

9. True or false
Given $f(x) = x^2$ (5.2)

(a) $\int_{-2}^{1} x^2\, dx = \int_{-2}^{0} x^2\, dx + \int_{0}^{1} x^2\, dx$

(b) $\int_{-2}^{1} x^2\, dx = \int_{1}^{2} x^2\, dx$

(c) $\int_{-100}^{100} x^2\, dx = 0$

(d) $\int_{1}^{-2} x^2\, dx = -\int_{-2}^{1} x^2\, dx$

(e) $F(x) = \int_{0}^{x} t^2\, dt,\ F'(x) = x^2$

10. Given $\int_{0}^{4} f(x)\, dx = 10,\ \int_{2}^{4} f(x)\, dx = 6,\ \int_{0}^{6} f(x)\, dx = 12$ (5.3)
find the following:

(a) $\int_{4}^{6} f(x)\, dx =$

(b) $\int_{4}^{2} f(x)\, dx =$

(c) $\int_{0}^{2} f(x)\, dx =$

(d) $\int_{2}^{2} f(x)\, dx =$

11. Given $F(x) = \int_{0}^{x} (t^2 + 2t - 1)\, dt$ (5.3)
find:
(a) $F'(0) =$ (b) $F'(1) =$ (c) $F'(-1) =$

12. Given $f(x) = \sin x$ find two different antiderivatives (5.3)
$F(x),\ G(x)$ for $f(x)$
(a) $F(x) =$ (b) $G(x) =$

13. True or false
The value of $\int_{0}^{2} x^2\, dx$ can be found using either of the following: (5.4)

(a) $F(x) = \dfrac{x^3}{3} - 1$ (b) $G(x) = \dfrac{x^3}{3} + 5$

14. For $f(x) = 2x$ find $\int_{2}^{5} f(x)\, dx$ using the Fundamental Theorem of Integral Calculus (5.4)

15. State the Fundamental Theorem of Integral Calculus (5.4)

16. Find:

(a) $\int_{0}^{1} x\, dx$

(b) $\int_{0}^{1} 2x\, dx$

(c) $\int_{0}^{1} 100x\, dx$

(d) $\int_{0}^{1} \alpha x\, dx,\ \alpha$ a constant (5.4)

17. True or false (5.4)
$f(x) = 1\quad g(x) = \cos x\quad h(x) = \sin x$

$\int_{0}^{\pi/2} f(x)\, dx = \int_{0}^{\pi/2} g(x)\, dx = \int_{0}^{\pi/2} \sin x\, dx$

18. True or false (5.4)
    If $f(x)$ is an even function

    (a) $\int_{-a}^{a} f(x)\,dx = 0$

    (b) $\int_{-a}^{a} f(x)\,dx = 2\int_{0}^{a} f(x)\,dx$

19. True or false (5.5)
    If $f(x)$ is an odd function then the area from $[-a,\ a]\ a > 0$ equals

    (a) $\int_{-a}^{a} f(x)\,dx$

    (b) $\int_{-a}^{a} f(x)\,dx = \int_{-a}^{a} |f(x)|\,dx$

    (c) $\int_{-a}^{a} f(x)\,dx = 2\int_{0}^{a} f(x)\,dx$

20. True or false (5.5)
    For any function $f(x)$, continuous $\left|\int_{a}^{b} f(x)\,dx\right| = \int_{a}^{b} |f(x)|\,dx$

21. Find the limits of integration that will determine the area bounded by $y = 4$ and $y = x^2 + 3x$. (5.6)

22. Find:

    (a) $\int x\,dx$ (d) $\int \cos x\,dx$

    (b) $\int (x^2 + 2x - 4)\,dx$ (e) $\int \sec x \tan x\,dx$

    (c) $\int \sin x\,dx$

23. Given $a(t) = t,\ v(1) = 2,\ s(1) = 6\dfrac{1}{6}$ find the equation of motion. (5.6)

24. For $\int (2x + 3)^4\,dx$ let $u = $ _____ (5.7)

    then $du = $ _____

25. For $\int \dfrac{x}{\sqrt{1+x^2}}\,dx$ let $u = $ _____ (5.7)

    $du = $ _____

26. For $\int \sin(5x - \pi)\,dx$ let $u = $ _____ (5.7)

    $du = $ _____

27. For $\int (a + x^n)^b$ we need $u = $ _____ (5.7)

    For $\int (a + x^u)^b$ _____ dx $du = $ _____

    a, b constants $b \neq 0$

28. If $f(x) \leq g(x)$ for all $x \in [a, b]$ then (5.7)

    $\int_{a}^{b} f(x)\,dx \leq \int_{a}^{b} g(x)\,dx$ is true iff

    $f(x)$ and $g(x) \geq 0$

29. True or false
v(t) is the speed of an object at any time t. (5.6)

30. The equation of motion for an object that moves along a straight line with constant acceleration, a, from an initial position, x_0, with initial velocity, V_0, is $x(t) = \frac{1}{2}at^2 + V_0 t + x_0$.
If an object is free falling explain the values (5.6)
$a =$
$V_0 =$
$x_0 =$

31. Can you find (5.7)
   (a) $\int \dfrac{dx}{(2 - 3x)^{1/2}}$ how ? let $u =$
   (b) $\int \dfrac{\sin x}{\cos 5x} dx$ how ? let $u =$
   (c) $\int \dfrac{17 dx}{(2x - 1)^{3/2}}$ how ?
   (d) $\int \dfrac{3x^2 - 6}{(x^3 - 3x^2)^{10}} dx$ how ?

32. True or false
Given $\int_a^b f(x)\, dx > \int_a^b g(x)\, dx$ (5.8)
   (a) $f(x) > g(x)$ for all $x \in [a, b]$
   (b) $\int_a^b |f(x)|\, dx > \int_a^b |g(x)|\, dx$

33. True or false
   (a) $\int_a^b f(x) = 0$ iff $f(x) = 0$ (5.8)
   (b) $\int_a^b |f(x)|\, dx = 0$ iff $f(x) = 0$

34. State the Mean-Value Theorem for Integrals (5.9)

35. $\dfrac{\int_a^b f(x)\, dx}{b - a}$ is called the _____ of f on $[a, b]$ (5.9)

36. for $f(x) = x$ on $[0, 4]$
find c such that $\int_0^4 f(x)\, dx = f(c)(4 - 0)$ (5.9)

Answers to Chapter 5, Part A Questions

1. b

2. (a) T   (b) F

3. The unique number I that satisfies the inequality $L_f(P) \leq I \leq U_1(P)$ for all partitions P of $[a,b]$

4. T

5. $\int_0^4 f(x) = \frac{1}{2}x^2 \Big|_0^4 = \frac{1}{2}(4)(4)$

6. $\int_a^b f(x)\,dx = \lim_{\|p\|\to 0} [f(x_1^*)\Delta x_1 + f(x_2^*)\Delta x_2 + \cdots + f(x_n^*)\Delta x_n)]$

7. (a) $L_f(P) = 3 \cdot 1 + 5 = 8$
   (b) $U_f(P) = 5 + 7 = 12$
   (c) $(1.5 \times 2) + 1 + (2.5 \times 2) + 1 = 10$
   (d) no
   (e) yes
   (f) $L_f(P) \leq S^*(P) \leq U_f(P)$

8. (a) $L_f(P) = 1 + 2 + 3 = 6$
   (b) $L_f(Q) = (1 + 1.5 + 2 + 2.5 + 3 + 3.5)/2 = 6.75$
   (c) $L_f(P) \leq L_f(Q)$
   (d) $U_f(Q) = (1.5 + 2 + 2.5 + 3 + 3.5 + 4)/2 = 8.25$
   (e) $U_f(P) = 2 + 3 + 4 = 9$
   (f) $L_f(P) < L_f(Q) < U_f(Q) < U_f(P)$

9. (a) T   (d) T
   (b) F   (e) T
   (c) F

10. (a) 2   (c) +4
    (b) −6   (d) 0

11. (a) −1   (b) 2   (c) −2

12. (a) $-\cos x$
    (b) $1 - \cos x$ other answer $-\cos x + c$

13. (a) T   (b) T

14. 21

15. Let f be continuous on $[a,b]$. If G is any anti-derivative for f on $[a,b]$, then
    $$\int_a^b f(t)\,dt = G(b) - G(a)$$

16. (a) $\frac{1}{2}$ (c) 50
    (b) 1   (d) $\frac{\alpha}{2}$

17. T

18. (a) F   (b) T

19. (a) F   (b) F   (c) T

20. F

21. $\int_{-4}^{1} (4 - (x^2 + 3x)\,dx$

22. (a) $\frac{x^2}{2} + C$ (c) $-\cos x + c$
    (b) $\frac{x^3}{3} + x^2 - 4x + c$ (d) $\sin x + c$
         (e) $\sec x + c$

23. $x(t) = \frac{1}{6}t^3 + \frac{3}{2}t + \frac{9}{2}$

24. $u = 2x + 3$
    $du = 2\,dx$

25. $u = x^2$
    $du = 2x\,dx$

26. $u = 5x - \pi$
    $du = 5\,dx$

27. $u = a + x^n$
    $du = nx^{n-1}\,dx$

28. T

29. F

30. $a = -g$
    $V_0 = 0$
    $X_0 = h$

31. (a) $\int \frac{dx}{(2-3x)^{1/2}} \quad \begin{array}{l} u = 2 - 3x \\ du = -3dx \end{array} -\frac{1}{3} \int \frac{du}{u^{1/2}}$
    $= -\frac{2}{3} u^{1/2} + c$
    $= -\frac{2}{3}(2 - 3x)^{1/2} + c$

    (b) $\int \frac{\sin x}{\cos^5 x}\,dx \quad \begin{array}{l} u = \cos x \\ du = -\sin x\,dx \end{array} -\int \frac{du}{u^5}$
    $= \frac{1}{4}u^{-4} + c = \frac{1}{4}\cos^4 x + c$

(c) $\displaystyle\int \frac{17\,dx}{(2x-1)^{3/2}} \quad \begin{array}{l} u = 2x-1 \\ du = 2dx \end{array} \quad \frac{1}{2}\int \frac{17\,du}{u^{3/2}}$

$\displaystyle = \frac{1}{2} \times \frac{2}{5} \times u^{5/2} \times 17 + c$

$\displaystyle = \frac{17}{5}(2x-1)^{5/2} + c$

(d) $\displaystyle\int \frac{3x^2 - 6x}{(x^3 - 3x^2)^{10}}\,dx \quad \begin{array}{l} u = x^3 - 3x^2 \\ du = (3x^2 - 6x)9 \end{array} \int \frac{du}{u^{10}}$

$\displaystyle = -\frac{1}{9}u^{-9} + c = -\frac{1}{9}(x^3 - 3x^2)^{-9} + c$

32. (a) F (b) F

33. (a) F (b) T

34. If f is continuous on $[a, b]$ then there is at least one number c in (a, b) for which

$$\int_a^b f(x)\,dx = f(c)(b - a)$$

35. the average value of f on $[a, b]$

36. $\displaystyle f(c) = \frac{\int_0^4 f(x)dx}{4} = \frac{\left[\frac{1}{2}x^2\right]_0^4}{4} = 2 = x$

Part B

5.2 The Definite Integral of a Continuous Function

1. Find $L_f(P)$ and $U_f(P)$ for $f(x) = 3x$, $x \in [-1, 0]$; $P = \left\{-1, \frac{-3}{4}, \frac{-1}{2}, \frac{-1}{4}, 0\right\}$.

2. Find $L_f(P)$ and $U_f(P)$ for $f(x) = 1 + x$, $x \in [0, 1]$; $P = \left\{0, \frac{1}{4}, \frac{1}{2}, 1\right\}$.

3. Find $L_f(P)$ and $U_f(P)$ for $f(x) = 1 + x^2$, $x \in [0, 2]$; $P = \left\{0, \frac{1}{2}, 1, \frac{3}{2}, 2\right\}$.

4. Find $L_f(P)$ and $U_f(P)$ for $f(x) = \sqrt{x-1}$, $x \in [1, 2]$; $P = \left\{1, \frac{5}{4}, \frac{3}{2}, \frac{7}{4}, 2\right\}$.

5. Find $L_f(P)$ and $U_f(P)$ for $f(x) = |x|$, $x \in [0, 1]$; $P = \left\{0, \frac{1}{4}, \frac{3}{4}, 1\right\}$.

6. Find $L_f(P)$ and $U_f(P)$ for $f(x) = x^2$, $x \in [0, 1]$; $P = \left\{0, \frac{1}{3}, \frac{1}{2}, \frac{2}{3}, \frac{3}{4}, 1\right\}$.

5.3 The Function $F(x) = \int_a^x f(t)\,dt$

7. Given that $\int_0^1 f(x)\,dx = 5$, $\int_0^3 f(x)\,dx = 3$, $\int_3^7 f(x)\,dx = 1$ find each of the following:
   (a) $\int_0^7 f(x)\,dx$
   (b) $\int_1^3 f(x)\,dx$
   (c) $\int_1^7 f(x)\,dx$
   (d) $\int_3^0 f(x)\,dx$

8. Given that $\int_1^5 f(x)\,dx = 7$, $\int_3^5 f(x)\,dx = 3$, $\int_1^7 f(x)\,dx = 12$ find each of the following:
   (a) $\int_5^7 f(x)\,dx$
   (b) $\int_1^3 f(x)\,dx$
   (c) $\int_3^7 f(x)\,dx$
   (d) $\int_5^3 f(x)\,dx$

9. For $x > -1$, set $F(x) = \int_0^x \sqrt{2t+2}\,dt$
   (a) Find $F(0)$
   (b) Find $F'(x)$
   (c) Find $F'(1)$

10. For $x > 0$, set $F(x) = \int_1^x \frac{dt}{t+1}$
    (a) Find $F(1)$
    (b) Find $F'(x)$
    (c) Find $F'(1)$

11. For the function $F(x) = \int_0^x \frac{dt}{t^2+4}$
    (a) Find $F'(-1)$
    (b) Find $F'(0)$
    (c) Find $F'(1)$
    (d) Find $F''(x)$

12. For the function $F(x) = \int_0^x t\sqrt{t^2+9}\,dt$
    (a) Find $F'(-1)$
    (b) Find $F'(0)$
    (c) Find $F'(1)$
    (d) Find $F''(x)$

13. For the function $F(x) = \int_0^x t\sqrt{t^2+4}\,dt$
    (a) Find $F'(-1)$
    (b) Find $F'(0)$
    (c) Find $F'(1)$
    (d) Find $F''(x)$

14. For the function $F(x) = \int_0^x (t+2)^2\,dt$
    (a) Find $F'(-1)$
    (b) Find $F'(0)$
    (c) Find $F'(1)$
    (d) Find $F''(x)$

5.4 The Fundamental Theorem of Integral Calculus

15. Evaluate $\int_0^1 (3x-2)\,dx$.

16. Evaluate $\int_{-2}^{-1} 3x^5\,dx$.

17. Evaluate $\int_0^2 (x^2+2x+5)\,dx$.

18. Evaluate $\int_{-1}^1 (x^3-x+5)\,dx$.

19. Evaluate $\int_1^2 \left(\sqrt{x}+\frac{1}{\sqrt{x}}\right)dx$.

20. Evaluate $\int_1^2 \frac{7}{\sqrt{x}}\,dx$.

21. Evaluate $\int_1^2 \frac{x^4+1}{x^3}\,dx$.

22. Evaluate $\int_1^2 \left(x^2+8x+\frac{3}{x^2}\right)dx$.

23. Evaluate $\int_0^1 (3\sqrt{x}+1)\,dx$.

24. Evaluate $\int_1^2 \left(x^2-\frac{3}{x^4}\right)dx$.

25. Evaluate $\int_1^2 \frac{7}{t^4}\,dt$.

26. Evaluate $\int_0^1 (x+1)^{11}\,dx$.

27. Evaluate $\int_0^1 x^3|2x-1|\,dx$.

28. Evaluate $\int_0^1 (x^2+1)^2\,dx$.

29. Evaluate $\int_{\pi/3}^{3\pi/2} \cos x \, dx$.

30. Evaluate $\int_0^{\pi/4} \sin x \, dx$.

31. Evaluate $\int_0^{\pi/4} \sec^2 x \, dx$.

32. Evaluate $\int_0^{\pi/6} \sec x \tan x \, dx$.

33. Evaluate $\int_1^3 f(x) \, dx$, where $f(x) = \begin{cases} (x+1)^2, & 1 \leq x \leq 2 \\ 3 - x^2, & 2 < x \leq 3 \end{cases}$.

34. Evaluate $\int_0^{\pi} f(x) \, dx$, where $f(x) = \begin{cases} x, & 0 \leq x \leq \pi/3 \\ \sin x, & \pi/3 < x \leq \pi \end{cases}$.

35. Find the area between the graph of $f(x) = x^2$ and the x-axis for $x \in [0, 2]$.

36. Find the area between the graph of $f(x) = \dfrac{1}{x^2}$ and the x-axis for $x \in [1, 2]$.

37. Find the area between the graph of $f(x) = x^3 + 2$ and the x-axis for $x \in [1, 4]$.

38. Find the area between the graph of $f(x) = x^2 - x$ and the x-axis for $x \in [3, 8]$.

39. Find the area between the graph of $f(x) = x^2 - x - 6$ and the x-axis for $x \in [0, 2]$.

40. Find the area between the graph of $f(x) = \dfrac{1}{x^2}$ and the x-axis for $x \in [1, 4]$.

41. Find the area between the graph of $f(x) = (2x+1)^{-2}$ and the x-axis for $x \in [0, 2]$.

42. Find the area between the graph of $f(x) = \sin x$ and the x-axis for $x \in [\pi/6, 2\pi/3]$.

43. Find the area between the graph of $f(x) = \sqrt{x+3}$ and the x-axis for $x \in [1, 6]$.

44. Find the area between the graph of $f(x) = 2(x+5)^{-1/2}$ and the x-axis for $x \in [-1, 4]$.

45. Sketch the region bounded by the curves $y = \dfrac{1}{2} x^2$ and $y = x + 4$, and find its area.

46. Sketch the region bounded by the curves $y = x^2$ and $2x - y + 3 = 0$, and find its area.

47. Sketch the region bounded by the curves $x^2 = 8y$ and $x = 2y - 8$, and find its area.

48. Sketch the region bounded by the curves $y = x^2 - 4x + 4$ and $y = x$, and find its area.

49. Sketch the region bounded by the curves $y = x + 5$ and $y = x^2 - 1$, and find its area.

50. Sketch the region bounded by the curves $y = x^3$, $x = -1$, $x = 2$, and $y = 0$, and find its area.

51. Sketch the region bounded by the curves $y = 2 - x^2$ and $y = -x$, and find its area.

52. Sketch the region bounded by the curves $y = x^3 + 1$, $x = -1$, $x = 2$, and the x-axis, and find its area.

5.6 Indefinite Integrals

53. Calculate $\displaystyle\int \frac{dx}{x^5}$.

54. Calculate $\displaystyle\int (3x+2)^2\, dx$.

55. Calculate $\displaystyle\int (2x^2+5)\, dx$.

56. Calculate $\displaystyle\int \frac{2dx}{\sqrt{4x+5}}$.

57. Calculate $\displaystyle\int \frac{x^5+2}{x^7}\, dx$.

58. Calculate $\displaystyle\int \frac{3x^3+2x}{x^2}\, dx$.

59. Calculate $\displaystyle\int \frac{(1+x)^2}{x^{1/2}}\, dx$.

60. Calculate $\displaystyle\int (x^3+2)^2\, dx$.

61. Calculate $\displaystyle\int \frac{x^2-4}{\sqrt[3]{x^2}}\, dx$.

62. Calculate $\displaystyle\int (x+1)\sqrt{x}\, dx$.

63. Calculate $\displaystyle\int (\sqrt{x}+2)^2\, dx$.

64. Find f given that $f'(x) = 3x+1$ and $f(2) = 3$.

65. Find f given that $f'(x) = 2x^2+3x+1$ and $f(0) = 2$.

66. Find f given that $f'(x) = \sin x$ and $f(\pi/2) = 2$.

67. Find f given that $f''(x) = 4x-1$, $f'(1) = 3$, and $f(0) = 1$.

68. Find f given that $f''(x) = x^2+2x$, $f'(0) = 3$, and $f(2) = 3$.

69. Find f given that $f''(x) = \cos x$, $f'(\pi) = 2$, and $f(\pi) = 1$.

70. An object moves along a coordinate line with velocity $v(t) = 2t^2 - 6t - 8$ units per second. Its initial position (position at time $t=0$) is 3 units to the left of the origin.
    (a) Find the position of the object 2 seconds later.
    (b) Find the total distance traveled by the object during those 2 seconds.

71. An object moves along a coordinate line with acceleration $a(t) = \dfrac{1}{2}(t+1)^3$ units per second per second.
    (a) Find the velocity function given that the initial velocity is 4 units per second.
    (b) Find the position function given that the initial velocity is 4 units per second and the initial position is the origin.

72. An object moves along a coordinate line with acceleration $a(t) = (2t+1)^{-1/2}$ units per second.
    (a) Find the velocity function given that the initial velocity is 2 units per second.
    (b) Find the position function given that the initial velocity is 2 units per second and the initial position is the origin.

73. An object moves along a coordinate line with velocity $v(t) = 3 - 2t^2$ units per second. Its initial position is 3 units to the left of the origin.
    (a) Find the position of the object 5 seconds later.
    (b) Find the total distance traveled by the object during those 5 seconds.

74. A ball is rolled across a level floor with an initial velocity of 28 feet per second. How far will the ball roll if the speed diminishes by 4 ft/sec² due to friction?

75. A particle, initially moving at 16 cm/sec, is slowing down at the rate of 0.8 m/sec². How far will the particle travel before coming to rest?

76. A jet plane moves with constant acceleration a from rest to a velocity of 300 ft/sec in a distance of 450 ft. Find a.

77. A particle which starts at the origin moves along the x-axis from time $t = 0$ to time $t = 3$ with velocity $v(t) = t^2 - t - 2$. Determine the final position of the particle and the total distance traveled.

78. A particle which starts at the origin moves along the x-axis from time $t = 0$ to time $t = 6$ with velocity $v(t) = 4 - t$. Determine the final position of the particle and the total distance traveled.

79. A particle which starts at the origin moves along the x-axis from time $t = 0$ to time $t = 5$ with velocity $v(t) = 8 - 2t$. Determine the final position of the particle and the total distance traveled.

80. A particle which starts at the origin moves along the x-axis from time $t = 0$ to time $t = 3$ with velocity $v(t) = t^2 - 3t + 2$. Determine the final position of the particle and the total distance traveled.

81. A particle which starts at the origin moves along the x-axis from time $t = 0$ to time $t = 4$ with velocity $v(t) = t^2 - 4t + 3$. Determine the final position of the particle and the total distance traveled.

82. A particle which starts at the origin moves along the x-axis from time $t = 0$ to time $t = 2$ with velocity $v(t) = t^2 + t - 2$. Determine the final position of the particle and the total distance traveled.

83. A particle which starts at the origin moves along the x-axis from time $t = 1$ to time $t = 3$ with velocity $v(t) = t - 8/t^2$. Determine the final position of the particle and the total distance traveled.

84. A rapid transit trolley moves with a constant acceleration and covers the distance between two points 300 feet apart in 8 seconds. Its velocity as it passes the second point is 50 ft/sec.
    (a) Find its acceleration.
    (b) Find the velocity of the trolley as it passes the first point.

5.7 The u-Substitution; Change of Variables

85. Calculate $\int 3x\sqrt{1 - 2x^2} \, dx$.

86. Calculate $\int t^2(2 - 3t^3)^3 \, dt$.

87. Calculate $\int \dfrac{4x}{\sqrt[3]{8 - x^2}} \, dx$.

88. Calculate $\int x^3 \sqrt{5x^4 - 18}\, dx$.

89. Calculate $\int x\sqrt{x - 5}\, dx$.

90. Calculate $\int \dfrac{dx}{(x+1)^2}$.

91. Calculate $\int (x^2 + 1)(x^3 + 3x)^{10}\, dx$.

92. Calculate $\int x^3 \sqrt{x - 2}\, dx$.

93. Calculate $\int \dfrac{x^2}{\sqrt{x+1}}\, dx$.

94. Calculate $\int \dfrac{x-2}{(x^2 - 4x + 4)^2}\, dx$.

95. Evaluate $\int_0^1 \dfrac{dx}{\sqrt{x+1}}$.

96. Evaluate $\int_1^2 (x^2 + 1)\sqrt{2x^3 + 6x}\, dx$.

97. Evaluate $\int_0^3 \sqrt{x^4 + 2x^2 + 1}\, dx$.

98. Evaluate $\int_0^1 x\sqrt{9x^2 + 16}\, dx$.

99. Evaluate $\int_0^1 \dfrac{x}{(1+x^2)^2}\, dx$.

100. Evaluate $\int_0^3 x\sqrt{9 - x^2}\, dx$.

101. Evaluate $\int_0^4 \dfrac{x}{\sqrt{9+x^2}}\, dx$.

102. Evaluate $\int_1^2 \dfrac{x^3}{\sqrt{3x^4 + 1}}\, dx$.

103. Evaluate $\int_1^2 x^2 \sqrt{x - 1}\, dx$.

104. Evaluate $\int_1^5 x\sqrt{2x - 1}\, dx$.

105. Calculate $\int \sin(5x + 3)\, dx$.

106. Calculate $\int \cos^2(2x + 1)\, dx$.

107. Calculate $\int \cos^2 2x \sin 2x \, dx$.

108. Calculate $\int x^{1/2} \sin x^{3/2} \, dx$.

109. Calculate $\int (2 + \cos 3x)^{3/2} \sin 3x \, dx$.

110. Calculate $\int \dfrac{\cos 2x}{(1 + \sin 2x)^2} \, dx$.

111. Calculate $\int \dfrac{dx}{\cos^2 x}$.

112. Calculate $\int (x^{-2} \sec^2 x + 3 \sin 2x) \, dx$.

113. Calculate $\int \dfrac{dx}{\sin^2 3x}$.

114. Calculate $\int (2 + \sin 3t)^{1/2} \cos 3t \, dt$.

115. Calculate $\int \csc 2t \cot 2t \, dt$.

116. Calculate $\int \tan^3 5x \sec^2 5x \, dx$.

117. Calculate $\int \dfrac{\sin x \, dx}{\cos^3 x}$.

118. Calculate $\int x \sec^2 x^2 \, dx$.

119. Calculate $\int x^3 \sin(x^4 + 2) \, dx$.

120. Evaluate $\int_{\pi/6}^{\pi/3} (\cos t - \csc t \cot t) \, dt$.

121. Evaluate $\int_0^{\pi/4} \sec^2 t \, dt$.

122. Evaluate $\int_{\pi/3}^{2/\pi} (t - \csc t \cot t) \, dt$.

123. Evaluate $\int_{4/\pi}^{2/\pi} \dfrac{\sin(1/t)}{t^2} \, dt$.

124. Evaluate $\int_0^{\pi/4} \cos^2 3t \sin 3t \, dt$.

125. Evaluate $\int_{\sqrt{\pi/3}}^{\sqrt{\pi/2}} \dfrac{t}{\sin^2 (t^2/2)} \, dt$.

126. Evaluate $\int_0^{\pi/8} (2x + \sec 2x \tan 2x)\, dx$.

127. Find the area bounded by $y = \cos \pi x$, $y = \sin \pi x$, $x = 1/4$, and $x = 1/2$.

128. Find the area bounded by $y = \sin x$, $y = 2x/\pi$, and $x = 0$.

129. Find the area bounded by $y = \frac{1}{2} \cos^2 \pi x$, $y = -\sin^2 \pi x$, $x = 0$, and $x = 1/2$.

5.8 Additional Properties of the Definite Integral

130. Calculate $\dfrac{d}{dx}\left[\displaystyle\int_1^x (t^3 + 1)\, dt\right]$.

131. Calculate $\dfrac{d}{dx}\left[\displaystyle\int_0^x (t + 1)^{1/2}\, dt\right]$.

132. Calculate $\dfrac{d}{dx}\left[\displaystyle\int_2^{3x^2} (t^2 - 4)^{2/3}\, dt\right]$.

133. Calculate $\dfrac{d}{dx}\left(\displaystyle\int_x^{x^2} \dfrac{1}{\sqrt{1 - 3t^2}}\, dt\right)$.

134. Calculate $\dfrac{d}{dx}\left(\displaystyle\int_{1-x}^{1+x} \dfrac{dt}{\sqrt{2t + 5}}\right)$.

135. Calculate $\dfrac{d}{dx}\left(\displaystyle\int_{1/x}^{1/\sqrt{x}} \sin t^2\, dt\right)$.

136. Find $H'(2)$ given that $H(x) = \displaystyle\int_{\sqrt{x}}^{x^2+x} \dfrac{3}{2 + \sqrt{2t}}\, dt$.

137. Find $H'(2)$ given that $H(x) = \dfrac{1}{x}\displaystyle\int_2^x [2t + H'(t)]\, dt$.

138. Suppose f is continuous and $\displaystyle\int_a^b |f(x)|\, dx = 0$.

    (a) Does it necessarily follow that $\displaystyle\int_a^b f(x)\, dx = 0$?

    (b) What can you conclude about $f(x)$ on $[a, b]$?

139. Let Ω be the region below the graph of $f(x) = 2x + 3$, $x \in [0, 2]$. Draw a figure showing the Riemann sum $S^*(P)$ as an estimate for this area. Take $P = \left\{0, \dfrac{1}{4}, \dfrac{3}{4}, 1, \dfrac{3}{2}, 2\right\}$ and let the x_i^* be the midpoints of the subintervals. Evaluate the Riemann sum.

140. Set $f(x) = 3x + 1$, $x \in [0, 1]$. Take $P = \left\{0, \dfrac{1}{8}, \dfrac{1}{4}, \dfrac{3}{8}, \dfrac{1}{2}, 1\right\}$ and set $x_1^* = \dfrac{1}{16}$, $x_2^* = \dfrac{3}{16}$, $x_3^* = \dfrac{3}{8}$, $x_4^* = \dfrac{5}{8}$, $x_5^* = \dfrac{3}{4}$. Calculate the following:

    (a) $\Delta x_1, \Delta x_2, \Delta x_3, \Delta x_4, \Delta x_5$
    (b) $\|P\|$

(c) m_1, m_2, m_3, m_4, m_5
(d) $f(x_1^*), f(x_2^*), f(x_3^*), f(x_4^*), f(x_5^*)$
(e) M_1, M_2, M_3, M_4, M_5
(f) $L_f(P)$
(g) $S^*(P)$
(h) $U_f(P)$
(i) $\int_0^1 f(x)\,dx$

5.9 Mean-Value Theorems for Integrals; Average Values

141. Determine the average value of $f(x) = \sqrt{4x+1}$ on the interval $[0, 2]$ and find a point c in this interval at which the function takes on this average value.

142. Determine the average value of $f(x) = x^3\sqrt{3x^4+1}$ on the interval $[-1, 2]$.

143. Determine the average value of $f(x) = x^3 + 1$ on the interval $[0, 2]$ and find a point c in this interval at which the function takes on this average value.

144. Determine the average value of $f(x) = x \cos x^2$ on the interval $[0, \pi/2]$.

145. Determine the average value of $f(x) = \cos x$ on the interval $[0, \pi/2]$ and find a point c in this interval at which the function takes on this average value.

146. Find the average distance of the parabolic arc $y = 2(x+1)^2$, $x \in [0, \sqrt{2}]$ from (a) the x-axis, and (b) the y-axis.

147. A rod lies on the x-axis from $x = 1$ to $x = L > 1$. If the density at any point x on the rod is $3/x^3$, find the center of mass of the rod.

Answers to Chapter 5, Part B Questions

1. $L_f(P) = -15/8$; $U_f(P) = -9/8$
2. $L_f(P) = 21/16$; $U_f(P) = 27/16$
3. $L_f(P) = -15/4$; $U_f(P) = 23/4$
4. $L_f(P) = \dfrac{1+\sqrt{2}+\sqrt{3}}{8} = 0.5183$
   $U_f(P) = \dfrac{3+\sqrt{2}+\sqrt{3}}{8} = 0.7683$
5. $L_f(P) = 13/8$; $U_f(P) = 19/8$
6. $L_f(P) = \dfrac{137}{576} = 0.2378$
   $U_f(P) = \dfrac{259}{576} = 0.4497$
7. (a) 4 (c) -1
   (b) -2 (d) -3
8. (a) 5 (c) 8
   (b) 4 (d) -3
9. (a) 0 (c) 2
   (b) $\sqrt{2x+2}$
10. (a) 0 (c) $\frac{1}{2}$
    (b) $\dfrac{1}{x+1}$
11. (a) 1/5 (c) 1/5
    (b) 1/4 (d) $\dfrac{-2x}{(x^2+4)^2}$
12. (a) $\sqrt{10}$ (c) $\sqrt{10}$
    (b) 3 (d) $\dfrac{x}{x^2+9}$
13. (a) $-\sqrt{5}$ (c) $\sqrt{5}$
    (b) 0 (d) $\dfrac{2x^2+4}{\sqrt{x^2+4}}$
14. (a) 1 (c) 9
    (b) 4 (d) $2(x+2)$
15. $-\frac{1}{2}$
16. $-63/2$
17. 50/3
18. 10
19. $\dfrac{1}{3}(10\sqrt{2}-8)$
20. $14(\sqrt{2}-1)$
21. 15/8
22. 95/6
23. 3
24. 35/24
25. 49/24
26. 1024/3
27. 5/32
28. 28/15
29. -2
30. $\dfrac{1}{2}(2-\sqrt{2})$
31. 1
32. $\dfrac{2-\sqrt{3}}{\sqrt{3}} = \dfrac{2\sqrt{3}-3}{3}$
33. 3
34. $\dfrac{\pi^2+27}{18}$
35. 8/3
36. $\frac{1}{2}$
37. 279/4
38. 805/6
39. 34/3
40. $\frac{3}{4}$
41. 2/5
42. $\dfrac{1+\sqrt{3}}{2}$
43. 38/3
44. 4

45. 18

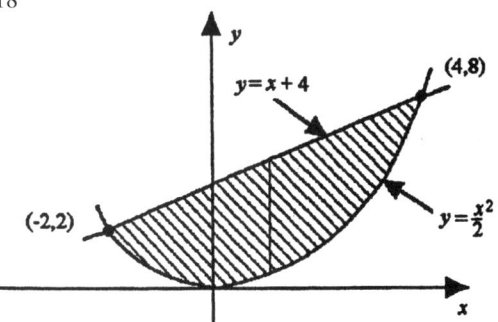

46. 32/3

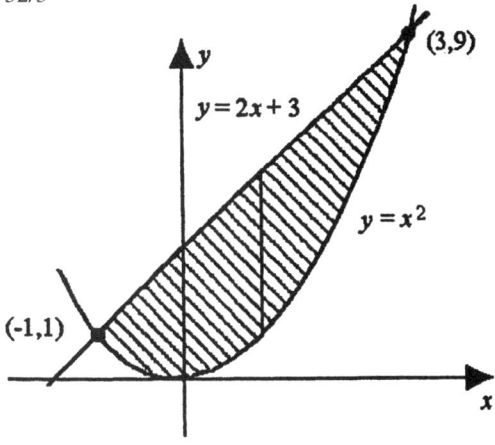

47. 36

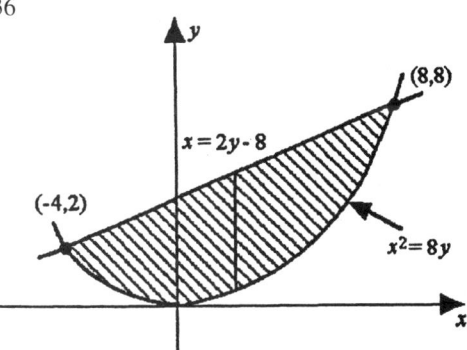

48. 9/2

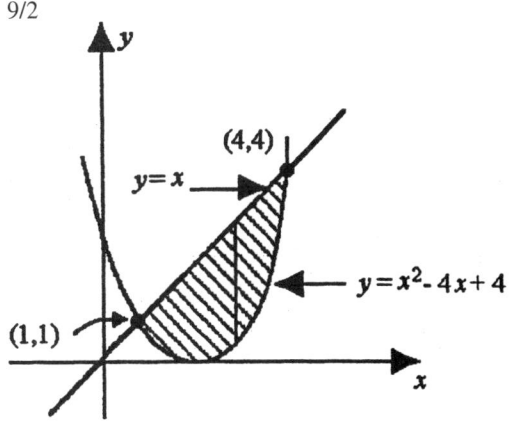

49. 125/6

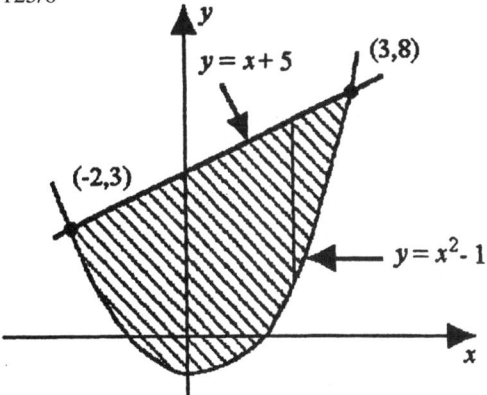

50. 17/4

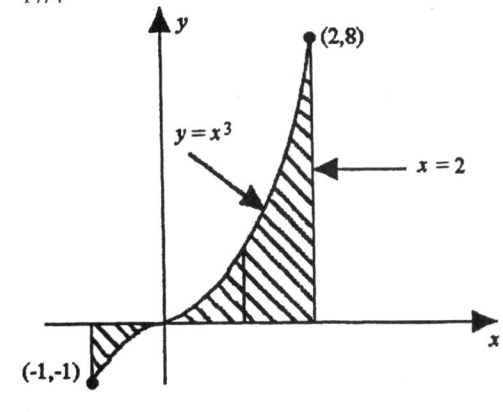

51. 9/2

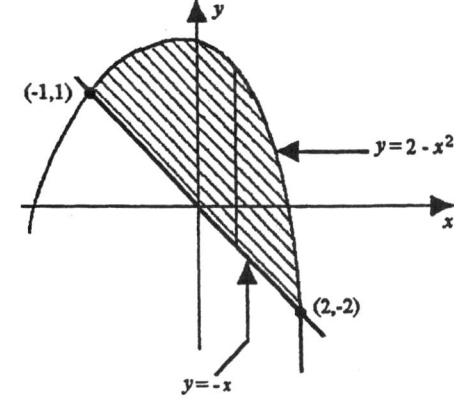

52. 27/4

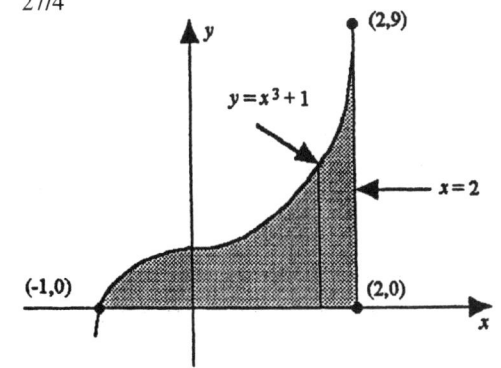

53. $-\dfrac{1}{4x^4} + C$

54. $\dfrac{1}{9}(3x+2)^3 + C$

55. $\dfrac{2}{3}x^3 + 5x + C$

56. $\sqrt{4x+5} + C$

57. $-\dfrac{1}{x} - \dfrac{1}{3x^6} C$

58. $\dfrac{3}{2}x^2 + 2\ln x + C$

59. $2\sqrt{x} + \dfrac{4}{3}x^{3/2} + \dfrac{2}{5}x^{5/2} + C$

60. $\dfrac{x^7}{7} + x^4 + 4x + C$

61. $\dfrac{3}{7}x^{7/3} - 12x^{1/3} + C$

62. $\dfrac{2}{5}x^{5/2} + \dfrac{2}{3}x^{3/2} + C$

63. $\dfrac{x^2}{2} + \dfrac{8}{3}x^{3/2} + 4x + C$

64. $\dfrac{3}{2}x^2 + x - 5$

65. $\dfrac{2}{3}x^3 + \dfrac{3}{2}x^2 + x + 2$

66. $-\cos x + 2$

67. $\dfrac{2}{3}x^3 + \dfrac{x^2}{2} + 2x + 1$

68. $\dfrac{x^4}{12} + \dfrac{x^3}{3} + 3x - 7$

69. $-\cos x + 2x - 2\pi$

70. (a) $-77/3$ (b) $68/3$

71. (a) $\dfrac{1}{8}(t+1)^4 + \dfrac{31}{8}$

    (b) $\dfrac{1}{40}(t+1)^5 + \dfrac{31}{8}t - \dfrac{1}{40}$

72. (a) $\sqrt{2t+1} + 1$

    (b) $\dfrac{1}{3}(2t+1)^{3/2} + t - \dfrac{1}{3}$

73. (a) $-71\ 1/3$ units to the left of the origin

    (b) $\dfrac{205 + 3\sqrt{6}}{3} = 70.78$ units

74. 98 ft

75. 160 cm

76. 100 ft/sec²

77. $-3/2;\ 31/6$

78. 6; 10

79. 15; 17

80. 3/2; 11/6

81. 4/3; 4

82. 2/3; 3

83. $-4/3;\ 11/3$

84. (a) 25/8 ft/sec²   (b) 25 ft/sec

85. $-\dfrac{1}{2}(1-2x^2)^{3/2} + C$

86. $-\dfrac{1}{36}(2-3t^3)^4 + C$

87. $-3(8-x^2)^{2/3} + C$

88. $\dfrac{1}{30}(5x^4 - 18)^{3/2} + C$

89. $\dfrac{2}{5}(x-5)^{5/2} + \dfrac{10}{3}(x-5)^{3/2} + C$

90. $-\dfrac{1}{x+1} + C$

91. $\dfrac{1}{33}(x^3 + 3x)^{11} + C$

92. $\dfrac{2}{9}(x-2)^{9/2} + \dfrac{12}{7}(x-2)^{7/2} + \dfrac{24}{5}(x-2)^{5/2}$
    $+ \dfrac{16}{3}(x-2)^{3/2} + C$

93. $\dfrac{2}{5}(x+1)^{5/2} - \dfrac{4}{3}(x+1)^{3/2} + 2(x+1)^{1/2} + C$

94. $\dfrac{-1}{2(x^2 - 4x + 4)} + C$

Answers

95. $2(\sqrt{2} - 1)$

96. $\frac{1}{9}(28^{3/2} - 8^{3/2}) = 13.95$

97. 12

98. 61/27

99. ¼

100. 9

101. 2

102. 5/6

103. 184/105

104. 428/15

105. $-\frac{1}{5}\cos(5x + 3) + C$

106. $\frac{x}{2} + \frac{1}{8}\sin 2(2x + 1) + C$

107. $-\frac{1}{6}\cos^3 2x + C$

108. $-\frac{2}{3}\cos x^{3/2} + C$

109. $-\frac{2}{15}(2 + \cos 3x)^{5/2} + C$

110. $-\frac{1}{2(1 + \sin 2x)} + C$

111. $\tan x + C$

112. $-\frac{1}{x} + \tan x - \frac{3}{2}\cos 2x + C$

113. $-\frac{1}{3}\cot 3x + C$

114. $\frac{2}{9}(2 + \sin 3t)^{3/2} + C$

115. $-\frac{1}{2}\csc 2t + C$

116. $\frac{1}{20}\tan^4 5x + C$

117. $\frac{1}{2}\sec^2 x + C$

118. $\frac{1}{2}\tan x^2 + C$

119. $-\frac{1}{4}\cos(x^4 + 2) + C$

120. $\frac{7\sqrt{3} - 15}{6} = -0.479$

121. 1

122. $1 + \frac{5\pi^2}{72} - \frac{2}{\sqrt{3}} = -0.5307$

123. $-\frac{\sqrt{2}}{2}$

124. 7/72

125. $\sqrt{3} - 1$

126. $\frac{\pi^2}{64} + \frac{\sqrt{2}}{2} - \frac{1}{2} = -0.3613$

127. $\frac{1}{\pi}(\sqrt{2} - 1) = 0.1318$

128. $1 - \pi/4$

129. 3/8

130. $x^3 + 1$

131. $(x + 1)^{1/2}$

132. $6x(9x^4 - 4)^{2/3}$

133. $\frac{2x}{\sqrt{1 - 3x^4}} - \frac{1}{\sqrt{1 - 3x^2}}$

134. $\frac{1}{\sqrt{2x + 7}} - \frac{1}{\sqrt{-2x + 7}}$

135. $\frac{\sin(1/x^2)}{x^2} - \frac{\sin(1/x)}{2x^{3/2}}$

136. $\frac{15}{2 + 2\sqrt{3}} - \frac{3}{4\sqrt{2} + 4(2^{1/4})}$

137. 4

138. (a) yes
    (b) $f(x) = 0$ for all $x \in [a, b]$

139.

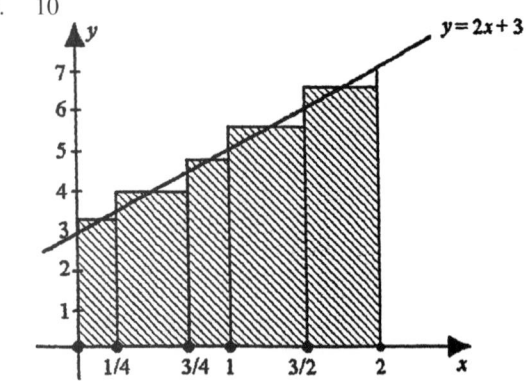

140. (a) $\Delta x_1 = 1/8, \Delta x_2 = 1/8, \Delta x_3 = 1/8,$
$\Delta x_4 = 1/8, \Delta x_5 = 1/2$
(b) $\|P\| = 1/2$
(c) $m_1 = 1, m_2 = 11/8, m_3 = 7/4, m_4 = 17/8,$
$m_5 = 5/2$
(d) $f(x_1^*) = 19/16, f(x_2^*) = 25/16,$
$f(x_3^*) = 17/8, f(x_4^*) = 23/8, f(x_5^*) = 13/4$
(e) $M_1 = 11/8, M_2 = 7/4, M_3 = 17/8,$
$M_4 = 5/2, M_5 = 4$
(f) $L_f(P) = 65/32$
(g) $S^*(P) = 83/32$
(h) $U_f(P) = 95/32$
(i) $\int_0^1 f(x)\,dx = 5/2 = 80/32$

141. $13/6; c = 133/144$

142. $335/54$

143. $3; c = \sqrt[3]{2} = 1.26$

144. $\dfrac{1}{\sqrt{2\pi}}$

145. $2/\pi; c = 0.8807$ rad

146. (a) $\dfrac{10 + 6\sqrt{2}}{3} = 6.1618$
(b) $2/3$

147. $2L(L+1)$

Part C

1. Let $P = \{x_0, x_1, x_2, \ldots, x_{n-1}, x_n\}$ be a regular partition of $[0, b]$ and let $f(x) = x^2$. (5.2)
   (a) Show that $L_f(P) = \dfrac{b^3}{n^3}[0^2 + 1^2 + 2^2 + \cdots + (n-1)^2]$
   (b) Show that $U_f(P) = \dfrac{b^3}{n^3}[1^2 + 2^2 + 3^2 + \cdots + n^2]$
   (c) Show that $L_f(P), U_f(P) \to \dfrac{b^3}{3}$ as $n \to \infty$

2. A function f is piecewise continuous on $[a, b]$ if it is continuous except, possibly, for a finite set of points at which it has a jump discontinuity. It can be shown that if f is piecewise continuous on $[a, b]$ then $\int_a^b f(x)\,dx$ exists
   Find:
   $$\int_{-2}^{2} f(x)\,dx \qquad f(x) = \begin{cases} -x & -2 \leq x < 0 \\ 1 & 0 \leq x < 2 \end{cases}$$
   Sketch the graph of f. (5.3)

3. Consider the discontinuous function defined for all x in $[0, 2]$ $f(x) = \begin{cases} 0 & x \neq 1 \\ 1 & x = 1 \end{cases}$ show that f is integrable on the interval and that $\int_0^2 f(x)\,dx = 0$. (5.2)

4. Let f be continuous and define F by $F(x) = \int_0^x \left[t \int_1^t f(u)\,du \right] dt$
   Find:
   (a) $F'(x)$ (b) $F'(1)$ (c) $F''(x)$ (d) $F''(1)$ (5.3)

5. Evaluate $\displaystyle\int_0^{3\pi/2} f(x)\,dx$ where $f(x) = \begin{cases} 2\sin x & 0 \leq x \leq \pi/2 \\ \dfrac{1}{2}\cos x & \pi/2 < x \leq 3\pi/2 \end{cases}$ (5.4)

6. Compare $\dfrac{d}{dx}\left[\int_a^x f(t)\,dt\right]$ to $\int_a^x \dfrac{d}{dt}[f(t)]\,dt$ (5.4)

7. Let $f(x) = \cos x + \sin x$ $\quad x \in [-\pi, \pi]$ (5.5)
   (a) Evaluate $\displaystyle\int_{-\pi}^{\pi} f(x)\,dx$
   (b) Sketch the graph of f and find the area bounded by the graph and the x-axis for $x \in [-\pi, \pi]$.

8. A particle moves along the x-axis with velocity $v(t) = At^2 + 1$. Determine A given that $x(1) = x(0)$. Compute the total distance traveled during the first second. (5.6)

9. Derive the formula $\displaystyle\int \cos^2 x\,dx = \dfrac{1}{2}x + \dfrac{1}{4}\sin 2x + C$ $\quad$ Hint: $\cos^2\theta = \dfrac{1}{2}(1 + \cos 2\theta)$ (5.7)

10. Given $\displaystyle\int \sin x \cos x\,dx$ (5.7)
    (a) let $u = \sin x$ and find the integral
    (b) let $u = \cos x$ and find the integral
    (c) reconcile the answers

11. Let f be a continuous function and show that (5.7)
    (a) $\displaystyle\int_{a+c}^{b+c} f(x-c)\,dx = \int_a^b f(x)\,dx$ and if $C = 0$

(b) $\dfrac{1}{c}\displaystyle\int_{ac}^{bc} f(x/c)\,dx = \int_a^b f(x)\,dx$

True or false

(c) Is $\displaystyle\int_2^4 (x-2)^2\,dx = \int_0^2 x^2\,dx$

(d) $\displaystyle\int_2^4 (x-2)\,dx = \int_\square^\square \square\,dx$

12. Calculate the derivative $\dfrac{d}{dx}\left(\displaystyle\int_{\sqrt{x}}^{x^2+x} \dfrac{dt}{2+\sqrt{t}}\right)$. (5.8)

13. Given f is continuous on $[-a, a]$ prove $\displaystyle\int_{-a}^{a} f(x)\,dx = 0$, if f is odd. (5.8)

14. Given if f is continuous on $[a, b]$ and $\displaystyle\int_a^b |f(x)|\,dx = 0$ then $f(x) = 0$ for all $x \in [a, b]$. (5.8)

15. Show that $\dfrac{d}{dx}\displaystyle\int_x^b f(t)\,dt = -f(x)$.

16. Show that $\dfrac{d}{dx}\displaystyle\int_a^{u(x)} f(t)\,dt = f(u(x))\dfrac{du}{dx}$

17. Buffon's Needle Expirement
A horizontal plane is ruled in 2-inch parallel strips if a 2-inch needle is dropped randomely onto the plane, the probability that the needle will intersect a line is;

$$P = \dfrac{2}{\pi}\int_0^{\pi/2} \sin\theta\,d\theta$$

where θ is the angle between the needle and any line. Determine P for the above.

18. Given $F(x) = \displaystyle\int_{-2}^{x} (t^2 - 2t)\,dt$ find $F'(x)$

19. Given $F(x) = \displaystyle\int_0^x t\cos t\,dt$ find $F'(x)$

20. Given $F(x) = \displaystyle\int_x^{x+2} (4t+1)\,dt$ find $F'(x)$

21. Archimedes formula for the area of a parabolic arc: $A = \dfrac{2}{3}bh$ where $b =$ length of the base of the arch and h is the height of the arch.
(a) Use Archimedes formula on the area of the parabolic arch formed by $f(x) = 4 - x^2$
(b) Verify the answer to part a by integration

22. Let $f(x) = \displaystyle\int_0^{1/x} \dfrac{1}{t^2+1}\,dt + \int_0^x \dfrac{1}{t^2+1}\,dt$ if $x > 0$, then $f(x)$ is _____.

Answers to Chapter 5, Part C Questions

1. (a) Note $x_i = \dfrac{ih}{n}$ $\Delta x = \dfrac{b}{n}$, since $f(x)$ increases on $[0, b]$

$$L_f(P) = \left(\dfrac{0b}{n}\right)^2 \left(\dfrac{b}{n}\right) + \left(\dfrac{b}{n}\right)^2 \left(\dfrac{b}{n}\right)$$
$$+ \left(\dfrac{2b}{n}\right)^2 \left(\dfrac{b}{n}\right) \cdots$$
$$+ \left(\dfrac{(n-1)b}{n}\right)^2 \left(\dfrac{b}{n}\right)$$
$$= \dfrac{b^3}{n^3}(0^2 + 1^2 + 2^2 + \cdots (n-1)^2)$$

(Riemen sum)

(b) Similar to (a)

(c) Since
$$0^2 + 1^2 + \cdots (n-1)^2 = \dfrac{(n-1)n(2n-1)}{6}$$
$$\therefore L_f(P) = \dfrac{b^3}{n^3} \dfrac{(n-1)n(2n-1)}{6}$$
$$Uf(P) = \dfrac{b^3}{n^3} \dfrac{n(n+1)(2n+1)}{6}$$

As $n \to \infty$ $L_f(p) \to \dfrac{b^3}{3}$ $Uf(P) \to \dfrac{b^3}{3}$

2. $\therefore \displaystyle\int_{-2}^{2} f(x)\,dx = \int_{-2}^{0} -x\,dx + \int_{0}^{2} dx = 2 + 2 = 4.$

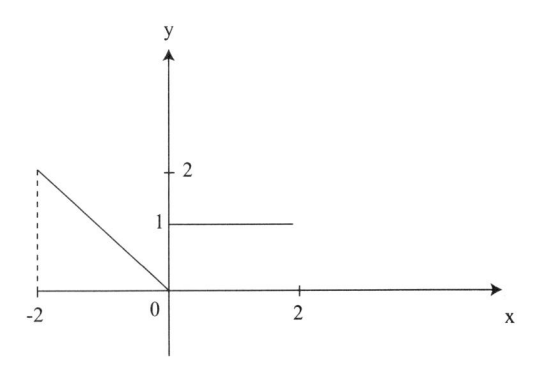

[graph showing piecewise function]

3. $f(x) = \begin{cases} 0 & x \neq 1 \\ 1 & x = 1 \end{cases}$

so $\displaystyle\int_0^2 f(x)\,dx = \lim_{b \to 1^-} \left(\int_0^b 0\,dx\right)$
$+ \displaystyle\lim_{(a \to 1^+)} \left(\int_b^2 0\,dx\right) = 0 + 0 = 0$ so it is

integrable and $\displaystyle\int_0^2 f(x)\,dx = 0$

4. (a) $F'(x) = x \displaystyle\int_1^x f(u)\,du$
   (b) $F'(1) = 0$

(c) $F''(x) = x\,f(x) + \displaystyle\int_1^x f(u)\,du$
(d) $F''(1) = f(1)$

5. $\displaystyle\int_0^{3\pi/2} f(x)\,dx = \int_0^{\pi/2} 2\sin x\,dx$
$+ \displaystyle\int_{\pi/2}^{3\pi/2} \dfrac{1}{2}\cos x\,dx$
$= [-2\cos x]_0^{\pi/2} + \left[\dfrac{1}{2}\sin x\right]_{\pi/2}^{3\pi/2}$
$= 1$

6. $\dfrac{d}{dx}\left[\displaystyle\int_a^x f(t)\,dt\right] = f(x)$

$\displaystyle\int_a^x \dfrac{d}{dx}[f(t)]\,dt = f(x) - f(a)$

7. (a) $\displaystyle\int_{-\pi}^{\pi} (\cos x + \sin x)\,dx = [\sin x - \cos \pi]_{-\pi}^{\pi}$
$= 0$

(b) Area $= -\displaystyle\int_{-\pi}^{-\pi/4} f(x)\,dx$
$+ \displaystyle\int_{-\pi/4}^{3\pi/4} f(x)\,dx - \int_{3\pi/4}^{\pi} f(x)\,dx = 4\sqrt{2}$

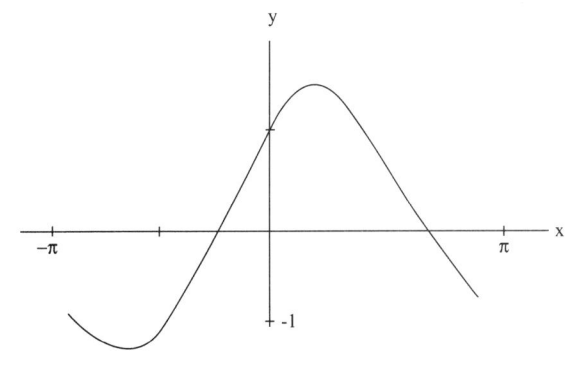

[graph of function on $[-\pi, \pi]$]

8. $x(t) = \displaystyle\int (At^2 + 1)\,dt = \dfrac{1}{3}At^3 + t + c$

$x(1) - x(0) = \left(\dfrac{1}{3}A + 1 + C\right) - C = \dfrac{A}{3} + 1$
$= 0 \therefore A = -3$

Distance traveled
$\displaystyle\int_0^{1/\sqrt{3}} (1 - 3t^2)\,dt + \int_{1/\sqrt{3}}^{1} (3t^2 - 1)\,dt$
$= 4\sqrt{3}/9.$

9. $\int \cos^2 x\,dx = \int \frac{1}{2}(1+\cos^2 x)\,dx$
$= \frac{1}{4}\int (1+\cos 2x)\,d(2x)$
$= \frac{1}{4}(2x + \sin 2x) + C$
$= \frac{1}{2}x + \frac{1}{4}\sin 2x + C$

10. (a) $u = \sin x \quad du = -\cos x\,dx$
$\int \sin x \cos dx = -\int u\,du$
$= -\frac{1}{2}u^2 + C = -\frac{1}{2}\cos^2 x + C$
(b) $u = \cos x \quad du = \sin x\,dx$
$\int \sin x \cos dx = \int u\,du = \frac{1}{2}u^2 + c$
$= \frac{1}{2}\sin^2 x + c'$
(c) since $\frac{1}{2}(\sin^2 x + \cos^2 x) = \frac{1}{2}$
$\therefore c' = c + \frac{1}{2}$

11. (a) $\begin{cases} u = x-c & x = a+c \to u = a \\ du = dx & x = b+c \to u = b \end{cases}$
$\int_{a+c}^{b+c} f(y-c)\,dx = \int_a^b f(u)\,du = \int_a^b f(x)\,dx$

(b) $c \neq 0$
$\begin{cases} u = \frac{x}{c} & x = a \cdot c \Rightarrow u = a \\ du = \frac{dx}{c} & x = b \cdot c \Rightarrow u = b \end{cases}$
$\int_{x/c}^{b/c} f\left(\frac{x}{c}\right)dx = \int_a^b f(u)\,du = \int_a^b f(x)\,dx$

(c) T

(d) $\int_0^2 x\,dx$

12. $\frac{d}{dx}\left(\int_{\sqrt{x}}^{x^2+x} \frac{dt}{2+\sqrt{t}}\right) = \frac{2x+1}{2+\sqrt{x^2+x}}$
$-\frac{1}{2+\sqrt[4]{x}} \cdot \frac{1}{2\sqrt{x}}$

13. Pf.
$\int_{-a}^a f(x)\,dx = \int_{-a}^0 f(x)\,dx + \int_0^a f(x)\,dx$
$= -\int_a^0 f(-x)\,dx + \int_0^a f(x)\,dx$
since f is odd, $f(-x) = -f(x)$

$\therefore \int_{-a}^a f(x)\,dx = \int_a^0 f(x)\,dx + \int_0^a f(x)\,dx$
$= -\int_0^a f(x)\,dx + \int_0^a f(x)\,dx$
$= 0$.

14. Suppose $f(c) > 0$ for some $c \in (a, b)$ then their exist $\delta > 0$ such that $f(x) > 0$ for all $x \in (c-\delta, c+\delta)$. By choosing δ we can make $(c-\delta, c+\delta) \subset (a, b)$
thus
$\int_a^b |f(x)|\,dx \geq \int_{c-\delta}^{c+\delta} |f(x)|\,dx > 0$

contradiction same argument holds for $f(c) < 0$
hence $f(x) = 0$ for all $x \in [a, b]$.

15. $\frac{d}{dx}\int_x^b f(x)\,dt = \frac{d}{dx}(F(b) - F(x))$
$= 0 - F'(x) = -f(x)$
Let $F(x)$ be anti derivative of or
$\frac{d}{dx}\int_x^b f(t)\,dt = \frac{d}{dx}\left(-\int_b^x f(t)\,dt\right)$
$= -\frac{d}{dx}\int_b^x f(t)\,dt = -f(x)$

16. $\frac{d}{dx}\int_a^{u(x)} f(t)\,dt = f(u(x))\frac{du}{dx}$

17. $P = \frac{2}{\pi}\int_0^{\pi/2} \sin\theta\,d\theta$
$= \frac{2}{\pi}(-\cos\theta)|_0^{\pi/2} = \frac{2}{\pi}(1-0) = \frac{2}{\pi}$

18. $F'(x) = x^2 - 2x$

19. $F'(x) = x \cos x$

20. $4(x+2) + 1 - (4x+1) = 8$

21. (a) $b = 4 \quad h = 4$
$A = \frac{2}{3}bh = \frac{32}{3}$
(b) $\left|\int_{-2}^2 (4-x^2)\,dx\right| = \left|\left(4x - \frac{x^3}{3}\right)\right|_{-2}^2$
$= \frac{32}{3}$.

22. constant

CHAPTER 6 Some Applications of the Interval

Part A

1. When does the $\int_a^b f(x)\,dx$, $f(x)$ continuous, represent the area bounded by $x = a$, $x = b$, and $f(x)$?

2. True or false

   If $\int_a^b [f(x) - g(x)]\,dx = A$ then $\int_a^b [g(x) - f(x)]\,dx = -A$.

3. When does the $\int_c^d f(y)\,dy$, $f(y)$ continuous, represent the area bounded by $y = c$, $y = d$, and $f(y)$?

4. $V = \pi \int_a^b [f(x)]^2\,dx$, the disc method for finding volumes $f(x)$ is _____ of each disc.

5. $V = \pi \int_a^b [(f(x))^2 - (g(x))^2]\,dx$, $g(x) \le f(x)$, the washer method for finding volume $f(x)$ is _____ of the washer and $g(x)$ is _____ of the washer.

6. True or false

   $\int_{-1}^5 x\,dx$ is the area bounded by $f(x) = x$ $x = -1$, $x = 5$.

7. True or false

   Given $g(x) \le f(x)$ for all $x \in [a, b]$ then the volume generated by revolving the region bounded by $x = a$, $f(x)$, $g(x)$, $x = b$ is $V = \pi \int_a^b [(f(x))^2 - (g(x))^2]\,dx$ will not work if $g(x) < 0$.

8. In the formula $V = 2\pi \int_a^b x\,f(x)\,dx$, the shell method formula, x represents _____.

9. The center of mass of a uniformly dense plate, density constant is called _____.

10. If $f(-x) = f(x)$ for all $x \in [a, b]$, $f(x) = 4 - x^2$ as an example on $[-2, 2]$. Then the centroid will be found _____.

11. State the first theorem of Pappus.

12. State Hooke's law.

13. If an object moves from $x = a$ to $x = b$ subject to a constant force F then the work W, done by F:

    (a) $W = $ _____.

    (b) if F is a variable force

    $W = $ _____.

14. If $A = $ area of a horizontal surface

    $h = $ depth of the liquid to A

    $\sigma = $ weight density of the liquid

    $F = $ force against A

    then $F = $

15. Given the centroid of a region that has an area 4 sq units, not crossing centroid (1, 2) find the volume formed by revolving the region about the axes.

    (a) the x-axis   (b) the y-axis

True or False

16. If the region bounded by $y = x^2$, y axis, $y = 1$ in the 1st Quadrant is revolved about x axis the volume generated is: $V = \pi \int_0^1 (1 - x^2)^2 \, dx$. (6.2)

17. The area in the first Quadrant bounded by $f(x) = \tan x$ and $g(x) = \cot x$, and the x axis is $A = 2 \int_0^{\pi/4} \tan x \, dx$. (6.1)

Answers to Chapter 6, Part A Questions

1. $f(x) \geq 0$ for $x \in [a, b]$

2. T

3. $f(y) \geq 0$ for $y \in [c, d]$

4. radius

5. outer radius, inner radius

6. F

7. T

8. radius of the partition

9. centroid

10. y axis $\left(0, \dfrac{8}{5}\right)$

11. A plane region is revolved about an axis that lies in its plane. If the region does not cross the axis, the volume of the resulting solid of revolution is the area of the region multiplied by the circumference of the circle descibled by the centroid of the region: $V = 2\pi \, kA$.

12. The force exerted by the spring is $F(x) = -kx$.

13. (a) $W = F(b - a)$ (b) $\displaystyle\int_a^b F \, dx$

14. $F = \sigma \, h \, A$

15. (a) 16π (b) 8π

16. F

17. T

Part B

6.1 More on Area

1. Sketch the region bounded by $y = x^2 - 4x + 5$ and $y = 2x - 3$. Represent the area of the region by one or more integrals (a) in terms of x; (b) in terms of y.

2. Sketch the region bounded by $x = y^2 - 4y + 2$ and $x = y - 2$. Represent the area of the region by one or more integrals (a) in terms of x; (b) in terms of y.

3. Sketch the region bounded by $y = 2x - x^2$ and $y = -3$. Represent the area of the region by one or more integrals (a) in terms of x; (b) in terms of y.

4. Sketch the region bounded by $y = x + 4/x^2$, the x-axis, $x = 2$, and $x = 4$.
   (a) Represent the area of the region by one or more integrals (b) Find the area.

5. Sketch the region bounded by $y = 4x - x^2$ and $y = 3$. Represent the area of the region by one or more integrals (a) in terms of x; (b) in terms of y. (c) Find the area.

6. Sketch the region bounded by $x = y^2 - 4y$ and $x = y$ and find its area.

7. Sketch the region bounded by $y = 3 - x^2$, $y = -x + 1$, $x = 0$, and $x = 2$, and find its area.

8. Sketch the region bounded by $x = 3y - y^2$ and $x + y = 3$, and find the area.

6.2 Volume by Parallel Cross Sections; Discs and Washers

9. Sketch the region Ω bounded by $x + y = 4$, $y = 0$, and $x = 0$, and find the volume of the solid generated by revolving the region about the x-axis.

10. Sketch the region Ω bounded by $y^2 = 4x$, $y = 2$, and $x = 4$, and find the volume of the solid generated by revolving the region about the x-axis.

11. Sketch the region Ω bounded by $y = 4 - x^2$ and $y = x + 2$, and find the volume of the solid generated by revolving the region about the x-axis.

12. Sketch the region Ω bounded by $y = x^2$, $y = 4$, and $x = 0$, and find the volume of the solid generated by revolving the region about the x-axis.

13. Sketch the region Ω bounded by $y^2 = x^3$, $x = 1$, and $y = 0$, and find the volume of the solid generated by revolving the region about the x-axis.

14. Sketch the region Ω bounded by $y^2 = x^2$, $x = 0$, and $y = 4$, and find the volume of the solid generated by revolving the region about the y-axis.

15. Sketch the region Ω bounded by $y = \sqrt{x}$, $y = 0$, and $x = 9$, and find the volume of the solid generated by revolving the region about the y-axis.

16. Sketch the region Ω bounded by $y^2 = 4x$, $x = 4$, and $y = 0$, and find the volume of the solid generated by revolving the region about the y-axis.

17. Sketch the region Ω bounded by $y^2 = x^3$, $x = 1$, and $y = 0$, and find the volume of the solid generated by revolving the region about the y-axis.

18. Sketch the region Ω bounded by $y = x^3$, $x = 2$, and $y = 0$, and find the volume of the solid generated by revolving the region about the line $x = 2$.

19. Sketch the region Ω bounded by $y = x^2$, $y = 0$, and $x = 2$, and find the volume of the solid generated by revolving the region about the line $x = 2$.

20. Sketch the region Ω bounded by $y = x^3$, $x = 1$, and $y = -1$, and find the volume of the solid generated by revolving the region about the line $y = -1$.

21. Sketch the region Ω bounded by $y = x^3/2$, $x = 2$, and $y = 0$, and find the volume of the solid generated by revolving the region about the line $y = 4$.

22. The base of a solid is a circle of radius 2. All sections that are perpendicular to the diameter are squares. Find the volume of the solid.

23. The steeple of a church is constructed in the form of a pyramid 45 feet high. The cross sections are all squares, and the base is a square of side 15 feet. Find the volume of the steeple.

6.3 Volume by the Shell Method

24. Sketch the region Ω bounded by $y = \sqrt{x}$, $y = 0$, and $x = 9$, and use the shell method to find the volume of the solid generated by revolving Ω about the y-axis.

25. Sketch the region Ω bounded by $y^2 = 4x$, $x = 4$, and $y = 0$, and use the shell method to find the volume of the solid generated by revolving Ω about the y-axis.

26. Sketch the region Ω bounded by $y^2 = 4x$, $y = 2$, and $x = 4$, and use the shell method to find the volume of the solid generated by revolving Ω about the y-axis.

27. Sketch the region Ω bounded by $y = 2x + 3$, $x = 1$, and $x = 4$, and $y = 0$, and use the shell method to find the volume of the solid generated by revolving Ω about the y-axis.

28. Sketch the region Ω bounded by $y = \sqrt{x+1}$, $x = 0$, $y = 0$, and $x = 3$, and use the shell method to find the volume of the solid generated by revolving Ω about the y-axis.

29. Sketch the region Ω bounded by $y = x^2$, $y = 4$, and $x = 0$, and use the shell method to find the volume of the solid generated by revolving Ω about the x-axis.

30. Sketch the region Ω bounded by $y = x^2$ and $x = y^2$, and use the shell method to find the volume of the solid generated by revolving Ω about the x-axis.

31. Sketch the region Ω bounded by $y = x^3$ and $y = x$, and use the shell method to find the volume of the solid generated by revolving Ω about the x-axis.

32. Sketch the region Ω bounded by the first quadrant of the circle $x^2 + y^2 = r^2$ and use the shell method to find the volume of the solid generated by revolving Ω about the x-axis.

33. Sketch the region Ω bounded by $x = 2y - y^2$ and $x = 0$, and use the shell method to find the volume of the solid generated by revolving Ω about the x-axis.

34. Sketch the region Ω bounded by $y = 2x$, $x = 0$, and $y = 2$, use the shell method to find the volume of the solid generated by revolving Ω about $x = 1$.

35. Sketch the region Ω bounded by $y^2 = 4x$, $y = x$, and use the shell method to find the volume of the solid generated by revolving Ω about $x = 4$.

36. Sketch the region Ω bounded by $y = x^2$, $y = 0$, and $x = 2$, and use the shell method to find the volume of the solid generated by revolving Ω about $x = 2$.

37. Sketch the region Ω bounded by $y = x^3$ and $x = 2$, and use the shell method to find the volume of the solid generated by revolving Ω about $x = 2$.

6.4 The Centroid of a Region; Pappus's Theorem on Volumes

38. Sketch the region bounded by $y = x^2$, $y = \sqrt{x}$. Determine the centroid of the region and the volume generated by revolving the region about each of the coordinate axes.

39. Sketch the region bounded by $y^2 = x^3$, $x = 1$, $y = 0$. Determine the centroid of the region and the volume generated by revolving the region about each of the coordinate axes.

40. Sketch the region bounded by $y^2 = 8x$, $y = 0$, and $x = 2$. Determine the centroid of the region and the volume generated by revolving the region about each of the coordinate axes.

41. Sketch the region bounded by $y = 2x$, $y = 4$, $y = 0$, and $x = 2$. Determine the centroid of the region and the volume generated by revolving the region about each of the coordinate axes.

42. Sketch the region bounded by $x^2 = 2y$ and $2x - y = 0$. Determine the centroid of the region and the volume generated by revolving the region about each of the coordinate axes.

43. Find the centroid of the bounded region determined by $y = 2x^2$ and $x - 2y + 3 = 0$.

44. Find the centroid of the bounded region determined by $y + 1 = 0$ and $x^2 + y = 0$.

45. Find the centroid of the bounded region determined by $y + x^2 + 2x$ and $y = 2x + 1$.

46. Locate the centroid of a solid cone of base radius 2 cm and height 4 cm.

47. Locate the centroid of a solid generated by revolving the region below the graph of $f(x) = 2 - x^2$, $x \in [0, 1]$.
    (a) about the x-axis.   (b) about the y-axis.

6.5 The Notion of Work

48. Find the work done by the force $F(x) = (x + 1)/\sqrt{x}$ pounds in moving an object from $x = 1$ foot to $x = 4$ feet along the x-axis.

49. Find the work done by the force $F(x) = x\sqrt{x + 3}$ Newtons in moving an object from $x = 1$ meter to $x = 6$ meters along the x-axis.

50. A spring exerts a force of 2 pounds when stretched 6 inches. How much work is required in stretching the spring from a length of 1 foot to a length of 2 feet?

51. A spring whose natural length is 10 feet exerts a force of 400 pounds when stretched 0.4 feet. How much work is required to stretch the spring from its natural length to 12 feet?

52. A spring exerts a force of 1 ton when stretched 10 feet beyond its natural length. How much work is required to stretch the spring 8 feet beyond its natural length?

53. A spring whose natural length is 18 inches exerts a force of 10 pounds when stretched 16 inches. How much work is required to stretch the spring from 4 inches beyond its natural length?

54. A dredger scoops a shovel full of mud weighing 2000 pounds from the bottom of a river at a constant rate. Water leaks uniformly at such a rate that half the weight of the contents is lost when the scoop has been lifted 25 feet. How much work is done by the dredger in lifting the mud this distance?

55. A 60-foot length of steel chain weighing 10 pounds per foot is hanging from the top of a building. How much work is required to pull half of it to the top?

56. A 50-foot chain weighing 10 pounds per foot supports a steel beam weighing 1000 pounds. How much work is done in winding 40 feet of the chain onto a drum?

57. A bucket weighing 1000 pounds is to be lifted from the bottom of a shaft 20 feet deep. The weight of the cable used to hoist it is 10 pounds per foot. How much work is done lifting the bucket to the top of the shaft?

58. A cylindrical tank 8 feet in diameter and 10 feet high is filled with water weighing 62.4 lbs/ft^3. How much work is required to pump the water over the top of the tank?

59. A cylindrical tank is to be filled with gasoline weighing 50 lbs/ft^3. If the tank is 20 feet high and 10 feet in diameter, how much work is done by the pump in filling the tank through a hole in the bottom of the tank?

60. A cylindrical tank 5 feet in diameter and 10 feet high is filled with oil whose density is 48 lbs/ft^3. How much work is required to pump the water over the top of the tank?

61. A conical tank (vertex down) has a diameter of 9 feet and is 12 feet deep. If the tank is filled with water of density 62.4 lbs/ft^3, how much work is required to pump the water over the top?

62. A conical tank (vertex down) has a diameter of 8 feet and is 10 feet deep. If the tank is filled to a depth of 6 feet with water of density 62.4 lbs/ft^3, how much work is required to pump the water over the top?

6.6 Fluid Force

63. A flat rectangular plate, 6 feet long and 3 feet wide, is submerged vertically in water (density 62.4 lbs/ft^3) with the 3-foot edge parallel to and 2 feet below the surface. Find the force against the surface of the plate.

64. A flat rectangular plate, 6 feet long and 3 feet wide, is submerged vertically in water (density 62.4 lbs/ft^3) with the 6-foot edge parallel to and 2 feet below the surface. Find the force against the surface of the plate.

65. A flat triangular plate whose dimensions are 5, 5, and 6 feet is submerged vertically in water (density 62.4 lbs/ft^3) so that its longer side is at the surface and parallel to it. Find the force against the surface of the plate.

66. A flat triangular plate whose dimensions are 5, 5, and 6 feet is submerged vertically in water (density 62.4 lbs/ft^3) so that its longer side is at the bottom and parallel to the surface, and its vertex is 2 feet below the surface. Find the force against the surface of the plate.

67. A flat plate, shaped in the form of a semicircle 6 feet in diameter, is submerged in water (density 62.4 lbs/ft^3) as shown. Find the force against the surface of the plate.

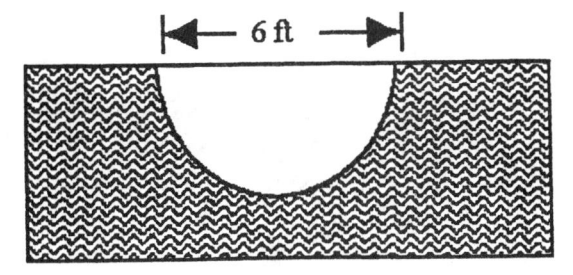

Answers to Chapter 6, Part B Questions

1. (a) $\int_{2}^{4} [(2x - 3) - (x^2 - 4x + 5)] dx$
   (b) $\frac{1}{2} \int_{1}^{5} \left(1 + 2\sqrt{y - 1} - y\right) dy$

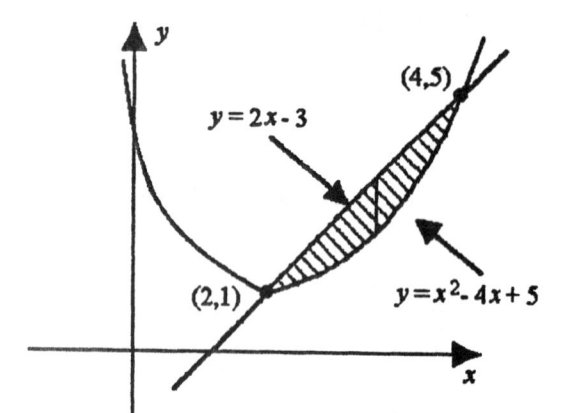

2. (a) $2 \int_{-2}^{-1} \sqrt{x + 2} \, dx + \int_{-1}^{2} \left(\sqrt{x + 2} - x\right) dx$
   (b) $\int_{1}^{4} [(y - 2) - (y^2 - 4y + 2)] \, dy$

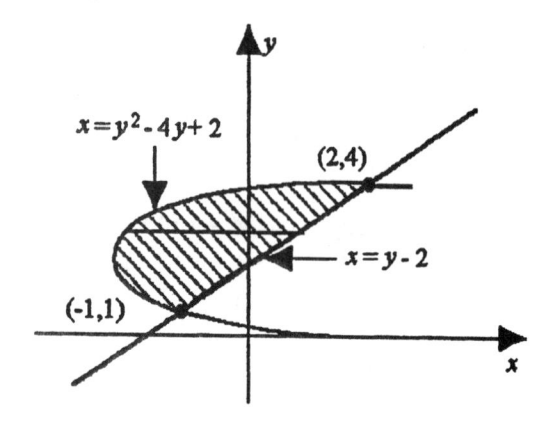

3. (a) $\int_{-1}^{3} [(2x - x^2) - (-3)] \, dx$
   (b) $2 \int_{-3}^{1} \sqrt{1 - y} \, dy$

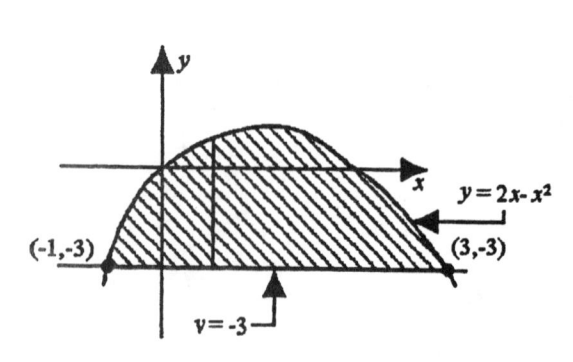

4. (a) $\int_{2}^{4} \left(x + \frac{4}{x^2}\right) dx$
   (b) 7

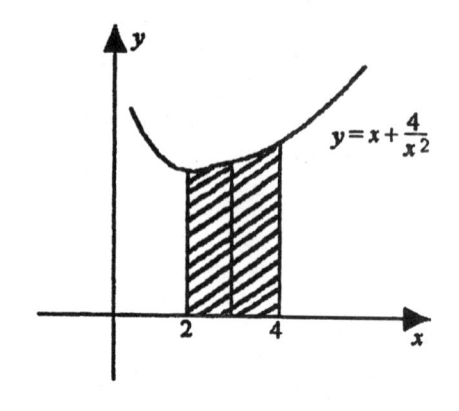

5. (a) $\int_{1}^{3} (4x - x^2 - 3) \, dx$
   (b) $2 \int_{3}^{4} \sqrt{4 - y} \, dy$
   (c) 4/3

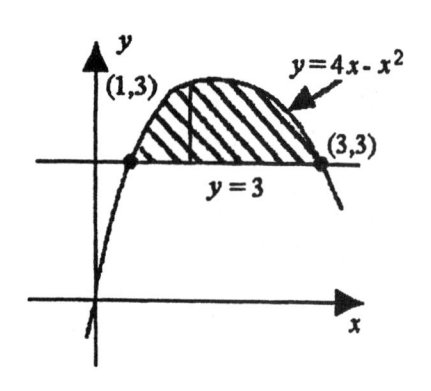

6. 125/6

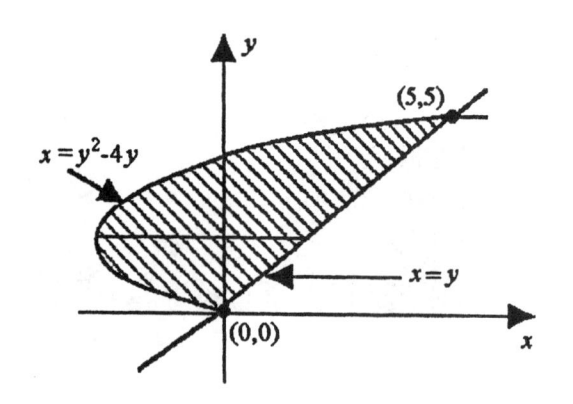

Answers

7. $10/3$

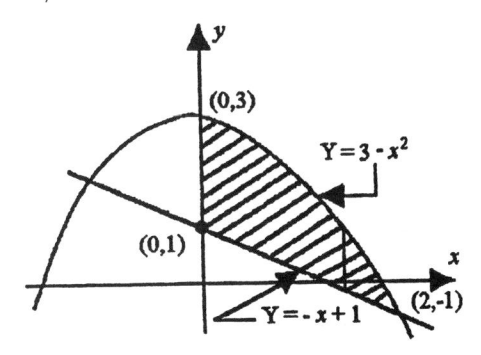

8. $4/3$

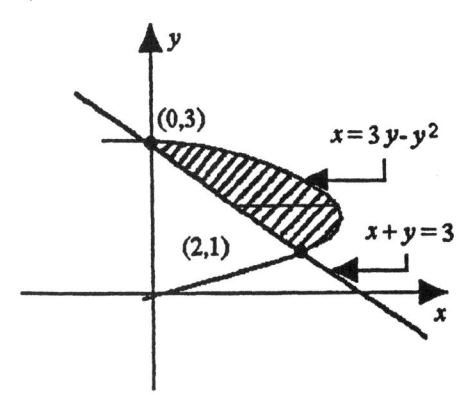

9. $64\pi/3$

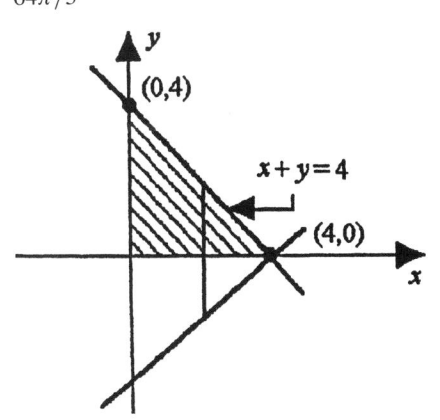

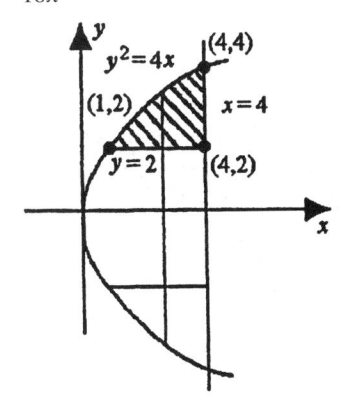

10. 18π

11. $108\pi/5$

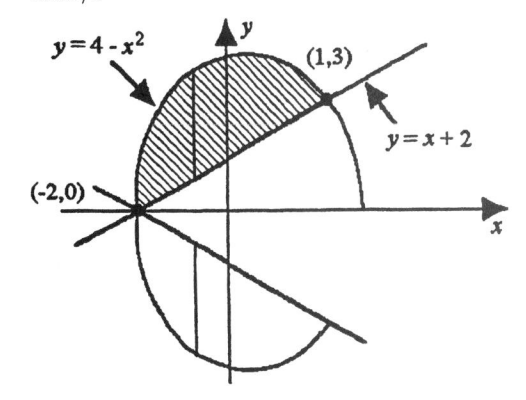

12. $128\pi/5$

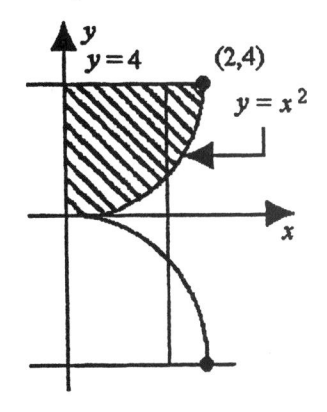

13. $\pi/4$

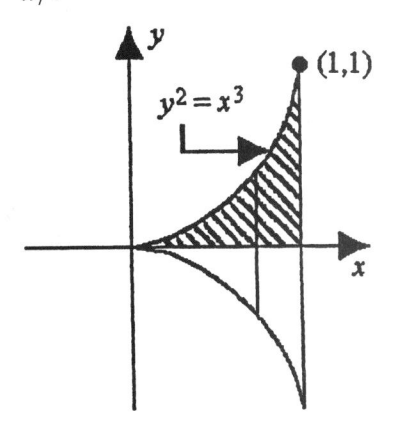

14. 8π

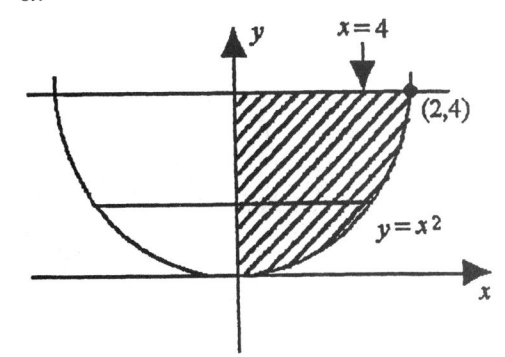

15. $972\pi/5$

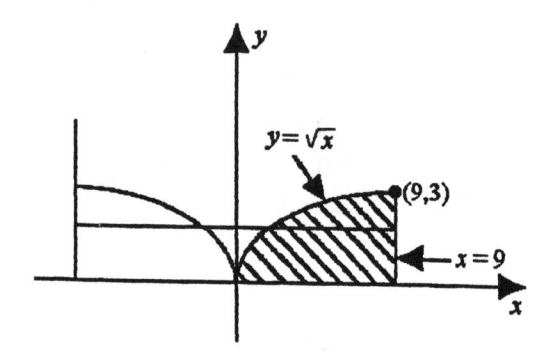

16. $256\pi/5$

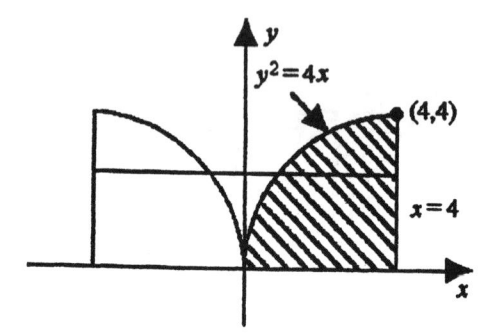

17. $4\pi/7$

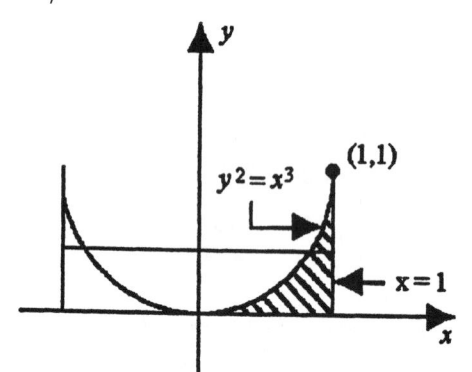

18. $16\pi/5$

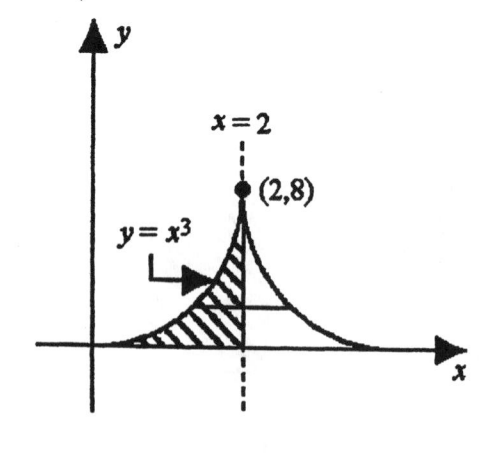

19. $8\pi/3$

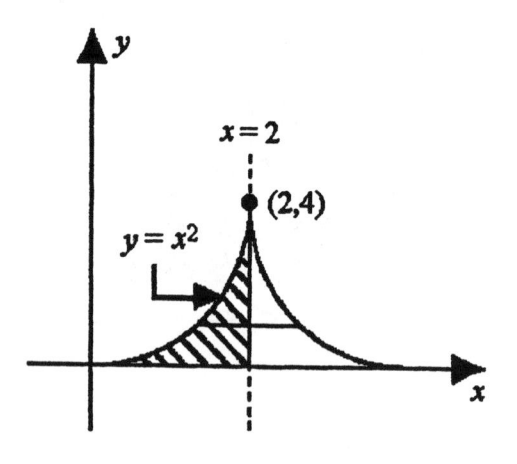

20. $16\pi/7$

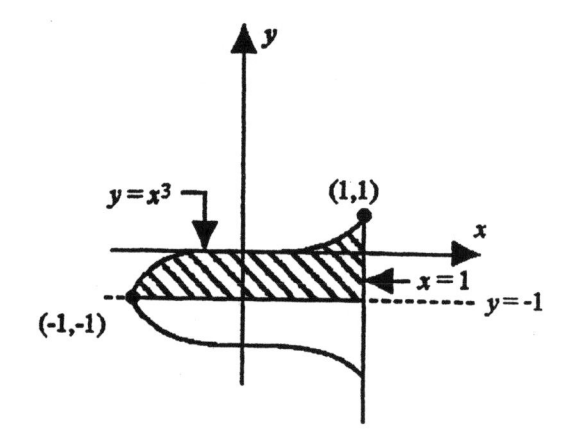

21. $80\pi/7$

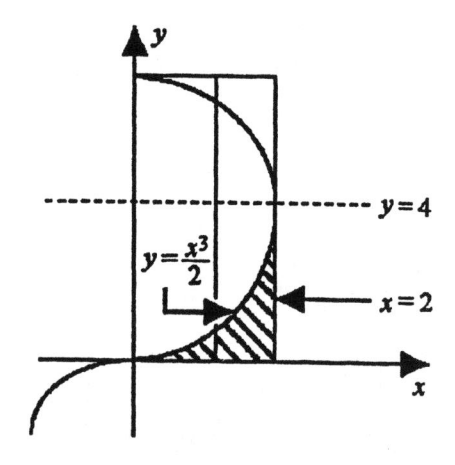

22. $128/3$

23. $3375\,ft^3$

Answers

24. $972\pi/5$

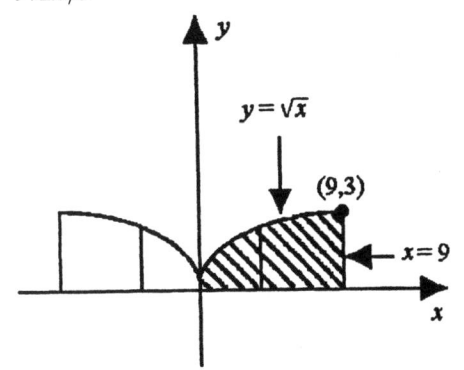

25. $256\pi/5$

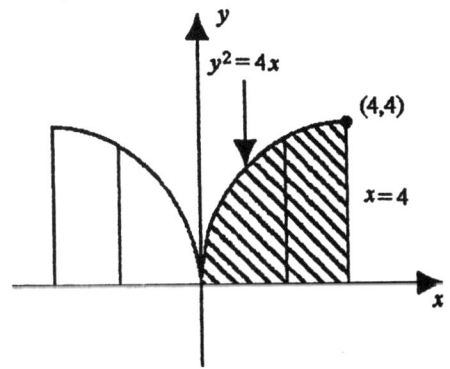

26. $98\pi/5$

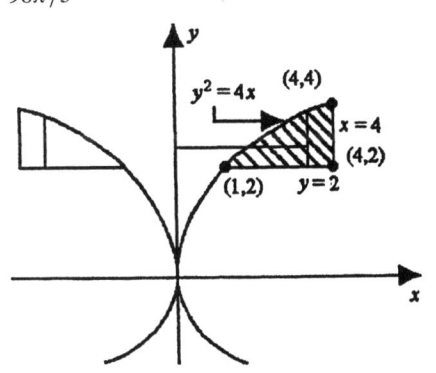

27. 129π

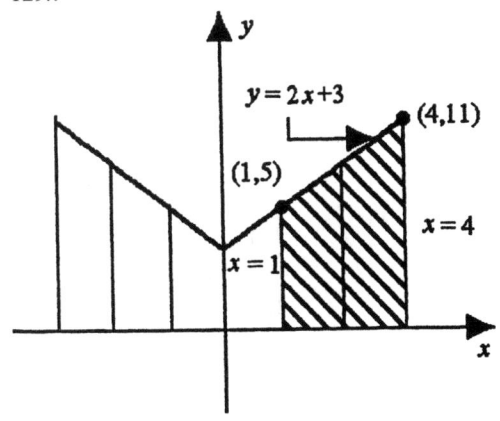

28. $232\pi/15$

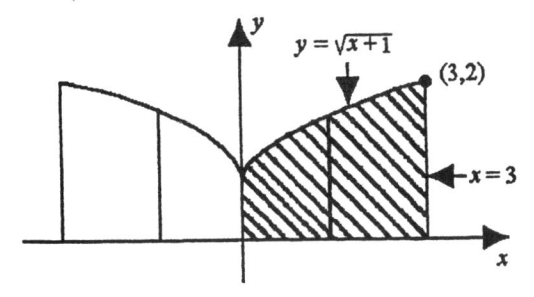

29. $128\pi/5$

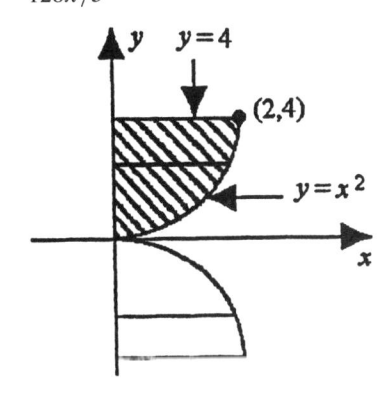

30. $3\pi/10$

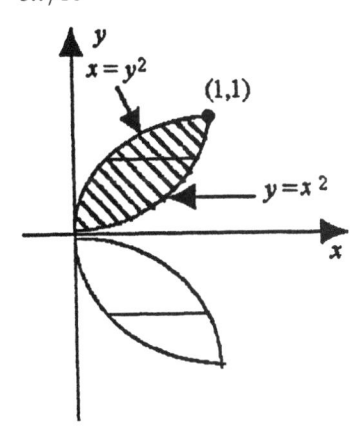

31. $4\pi/21$

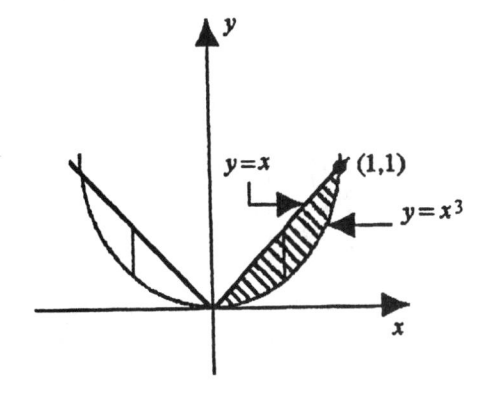

32. $2\pi r^3/3$

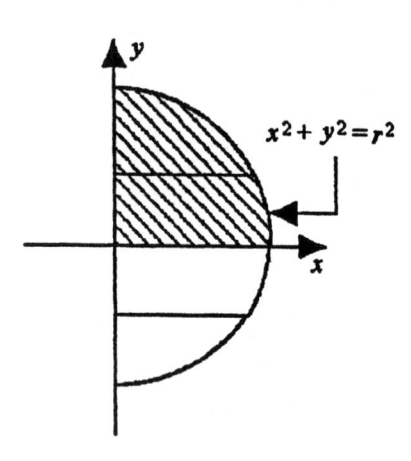

33. $8\pi/3$

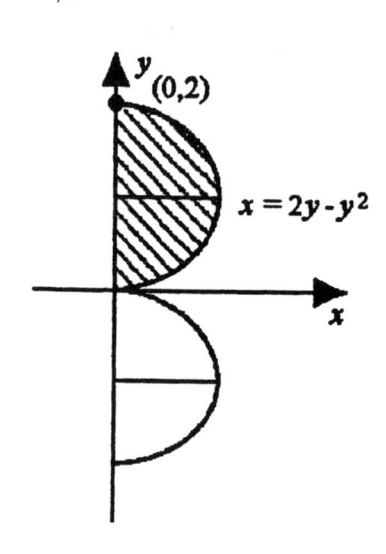

34. $4\pi/3$

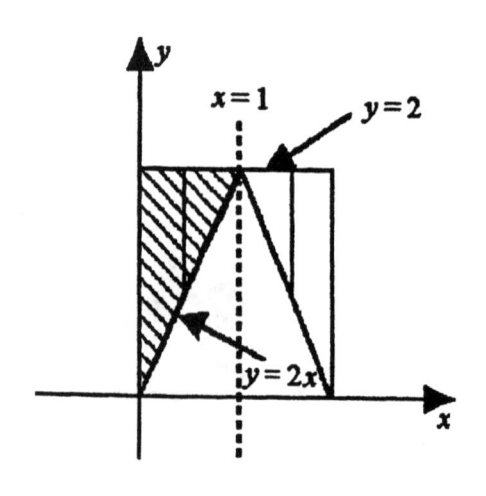

35. $64\pi/5$

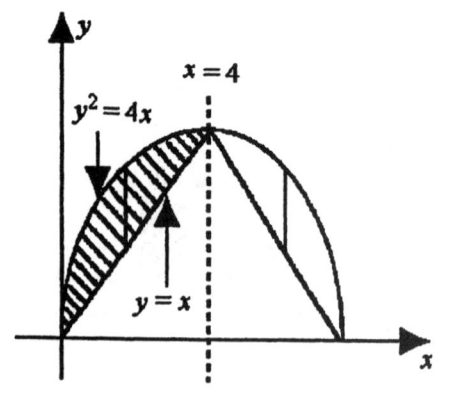

36. $8\pi/3$

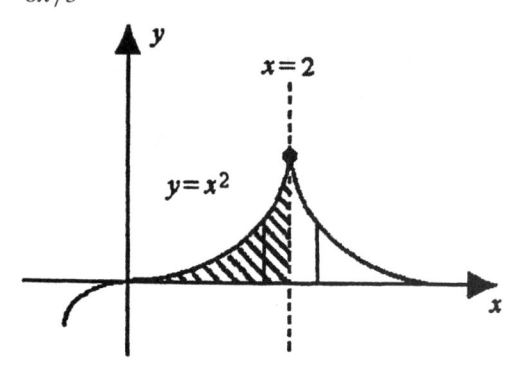

37. $16\pi/5$

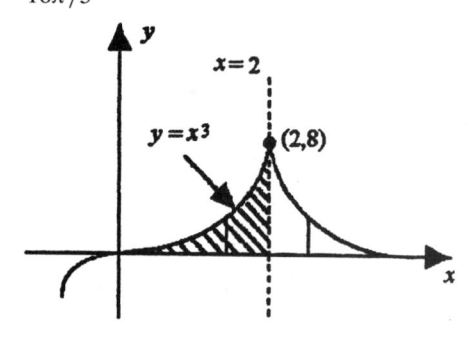

38. $(\bar{x}, \bar{y}) = \left(\dfrac{9}{20}, \dfrac{9}{20}\right)$; $V_x = V_y = \dfrac{3\pi}{10}$

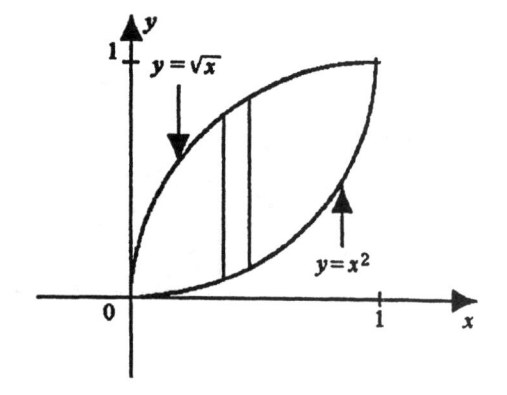

39. $(\bar{x}, \bar{y}) = \left(\dfrac{5}{7}, \dfrac{5}{16}\right)$; $V_x = \dfrac{\pi}{4}$, $V_y = \dfrac{4\pi}{7}$

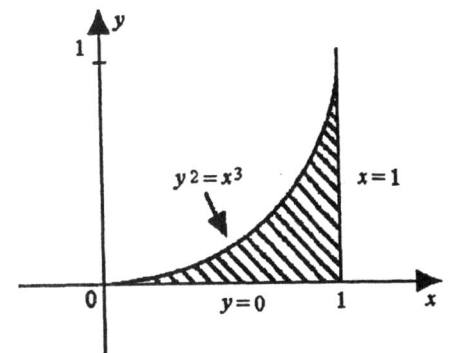

40. $(\bar{x}, \bar{y}) = \left(\dfrac{6}{5}, \dfrac{3}{2}\right)$; $V_x = 16\pi$, $V_y = \dfrac{64\pi}{5}$

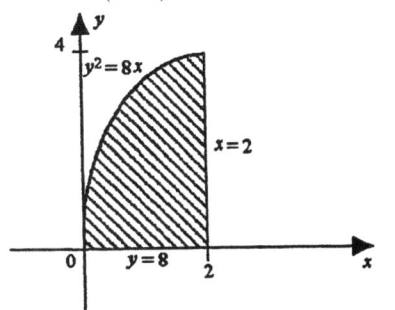

41. $(\bar{x}, \bar{y}) = \left(\dfrac{4}{3}, \dfrac{4}{3}\right)$; $V_x = V_y = \dfrac{32\pi}{3}$

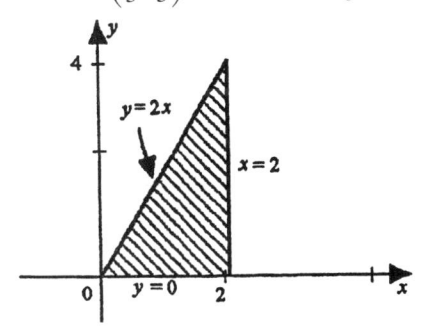

42. $(\bar{x}, \bar{y}) = \left(2, \dfrac{16}{5}\right)$; $V_x = \dfrac{512\pi}{15}$, $V_y = \dfrac{64\pi}{3}$

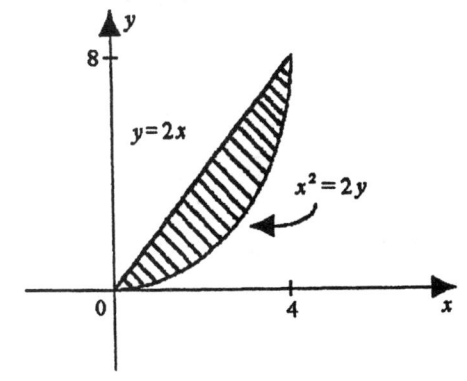

43. $(\bar{x}, \bar{y}) = \left(\dfrac{1}{8}, \dfrac{19}{20}\right)$

44. $(\bar{x}, \bar{y}) = \left(0, \dfrac{-3}{5}\right)$

45. $(\bar{x}, \bar{y}) = \left(0, \dfrac{3}{5}\right)$

46. $(\bar{x}, \bar{y}, \bar{z}) = (0, 0, 1)$

47. (a) $(\bar{x}, \bar{y}, \bar{z}) = \left(\dfrac{9}{20}, 0, 0\right)$

    (b) $(\bar{x}, \bar{y}, \bar{z}) = \left(0, \dfrac{43}{50}, 0\right)$

48. 20/3 ft-lbs

49. 232/5 joules

50. 6 ft-lbs

51. 2000 ft-lbs

52. 6400 ft-lbs

53. 5/12 ft-lbs

54. 37,500 ft-lbs

55. 13,500 ft-lbs

56. 52,000 ft-lbs

57. 22,000 ft-lbs

58. $49,920\pi$ ft-lbs

59. $250,000\pi$ ft-lbs

60. $15,000\pi$ ft-lbs

61. $15,163.2\pi$ ft-lbs

62. 3953.7π ft-lbs

63. 2995.2 lbs

64. 3931.2 lbs

65. 998.4 lbs

66. 3494.4 lbs

67. 1123.2 lbs

Part C

1. (a) Calculate the area of the region in the first quadrant bounded by the coordinate axes and the graph of the parabola $y = 1 + a - ax^2$ $a > 0$. (6.1)
   (b) Determine a so that the area found in part (a) is a minimum.

2. Let $p > 1$. Calculate the area A of the region bounded by the graph of $f(x) = 1/x^p$ and the x-axis for $x \in [1, b]$ and then calculate $\lim\limits_{b \to \infty} A(b)$. (6.1)

3. For $0 < p < 1$. Calculate the area A of the region bounded by the graph of $f(x) = 1/x^p$ and the x-axis for $x \in [1, b]$ and then calculate $\lim\limits_{b \to \infty} A(b)$. (6.1)

4. Find the volume of the solid that is generated by revolving the region bounded by $y = x^2$ and $y = 4x$ around
   (a) the line $x = 5$ (b) the line $x = -1$. (6.2)

5. A ball of radius r is cut into two pieces by a horizontal plane a units above the center of the ball. Determine the volume of the upper piece by using the shell method. (6.3)

6. The ellipse $b^2x^2 + a^2y^2 = a^2b^2$ encloses a region of area πab. Locate the centroid of the upper half of the region. (6.4)

7. A box that weighs w pounds is dropped to the floor from a height of d feet. (6.5)
   (a) What is the work done by gravity?
   (b) Show that the work is the same if the box slides to the floor along a smooth inclined plane.

8. The face of a rectangular dam in a river is 1000 feet wide by 100 ft deep, and it makes an angle of 30° with the vertical. Find the force of the water on the face of the dam if (6.6)
   (a) The water level is at the top of the dam.
   (b) The water level is 75 ft from the bottom of the river.

9. A hole of radius $\sqrt{3}$ is bored through the center of a sphere of radius 2. Find the volume removed. (6.2)+(6.3)

10. Two particles of matter of masses M_1 and M_2 are "a" units apart, how much work must be done to move them $2a$ units apart? (6.5)

11. Find the volume of the solid of revolution generated when the area bounded by $\sqrt{x} + \sqrt{y} = \sqrt{a}$, $x = 0$, $y = 0$ is revolved about the x-axis. (6.2)+(6.3)

12. The base of a solid is $x^2 + y^2 = 16$. Each plane perpendicular to the x-axis intersects the solid in a square cross section with one side in the base of the solid. Find the volume. (6.2)

13. A hemispherical tank of radius 8 ft is full of water. If a hole is made in the bottom, find the work done by gravity in emptying the tank (6.5)

14. Cavalieri's theorem states that if two solids have equal altitudes h and it all cross sections by planes parallel to their bases and at the same distances from their bases have equal areas, $A(x) = B(x)$ then the solids have the same volume. Prove the theorem. (6.1)

15. Let $r, h > 0$. The region bounded by $\dfrac{x}{r} + \dfrac{y}{h} = 1$ and the x-, y-axes is revolved about the y-axis. Find the volume. Do you recognize the formula? (6.2, 6.3, 6.4)

Answers to Chapter 6, Part C Questions

1. (a) $A = \int_0^{\sqrt{\frac{1+a}{a}}} (1 + a - ax^2)\, dx$

    $= \left[x + ax - \frac{ax^3}{3} \right]_0^{\sqrt{\frac{1+a}{a}}}$

    $= \frac{2(1+a)}{3} \sqrt{\frac{1+a}{a}}$

    (b) $A' = \frac{4}{3}\sqrt{\frac{1+a}{a}} - 2\frac{1+a}{3a^2\sqrt{\frac{1+a}{a}}} = 0$

    $\therefore a = \frac{1}{2}$ minimum

2. $A = \int_1^b \frac{1}{x}\, dx = \left[\frac{x^{1-p}}{1-p} \right]_1^b$

    $= \frac{1}{p-1}(1 - b^{1-p})$

    $\therefore \lim_{b\to\infty} A(b) = \frac{1}{p-1}$

3. If $0 < p < 1$ (same as #2)
    $\lim_{b\to\infty} A(b) \to -\infty$

4. (a) $V = \int_0^{16} \pi \left[\left(5 - \frac{y}{4}\right)^2 - (5 - \sqrt{y})^2 \right] dy$

    $= \pi \int_0^{16} \left(10\sqrt{y} - \frac{7}{2} y + \frac{y^2}{16} \right) dy$

    $= 64\pi$

    (b) $V = \int_0^{16} \pi \left[(\sqrt{y}+1)^2 - \left(\frac{y}{4}+1\right)^2 \right] dy$

    $= \pi \int_0^{16} \left(\frac{y}{2} + 2\sqrt{y} - \frac{y^2}{16} \right) dy$

    $= 64\pi$

5. $V = \int_0^{\sqrt{r^2-a^2}} 2\pi x \left(\sqrt{r^2 - x^2} - ax \right) dx$

    $= 2\pi \int_0^{\sqrt{r^2-a^2}} \left[x(r^2 - x^2)^{1/2} - ax^2 \right] dx$

    $= 2\pi \left[-\frac{1}{3}(r^2 - x^2)^{3/2} - \frac{a}{2} x^2 \right]_0^{\sqrt{r^2-a^2}}$

    $F(\sqrt{r^2-a^2}) = \left[-\frac{1}{3}(r^2 - r^2 + a^2)^{3/2} - \frac{a}{2}(r^2 - a^2) \right]$

    $= -\frac{1}{3}[a^3] - \frac{ar^2}{2} + \frac{a^3}{2} = \frac{1}{6}a^3 - \frac{ar^2}{2}$

    $F(0) = \left[-\frac{1}{3}(r^2)^{3/2} \right] = -\frac{1}{3}r^3$

    $= 2\pi \left[\frac{1}{6} a^3 - \frac{ar^2}{2} - \frac{1}{3}r^3 \right]$

    $= 2\pi \left[\frac{a^3 - 3ar^2 - 2r^2}{6} \right]$

    $= \frac{\pi}{3}(2r^3 + a^3 - 3ar^2)$

6. $A = \frac{\pi}{2} ab \quad \bar{x} = 0$ by symmetry

    $\bar{y} A = \int_{-a}^a \frac{1}{2} \left(\frac{b}{a} \sqrt{a^2 - x^2} \right) dx = \frac{2}{3} ab^2$

    $\Rightarrow \bar{y} = \frac{4}{3} \frac{b}{\pi}$

7. (a) $W = wd$ (ft-lb)
    (b) component of force along the inclined plane
    $= w \sin\theta$ and distance traveled $= \frac{d}{\sin\theta}$
    hence $W = wd$ same as (a)

8. (a) $F = \int_0^{100} 62.5 \times (1000) \sec(\pi/6)\, dx$

    $= \frac{125{,}000}{\sqrt{3}} \left[\frac{x^2}{2} \right]_0^{100}$

    $\cong 361{,}000{,}000$ lbs

    (b) $F = \int_0^{75} 62.5 \times (1000) \sec(\pi/6)\, dx$

    $\cong 203{,}000{,}000$ lbs

9. $\therefore 2y =$ height of shell cylindrical

    left overs of the sphere
    $dv = 2\pi \times (2y)\, dy = 4\pi \times \sqrt{4 - x^2}$
    so integrate
    $V = \int_{\sqrt{3}}^2 x\sqrt{4 - x^2}\, dx = 4\pi/3$

Since the volume of a sphere of radius 2 is

$$\frac{4}{3}\pi r^3 = \frac{4}{3}\pi(2)^3 = \frac{32}{3}\pi$$

hole = Sphere − Leftover after boring

$$= \frac{32}{3}\pi - \frac{4}{3}\pi$$

$$= \frac{28}{3}\pi$$

10. Let $F = G\dfrac{M_1 M_2}{a^2}$ where

$G = 6.667 \times 10^{-11}$ Nm²/kg², F be the universal gravitation

and the work is $W = F \cdot a = G\dfrac{M_1 M_2}{a}$

11. $\dfrac{\pi a^3}{15}$

12. The side of the square is $2\sqrt{16-x^2}$ hence the area is $4(16-x^2)$

the volume $= \displaystyle\int_{-4}^{4} 4(16-x^2)\,dx$

$$= 2\int_{0}^{4} 4(16-x^2)\,dx$$

$$= 8\int_{0}^{4}(16-x^2)\,dx$$

$$= 8\left[16x - \frac{x^3}{3}\right]_0^4$$

$$= \frac{1024}{3}$$

13. $r = \sqrt{8^2 - x^2}$

$dw = \pi\left(\sqrt{64-x^2}\right)^2 dx \cdot w \cdot (8-x)$

$$\int_0^8 \pi w\left(512 - 64x - 8x^2 + x^3\right)dx$$

$$= \frac{5120\,\pi(w)}{3}$$

14. $V = \displaystyle\int_0^h B(x)\,dx = \int_0^h A(x)\,dx$

Since $B(x) = A(x)$

15. $V = \displaystyle\int_0^r 2\pi \times \frac{h}{r}(r-x)\,dx$

$$= \frac{1}{3}\pi r^2 h \text{ volume of a cone}$$

CHAPTER 7 The Transcendental Functions
Part A

True or false (1–5)

1. Every even function is one to one (7.1)

2. Every odd function is one to one (7.1)

3. If $f(x)$ is a linear function then $f^{-1}(x)$ is a linear function (7.1)

4. If $f(x)$ is a quadratic function then $f^{-1}(x)$ is a quadratic function (7.1)

5. $\log_b 1 = 0$ (7.2)

Complete

6. $\log_b b =$ (7.2)

7. $\log_b x =$ (7.2)

8. $\int_1^e \dfrac{dt}{t} =$ (7.3)

9. (a) $\ln e^x =$ _____ for all real x (7.4)
   (b) $e^{\ln x} =$ _____ for $x > 0$

True or false

10. $\ln \dfrac{1}{2} = -\int_1^2 \dfrac{dt}{t}$ (7.4)

11. (a) $\int e^x \, dx =$ _____ (7.4)

    (b) $\int e^u \, du =$ _____

True or false

12. $\int \dfrac{2x \, dx}{x^2 + 1} = \ln(x^2 + 1) + c$ (7.4)

13. If u is a positive function of x and $r \in R$ then $\dfrac{d}{dx} u^r =$ (7.5)

14. (a) $\dfrac{d}{dx}(p^x) =$ _____ $p + 1$ (7.5)

    (b) $\int p^x \, dx =$ _____

15. $\dfrac{d}{dx}(\log_p x) =$ _____ $p > 0 \quad p \neq 1$ (7.5)

16. Give the Domain for each: (7.7)
    (a) $\sin^{-1} x$ Domain _____
    (b) $\cos^{-1} x$ Domain _____
    (c) $\tan^{-1} x$ Domain _____
    (d) $\sec^{-1} x$ Domain _____

17. (a) Define $\sinh x$ in terms of e^x (7.8)
    (b) Define $\cosh x$ in terms of e^x

18. True or false
    $$\frac{d}{dx}\sinh x = \cosh x \text{ and } \frac{d}{dx}\cosh x = \sinh x$$ (7.8)

19. Using $\sinh x$, $\cosh x$ (7.8)
    Define (a) $\tanh x =$
           (b) $\text{sech } x =$

True or false

20. $\lim_{n \to \infty} \left(1 + \frac{1}{n}\right)^n = e, n \neq 0$ (7.4)

True or false

21. Given f and f^{-1} both differentiable if a is in the domain of f such that $f(a) = b$ and if $f'(a) \neq 0$ then $\left(f^{-1}\right)'(b) = \dfrac{1}{f'(a)}$

22. Does $\cot x$, $0 < x < \pi$ have an inverse?

True or false (23–25)

23. $2^\pi = e^{2 \ln \pi}$

24. Does $f(x) = e^{-x^2}$ have an inverse?

25. $\ln x = k$ has a unique solution

Answers to Chapter 7, Part A Questions

1. F
2. F
3. T
4. F
5. T
6. 1
7. $b^y = x$
8. 1
9. (a) x (b) x
10. $\ln \dfrac{1}{2} = -\int_1^2 \dfrac{dt}{t}$
11. (a) $e^x + c$ (b) $e^u + c$
12. T
13. $ru^{r-1} \dfrac{du}{dx}$
14. (a) $p^x \ln p$ (b) $\dfrac{p^x}{\ln p} + c$
15. $\dfrac{1}{x \ln p}$
16. (a) $[-1, 1]$ (c) $(-\infty, \infty)$
    (b) $[-1, 1]$ (d) $(-\infty, \infty)$
17. (a) $\dfrac{1}{2}(e^x - e^{-x})$
    (b) $\dfrac{1}{2}(e^x + e^{-x})$
18. True
19. $\tanh x = \dfrac{e^x - e^{-x}}{e^x + e^{-x}}$
    $\operatorname{sech} x = \dfrac{2}{e^x + e^{-x}}$
20. $\ln\left(1 + \dfrac{1}{n}\right)^n = n \ln\left(1 + \dfrac{1}{n}\right)$
    $= \dfrac{\ln\left(1 + \dfrac{1}{n}\right)}{\dfrac{1}{n}} \quad \dfrac{\dfrac{1}{n} \to 0}{n \to \infty} 1$
    $\therefore \lim_{n \to \infty} \left(1 + \dfrac{1}{n}\right)^n = e$
21. T
22. Yes
23. F
24. F, No
25. T

Part B

7.1 One-to-One Functions; Inverses

1. Determine whether or not $f(x) = 4x + 3$ is one-to-one, and, if so, find its inverse.

2. Determine whether or not $f(x) = 5x - 7$ is one-to-one, and, if so, find its inverse.

3. Determine whether or not $f(x) = 2x^2 - 1$ is one-to-one, and, if so, find its inverse.

4. Determine whether or not $f(x) = 2x^3 + 1$ is one-to-one, and, if so, find its inverse.

5. Determine whether or not $f(x) = (1 - 3x)^3$ is one-to-one, and, if so, find its inverse.

6. Determine whether or not $f(x) = (x + 1)^4$ is one-to-one, and, if so, find its inverse.

7. Determine whether or not $f(x) = (x - 1)^5 + 1$ is one-to-one, and, if so, find its inverse.

8. Determine whether or not $f(x) = (4x + 5)^3$ is one-to-one, and, if so, find its inverse.

9. Determine whether or not $f(x) = 2 + (x + 1)^{5/3}$ is one-to-one, and, if so, find its inverse.

10. Determine whether or not $f(x) = \dfrac{1}{x - 2}$ is one-to-one, and, if so, find its inverse.

11. Determine whether or not $f(x) = \dfrac{2x - 3}{3x + 1}$ is one-to-one, and, if so, find its inverse.

12. Determine whether or not $f(x) = \dfrac{1}{x - 1} + 2$ is one-to-one, and, if so, find its inverse.

13. Determine whether or not $f(x) = \dfrac{2}{x^3 - 1}$ is one-to-one, and, if so, find its inverse.

14. Sketch the graph of f^{-1} given the graph of f shown below.

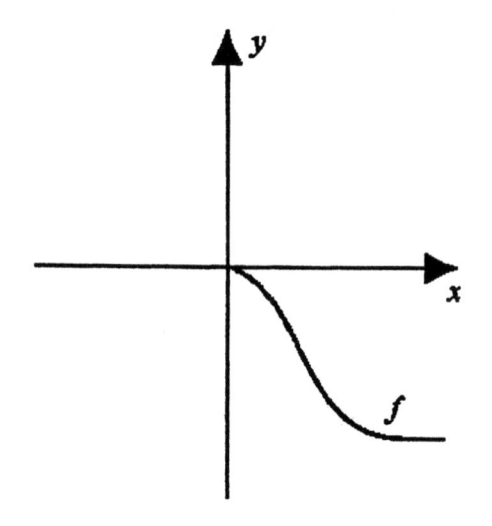

15. Sketch the graph of f^{-1} given the graph of f shown below.

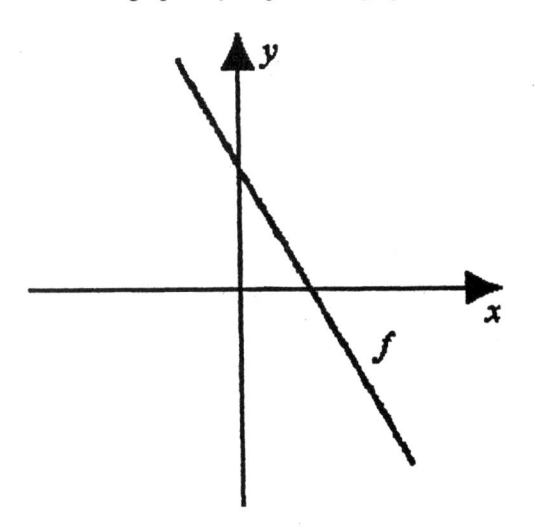

16. Given that f is one-to-one and $f(2) = 1$, $f'(2) = 3$, $f(3) = 2$, $f'(3) = 4$, deduce, if possible, $(f^{-1})'(2)$.

17. Show that $f(x) = x^3 + 2x - 5$ has an inverse and find $(f^{-1})'(7)$.

18. Show that $f(x) = \sin 2x - 4x$ has an inverse and find $(f^{-1})'(0)$.

19. Find a formula for $(f^{-1})'(x)$ given that f is one-to-one and satisfies $f'(x) = \dfrac{1}{f(x)}$.

20. Show that $f(x) = \displaystyle\int_{\pi/2}^{x} (1 + \sin^2 t)\,dt$ has an inverse and find $(f^{-1})'(0)$.

7.2 The Logarithm Function; Part I

21. Estimate ln 25 on the basis of the following table.

n	ln n	n	ln n
1	0.00	6	1.79
2	0.69	7	1.95
3	1.10	8	2.08
4	1.39	9	2.20
5	1.61	10	2.30

22. Estimate ln 2.2 on the basis of the following table.

n	ln n	n	ln n
1	0.00	6	1.79
2	0.69	7	1.95
3	1.10	8	2.08
4	1.39	9	2.20
5	1.61	10	2.30

23. Estimate $\ln 5^4$ on the basis of the following table.

n	$\ln n$	n	$\ln n$
1	0.00	6	1.79
2	0.69	7	1.95
3	1.10	8	2.08
4	1.39	9	2.20
5	1.61	10	2.30

24. Estimate $\ln \sqrt{425}$ on the basis of the following table.

n	$\ln n$	n	$\ln n$
1	0.00	6	1.79
2	0.69	7	1.95
3	1.10	8	2.08
4	1.39	9	2.20
5	1.61	10	2.30

25. Estimate $\ln 3.5 = \int_1^{3.5} \dfrac{dt}{t}$ using the approximation $\dfrac{1}{2}\left[L_f(P) + U_f(P)\right]$ with

$$P = \left\{1 = \dfrac{4}{4}, \dfrac{5}{4}, \dfrac{6}{4}, \dfrac{7}{4}, \dfrac{8}{4}, \dfrac{9}{4}, \dfrac{10}{4} = \dfrac{5}{2}\right\}$$

26. Taking $\ln 4 \approx 1.39$, use differentials to estimate (a) $\ln 4.25$; (b) $\ln 3.75$.

27. Solve $\ln x + \ln(x+2) = 0$ for x.

28. Solve $\ln(x+1) - \ln(x-2) = 1$ for x.

29. Solve $2 \ln x = \ln(x+2)$ for x.

30. Solve $(5 - \ln x)(2 \ln x) = 0$ for x.

7.3 The Logarithm Function; Part II

31. Determine the domain and find the derivative of $f(x) = \ln(x^3 - 1)$.

32. Determine the domain and find the derivative of $f(x) = \ln(1 - x^2)$.

33. Determine the domain and find the derivative of $f(x) = \ln 2x\sqrt{2+x}$.

34. Determine the domain and find the derivative of $f(x) = x \ln(2x + x^2)$.

35. Determine the domain and find the derivative of $f(x) = \ln 3x\sqrt{3 - x^2}$.

36. Determine the domain and find the derivative of $f(x) = \ln(\sec x)$.

37. Calculate $\displaystyle\int \dfrac{dx}{5 - 3x}$.

38. Calculate $\displaystyle\int \dfrac{(4x - 2)dx}{x^2 - x}$.

39. Calculate $\displaystyle\int \frac{x^2}{3x^3 - 5}\,dx$.

40. Calculate $\displaystyle\int \cot 3x\,dx$.

41. Calculate $\displaystyle\int \frac{\sin x + \cos x}{\sin x}\,dx$.

42. Calculate $\displaystyle\int_0^1 \frac{x}{6x^2 + 1}\,dx$.

43. Calculate $\displaystyle\int_{1/3}^{1/2} \frac{x}{1 - 3x^2}\,dx$.

44. Calculate $\displaystyle\int_0^{\pi/2} \frac{\sin x}{2 - \cos x}\,dx$.

45. Calculate $g'(x)$ by logarithmic differentiation if $g(x) = \dfrac{x\sqrt{x^2 + 1}}{(x + 1)^{2/3}}$.

46. Calculate $g'(x)$ by logarithmic differentiation if $g(x) = \sqrt[3]{\dfrac{3(x^2 + 5)\cos^4 2x}{(x^3 - 8)^2}}$.

47. Calculate $g'(x)$ by logarithmic differentiation if $g(x) = \sqrt[5]{\dfrac{\sin x}{(1 + x^5)^3}}$.

48. The region bounded by the graph of $f(x) = \dfrac{1}{5 - x^2}$ and the x-axis for $0 \le x \le 2$ is revolved around the y-axis. Find the volume of the solid that is generated.

49. Calculate $\displaystyle\int \tan 5x\,dx$.

50. Calculate $\displaystyle\int \sec \frac{2\pi x}{3}\,dx$.

51. Calculate $\displaystyle\int \csc\left(2\pi - \frac{x}{2}\right) dx$.

52. Calculate $\displaystyle\int \cot(2\pi x - 3)\,dx$.

53. Calculate $\displaystyle\int e^{2x} \sin e^{2x}\,dx$.

54. Calculate $\displaystyle\int \frac{\sec^2 2x\,dx}{2 + \tan 2x}$.

55. Calculate $\displaystyle\int \frac{3 \sec x \tan x\,dx}{(2 + \sec x)^3}$.

56. Evaluate $\displaystyle\int_0^{\ln \pi/3} e^{2x} \csc e^{2x}\,dx$.

57. Evaluate $\int_{7\pi/12}^{5\pi/6} \tan(2x - \pi)\,dx$.

58. Evaluate $\int_{\pi/4}^{\pi/2} \dfrac{3\csc x \cot x\,dx}{2 + \csc x}$.

7.4 The Exponential Function

59. Differentiate $y = e^{-3x/2}$.

60. Differentiate $y = e^{2x^2 - x}$.

61. Differentiate $y = e^{\sqrt[3]{x}+1}$.

62. Differentiate $y = e^{\sqrt[3]{x}} \ln \sqrt{x^3}$.

63. Differentiate $y = (e^{x^3+x} + 1)^2$.

64. Differentiate $y = \dfrac{e^{3x} - e^{2x}}{e^{3x} + e^{2x}}$.

65. Differentiate $y = e^{x \sin x}$.

66. Differentiate $y = e^{-2x} \sin 3x$.

67. Differentiate $y = e^{\sin 2x}$.

68. Differentiate $y = e^{1/x}$.

69. Differentiate $y = \dfrac{e^{2x}}{2x}$.

70. Calculate $\int \dfrac{e^x}{\sqrt{1 - e^x}}\,dx$.

71. Calculate $\int \dfrac{e^{2x} + e^{-2x}}{e^{2x} - e^{-2x}}\,dx$.

72. Calculate $\int e^{5x/2}\,dx$.

73. Calculate $\int \dfrac{e^{\sqrt[3]{x}}}{\sqrt[3]{x^2}}\,dx$.

74. Calculate $\int \left(\dfrac{1}{2}e^{2x} - e^{-2x}\right)^2 dx$.

75. Calculate $\int \dfrac{e^{3x}}{3e^{3x} - 2}\,dx$.

76. Calculate $\int \dfrac{2xe^{3x^2}}{e^{3x^2} + 2}\,dx$.

77. Evaluate $\int_0^1 e^{x/2}\,dx$.

78. Evaluate $\int_0^{\ln x/2} e^{-3x}\,dx$.

79. Evaluate $\int_0^1 \dfrac{e^{\sqrt{x}}}{\sqrt{x}}\,dx$.

80. Evaluate $\int_0^1 \dfrac{e^{2x}}{2+e^{2x}}\,dx$.

81. Sketch the region bounded by $y = e^{3x}$, $y = e^x$, and $x = 2$, and find its area.

82. Sketch the graph of $f(x) = xe^{-2x}$.

7.5 Arbitrary Powers; Other Bases

83. Find $\log_{10} 0.001$

84. Find $\log_9 243$

85. Find the derivative of $f(x) = 3^{\cos\sqrt{x}}$.

86. Find the derivative of $f(x) = 5^{2^x}$.

87. Calculate $\int 4^{2x}\,dx$.

88. Calculate $\int 3^{-x/2}\,dx$.

89. Calculate $\int 3x 5^{-x^2}\,dx$.

90. Calculate $\int \dfrac{\log_3 2x}{x}\,dx$.

91. Calculate $\int \dfrac{\log_4 \sqrt{2x+2}}{x}\,dx$.

92. Find $f'(e)$ if $f(x) = e^x + x^e$.

93. Find $f'(e)$ if $f(x) = \ln(\ln(\ln x^2))$.

94. Find $\dfrac{d}{dx}\left[(\sin x)^x\right]$ by logarithmic differentiation.

95. Find $\dfrac{d}{dx}(x^4 4^x)$ by logarithmic differentiation.

96. Find $\dfrac{d}{dx}\left[(\tan x)^x\right]$ by logarithmic differentiation.

97. Evaluate $\int_1^2 3^{-x}\,dx$.

138   Calculus: One and Several Variables

98. Evaluate $\int_0^1 2^{3x}\, dx$.

99. Evaluate $\int_0^5 \dfrac{4a^{\sqrt{2x-1}}}{\sqrt{2x-1}}\, dx$.

100. Evaluate $9^{(\ln 5)/(\ln 3)}$.

7.6   Exponential Growth and Decay

101. Find the amount of interest earned by $700 compounded continuously for 10 years (a) at 8% (b) at 9% (c) at 10%.

102. The population of a certain city increases at a rate proportional to the number of its inhabitants at any time. If the population of the city was originally 10,000 and it doubled in 15 years, in how many years will it triple?

103. A certain radioactive substance has a half-life of 1300 years. Assume an amount y_0 was initially present. (a) Find a formula for the amount of substance present at any time. (b) In how many years will only 1/10 of the original amount remain?

104. A tank initially contains 100 gal. of pure water. At time $t = 0$, a solution containing 4 lb. of dissolved salt per gal. flows into the tank at 3 gal./min. The well-stirred mixture is pumped out of the tank at the same rate. (a) How much salt is present at the end of 30 min? (b) How much salt is present after a very long time?

105. A tank initially contains 150 gal. of brine in which there is dissolved 30 lb. of salt. At $t = 0$, a brine solution containing 3 lb. of dissolved salt per gallon flows into the tank at 4 gal./min. The well-stirred mixture flows out of the tank at the same rate. How much salt is in the tank at the end of 10 min?

106. An object of unknown temperature is put in a room held at 30° F. After 10 minutes, the temperature of the object is 0° F; 10 minutes later it is 15° F. What was the object's initial temperature?

107. Show that $\dfrac{dy}{dx} - e^{-y} \sec^2 x = 0$ is a separable equation and then find its solution.

7.7   The Inverse Trigonometric Functions

108. Determine the exact value for $\tan^{-1}(\sqrt{3}/3)$.

109. Determine the exact value for $\cot\left[\sin^{-1}(-1/4)\right]$.

110. Determine the exact value for $\tan\left[\sin^{-1}(-1/4)\right]$.

111. Determine the exact value for $\tan\left[\sec^{-1}(3/2)\right]$.

112. Determine the exact value for $\sin^{-1}[\cot(\pi/4)]$.

113. Determine the exact value for $\sec\left[\sin^{-1}(3/4)\right]$.

114. Determine the exact value for $\cos\left[\sin^{-1}(-3/4)\right]$.

115. Determine the exact value for $\sin 2\left[\tan^{-1}(1/3)\right]$.

116. Determine the exact value for $\sin\left[\tan^{-1}(-2/3)\right]$.

117. Determine the exact value for $\cos\left[\tan^{-1}(-3/4)\right]$.

118. Taking $0 < x < 1$, calculate $\sin(\cos^{-1} x)$.

119. Taking $0 < x < 1/2$, calculate $\tan(\cos^{-1} 2x)$.

120. Differentiate $f(x) = x \tan^{-1} 3x$.

121. Differentiate $f(x) = x \sin^{-1} 2x$.

122. Differentiate $f(x) = \sec(\tan^{-1} x)$.

123. Differentiate $y = \cos^{-1}(\cos x)$.

124. Differentiate $f(x) = e^{3x} \sin^{-1} 2x$.

125. Differentiate $f(x) = \tan^{-1}\left(\dfrac{x-1}{x+1}\right)$.

126. Differentiate $f(x) = \sin^{-1} x + \sqrt{1 - x^2}$.

127. Differentiate $y = \sin^{-1} \dfrac{1}{\sqrt{1 + x^2}}$.

128. Differentiate $f(x) = \ln(x^2 + 4) - x \tan^{-1} \dfrac{x}{2}$.

129. Evaluate $\displaystyle\int_0^1 \dfrac{dx}{16 + 9x^2}$.

130. Evaluate $\displaystyle\int_0^{1/4} \dfrac{dx}{\sqrt{1 - 4x^2}}$.

131. Evaluate $\displaystyle\int_{-1/2}^0 \dfrac{dx}{\sqrt{1 + 4x^2}}$.

132. Evaluate $\displaystyle\int_1^3 \dfrac{dx}{x^2 - 6x + 13}$.

133. Calculate $\displaystyle\int \dfrac{x}{1 + x^4}\, dx$.

134. Calculate $\displaystyle\int \dfrac{dx}{1 + 9x^2}$.

135. Calculate $\displaystyle\int \dfrac{dx}{x\sqrt{1 - (\ln x)^2}}$.

7.8 The Hyperbolic Sine and Cosine Functions

136. Differentiate $y = \sinh x^3$.

137. Differentiate $y = \cosh x^2$.

138. Differentiate $y = e^{2 \ln \sinh 3x}$.

139. Differentiate $y = \cosh^2 3x$.

140. Differentiate $y = x \sinh \sqrt{x}$.

141. Differentiate $y = (\sinh 2x)^{3x}$.

142. Differentiate $y = \cosh(e^{-2x})$.

143. Differentiate $y = 2^{\cosh 3x}$.

144. Differentiate $y = \sinh 2x \cosh^2 2x$.

145. Calculate $\displaystyle\int \frac{\sinh x}{1 + \cosh x}\, dx$.

146. Calculate $\displaystyle\int \sqrt{\sinh 2x}\, \cosh 2x\, dx$.

147. Calculate $\displaystyle\int \sinh^5 \pi x \cosh \pi x\, dx$.

7.9 The Other Hyperbolic Functions

148. Differentiate $y = \operatorname{sech} x \, \tanh x$.

149. Differentiate $y = e^{3x} \tanh 2x$.

150. Differentiate $y = \coth(\sqrt{2x^2 + 1})$.

151. Differentiate $y = \operatorname{csch}(\tan e^{3x})$.

152. Differentiate $y = \tanh^2 x$.

153. Differentiate $y = \tanh^3(3x + 2)$.

154. Differentiate $y = x \operatorname{sech} x$.

155. Differentiate $y = \sqrt{1 + x^2} + \tanh x$.

156. Differentiate $y = (\tanh 3x)^2$.

157. Differentiate $y = \tanh(\sin 2x)$.

158. Calculate $\displaystyle\int \tanh 5x\, dx$.

159. Calculate $\displaystyle\int x \tanh x^2 \operatorname{sech}^2 x^2\, dx$.

Answers to Chapter 7, Part B Questions

1. one-to-one; $f^{-1}(x) = \dfrac{x-3}{4}$

2. one-to-one; $f^{-1}(x) = \dfrac{x+7}{5}$

3. not one-to-one

4. one-to-one; $f^{-1}(x) = \sqrt[3]{\dfrac{x-1}{2}}$

5. one-to-one; $f^{-1}(x) = \dfrac{1-\sqrt[3]{x}}{3}$

6. not one-to-one

7. one-to-one; $f^{-1}(x) = (x-1)^{1/5} + 1$

8. one-to-one; $f^{-1}(x) = \dfrac{\sqrt[3]{x}-5}{4}$

9. one-to-one; $f^{-1}(x) = (x-2)^{3/5} - 1$

10. one-to-one; $f^{-1}(x) = \dfrac{1+2x}{x}$

11. one-to-one; $f^{-1}(x) = \dfrac{x+3}{2-3x}$

12. one-to-one; $f^{-1}(x) = \dfrac{x-1}{x-2}$

13. one-to-one; $f^{-1}(x) = \sqrt[3]{\dfrac{2+x}{x}}$

14.

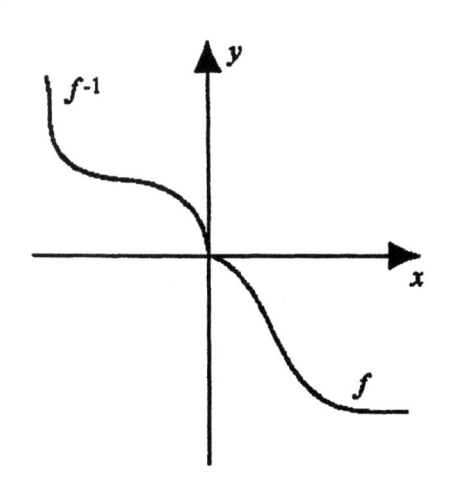

15.

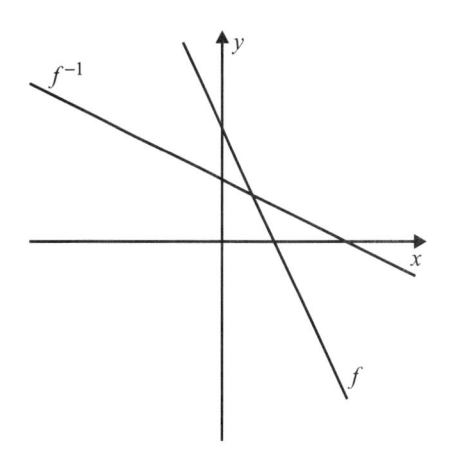

16. 1/4

17. $f'(x) = 3x^2 + 2 > 0$, so f is increasing and therefore, one-to-one. $(f^{-1})'(7) = 1/14$.

18. $f'(x) = 2\cos 2x - 4 < 0$, so f is decreasing and therefore, one-to-one. $(f^{-1})'(0) = -1/2$.

19. $(f^{-1})'(x) = x$

20. $f'(x) = 1 + \sin^2 x > 0$, so f is increasing and therefore, one-to-one. $(f^{-1})'(0) = 1/2$.

21. 3.22

22. 0.79

23. 6.44

24. 3.03

25. 1.3

26. (a) 1.45 (b) 1.33

27. $x = -1 + \sqrt{2}$ ($-1 - \sqrt{2}$ is not a solution because $h(-1 - \sqrt{2})$ is undefined.)

28. $x = \dfrac{2e+1}{e-1}$

29. $x = 2$ (-1 is not a solution because $\ln(-1)$ is undefined.)

30. $x = 1, e^5$

31. $\text{dom}(f) = (1, \infty)$; $f'(x) = \dfrac{3x^2}{x^3 - 1}$

32. $\text{dom}(f) = (-1, 1)$; $f'(x) = \dfrac{2x}{x^2 - 1}$

33. $\text{dom}(f) = (0, \infty)$;
$f'(x) = \dfrac{1}{x} + \dfrac{1}{2(2+x)} = \dfrac{4+3x}{2x(2+x)}$

34. $\text{dom}(f) = (-\infty, -2) \cup (0, \infty)$;
$f'(x) = \ln(2x + x^2) + \dfrac{2 + 2x}{2x + x^2}$

35. $\text{dom}(f) = (0, \sqrt{3})$; $f'(x) = \dfrac{3 - 2x^2}{x(3 - x^2)}$

36. $\text{dom}(f) = (2\pi k - \pi/2,\ 2\pi k + \pi/2)$, $k = 0, \pm1, \pm2, \ldots$; $f'(x) = \tan x$

37. $-\dfrac{1}{3}\ln|5 - 3x| + C$

38. $2\ln|x^2 - x| + C$

39. $\dfrac{1}{9}\ln|3x^3 - 5| + C$

40. $\dfrac{1}{3}\ln|\sin 3x| + C$

41. $x + \ln|\sin x| + C$

42. $\dfrac{1}{12}\ln 7$

43. $\dfrac{1}{6}\ln\dfrac{8}{3}$

44. $\ln 2$

45. $\dfrac{x\sqrt{x^2+1}}{(x+1)^{2/3}}\left(\dfrac{1}{x} + \dfrac{x}{x^2+1} - \dfrac{2}{3(x+1)}\right)$

46. $\dfrac{1}{3}\sqrt[3]{\dfrac{3(x^2+5)\cos^4 2x}{(x^3-8)^2}}$
$\times \left(\dfrac{2x}{x^2+5} - 8\tan 2x - \dfrac{6x^2}{x^3-8}\right)$

47. $\sqrt[5]{\dfrac{\sin x}{(1+x^5)^3}}\left(\dfrac{1}{5}\cot x - \dfrac{3x^4}{(1+x)^5}\right)$

48. $V = \pi \ln 5$

49. $\dfrac{1}{5}\ln|\sec 5x| + C$

50. $\dfrac{3}{2\pi}\ln\left|\sec\dfrac{2\pi x}{3} + \tan\dfrac{2\pi x}{3}\right| + C$

51. $-2\ln\left|\csc\left(2\pi - \dfrac{x}{2}\right) - \cot\left(2\pi - \dfrac{x}{2}\right)\right| + C$

52. $\dfrac{-1}{2\pi}\ln|\sin(2\pi x - 3)| + C$

53. $\dfrac{-1}{2}\cos e^{2x} + C$

54. $\dfrac{1}{2}\ln|2 + \tan 2x| + C$

55. $\dfrac{-3}{2}\dfrac{1}{(2 + \sec x)^2} + C$

56. $\dfrac{1}{2}\ln\left|\dfrac{\csc\dfrac{\pi^2}{9} - \cot\dfrac{\pi^2}{9}}{\csc 1 - \cot 1}\right|$

57. $\dfrac{1}{4}\ln 3$

58. $3\ln\dfrac{2 + \sqrt{2}}{3}$

59. $-\dfrac{3}{2}e^{-3x/2}$

60. $(4x - 1)e^{2x^2 - x}$

61. $\dfrac{1}{3}x^{-2/3}e^{\sqrt[3]{x}+1}$

62. $e^{\sqrt[3]{x}}\left(x^{-2/3}\ln x^2 + \dfrac{3}{2x}\right)$

63. $2(e^{x^3+x} + 1)(3x^2 + 1)e^{x^3+x}$

64. $\dfrac{2e^{5x}}{(e^{3x} + e^{2x})^2}$

65. $e^{x\sin x}(x\cos x + \sin x)$

66. $e^{-2x}(3\cos 3x - 2\sin 3x)$

67. $2e^{\sin 2x}\cos 2x$

68. $\dfrac{-e^{1/x}}{x^2}$

69. $\dfrac{1}{2}e^{2x}\dfrac{2x - 1}{x^2}$

70. $-2\sqrt{1 - e^x} + C$

Answers

71. $\frac{1}{2}\ln(e^{2x} + e^{-2x}) + C$

72. $\frac{2}{5}e^{5x/2} + C$

73. $3e^{\sqrt[3]{x}} + C$

74. $\frac{1}{16}e^{4x} - x - \frac{1}{4}e^{-4x} + C$

75. $\frac{1}{9}\ln|3e^{3x} - 2| + C$

76. $\frac{1}{3}\ln(e^{3x^2} + 2) + C$

77. $2(e^{1/2} - 1)$

78. $\frac{1}{3}\left[1 - \left(\frac{2}{\pi}\right)^3\right]$

79. $2(e - 1)$

80. $\frac{1}{2}\ln\frac{2 + e^2}{3}$

81. $\frac{1}{3}e^6 - e^2 + \frac{2}{3}$

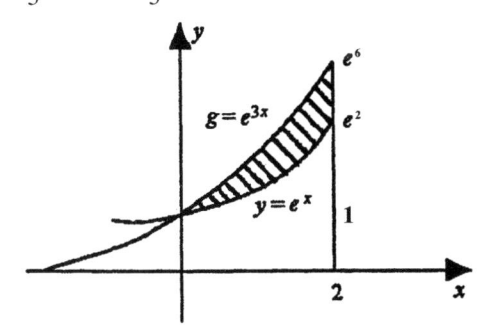

82.

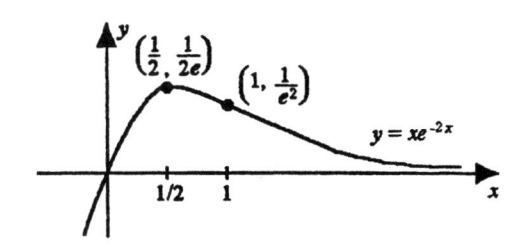

83. -3

84. $5/2$

85. $\frac{-3^{\cos\sqrt{x}}\sin\sqrt{x}(\ln 3)}{2\sqrt{x}}$

86. $5^{2^x} 2^x (\ln 5)(\ln 2)$

87. $\frac{1}{2\ln 4} 4^{2x} + C$

88. $\frac{-2}{\ln 3} 3^{-x/2} + C$

89. $\frac{-3}{2\ln 5} 5^{-x^2} + C$

90. $\frac{1}{2\ln 3}(\ln 2x)^2 + C$

91. $\frac{1}{4\ln 4}[\ln(2x)]^2 + 2\ln x + C$

92. $e^e\left(1 + \frac{1}{e}\right)$

93. $\frac{1}{e\ln 2}$

94. $(\sin x)^x[\ln \sin x + x \cot x]$

95. $x^4 4^x\left(\frac{4}{x} + \ln 4\right)$

96. $(\tan x)^x[\ln \tan x + x(\cot x + \tan x)]$

97. $\frac{2}{9\ln 3}$

98. $\frac{7}{3\ln 2}$

99. $\frac{4}{\ln a}(a^3 - a)$

100. 25

101. (a) $857.88 (c) $1202.80
     (b) $1021.72

102. 23.8 yrs

103. (a) $y(t) = y_0 e^{-0.0005332t}$
     (b) approximately 4319 yrs

104. (a) 237.4 lbs   (b) 400 lbs

105. approximately 128.3 lbs

106. $-30°$ F

107. The equation can be rewritten as
$e^y \, dy = \sec^2 x \, dx$; $y = \ln(\tan x + C)$

108. $\pi/6$

109. $-\sqrt{15}$

110. $\dfrac{-1}{\sqrt{15}} = \dfrac{-\sqrt{15}}{15}$

111. $\dfrac{\sqrt{5}}{2}$

112. $\pi/2$

113. $\dfrac{4}{\sqrt{7}} = \dfrac{4\sqrt{7}}{7}$

114. $\dfrac{\sqrt{7}}{4}$

115. $3/5$

116. $\dfrac{-2}{\sqrt{13}} = \dfrac{-2\sqrt{13}}{13}$

117. $\dfrac{\sqrt{7}}{4}$

118. $\sqrt{1-x^2}$

119. $\dfrac{\sqrt{1-4x^2}}{2x}$

120. $\dfrac{3x}{1+9x^2} + \tan^{-1} 3x$

121. $\dfrac{2x}{\sqrt{1-4x^2}} + \sin^{-1} 2x$

122. $\dfrac{x}{\sqrt{x^2+1}}$

123. 1

124. $\dfrac{2e^{3x}}{\sqrt{1-4x^2}} + 3e^{3x} \sin^{-1} 2x$

125. $\dfrac{1}{x^2+1}$

126. $\sqrt{\dfrac{1-x}{1+x}}$

127. $-\dfrac{1}{1+x^2}$

128. $-\tan^{-1} \dfrac{x}{2}$

129. $\dfrac{1}{12} \tan^{-1} \dfrac{3}{4}$

130. $\pi/12$

131. $\pi/8$

132. $\pi/8$

133. $\dfrac{1}{2} \tan^{-1} x^2 + C$

134. $\dfrac{1}{3} \tan^{-1} 3x + C$

135. $\sin^{-1}(\ln x) + C$

136. $3x^2 \cosh x^2$

137. $2x \sinh x^2$

138. $3 \sinh 6x$

139. $6 \cosh 3x \sinh 3x$

140. $\dfrac{\sqrt{x}}{2} \cosh \sqrt{x} + \sinh \sqrt{x}$

141. $(\sinh 2x)^{3x} (6x \coth 2x + 3 \ln \sinh 2x)$

142. $-2e^{-2x} \sinh(e^{-2x})$

143. $2^{\cosh 3x} (3 \ln 2 \sinh 3x)$

144. $4 \sinh^2 2x \cosh 2x + 2 \cosh^3 2x$

145. $\ln(1 + \cosh x) + C$

146. $\dfrac{1}{3}(\sinh 2x)^{3/2} + C$

147. $\dfrac{1}{6\pi} \sinh^6 \pi x + C$

148. $\text{sech}^2 x - \text{sech } x \tanh^2 x$

149. $2e^{3x} \text{sech}^2 2x + 3e^{3x} \tanh 2x$

150. $\dfrac{-1}{\sqrt{2x^2+1}} \text{csch}^2(\sqrt{2x^2+1})$

Answers

151. $-3e^{3x} \operatorname{csch}(\tan e^{3x}) \coth(\tan e^{3x}) \sec^2 e^{3x}$

152. $2 \tanh x \operatorname{sech}^2 x$

153. $9 \tanh^2(3x+2) \operatorname{sech}^2(3x+2)$

154. $\operatorname{sech} x - x \operatorname{sech} x \tanh x = \operatorname{sech} x(1 - x \tanh x)$

155. $\dfrac{x}{\sqrt{1+x^2}} + \operatorname{sech}^2 x$

156. $-6 \tanh 3x \operatorname{sech}^2 3x$

157. $2 \operatorname{sech}^2(\sin 2x) \cos 2x$

158. $\dfrac{1}{5} \ln(\cosh 5x) + C$

159. $\dfrac{1}{4} \tanh^2 x^2 + C$

Part C

1. Let $f(x) = \int_1^{2x} \sqrt{16 + t^4}\, dt$ (7.1)

   (a) Prove $f(x)$ has an inverse
   (b) Find $(f^{-1})'(0)$

2. Given $p(x) = a_n x^n + a_{n-1} x^{n-1} + \cdots + a_1 x + a_0$ a polynomial function. (7.1)
   (a) Can $p(x)$ have an inverse if its degree n is even? Explain.
   (b) Can $p(x)$ have an inverse if n is odd? Explain.

3. Use the mean-value theorem to show that $\dfrac{x-1}{x} < \ln x < x - 1$ for all $x > 0$ (7.2)

   Hint: Consider cases $x > 1$ and $0 < x < 1$ separately.

4. (a) Explain why $\int_0^{\pi} \sqrt{1 - \sin^2 x}\, dx > 0$ (7.3)
   (b) Evaluate the above integral

5. Find a formula for $\dfrac{d^n}{dx^n}\left(\ln \dfrac{1}{x}\right)$ (7.3)

6. Evaluate the limit numerically $\lim\limits_{x \to 1} \dfrac{e^x - e}{\ln x}$ recall $f'(c) = \lim\limits_{x \to c} \dfrac{f(x) - f(c)}{x - c}$ (7.4)

7. Evaluate $\int_0^1 \dfrac{5 p^{\sqrt{x+1}}}{\sqrt{x+1}}\, dx$ (7.5)

8. A boat moving in still water is subject to a retardation proportional to its velocity. Show that the velocity t seconds after the power is shut off is $v = ce^{-kt}$ where c is the velocity at the instant the power is shut off. (7.6)

9. Find the area of the region bounded by $y = \dfrac{3}{9 + x^2}$ and the x-axis between $x = -3$ and $x = 3$ (7.7)

10. The region bounded by $y = \dfrac{1}{x^2\sqrt{x^2 - 9}}$ and the x-axis between $x = 2\sqrt{3}$ and $x = 6$ is revolved around the y-axis. Find the volume of the solid that is generated. (7.7)

11. Evaluate the limit $\lim\limits_{x \to \infty} \dfrac{\sinh x}{e^x}$ (7.8)

12. Evaluate the limit $\lim\limits_{x \to \infty} \dfrac{\cosh x}{e^{ax}}$ when $0 < a < 1$ and when $a > 1$ (7.8)

13. Verify the formula, $a > 0$ $\int \dfrac{1}{\sqrt{x^2 - a^2}}\, dx = \cosh^{-1}\left(\dfrac{x}{a}\right) + C$ (7.9)

14. Show that $\log_a (x + \sqrt{x^2 - 1}) = -\log_a (x - \sqrt{x^2 - 1})$ (7.9)

15. Use the intermediate value theorem to show that $x + e^x = 0$ has a root. (7.6)

16. Show $\lim\limits_{x \to 0} \dfrac{a^x - 1}{x} = \ln a$ $a > 0$ (7.8)

17. Prove $\lim_{x \to \infty} \dfrac{e^x}{x^n} = \infty$, $n \in I^+$ (7.5)

18. If the number of bacteria doubles every hour how long will it take for 1000 bacteria to produce a billion? (7.6)

19. The top of a 15 ft ladder is sliding down a wall. When the base of the ladder is 9 ft from the wall, it is sliding away at a rate of 2 ft/s. What is the angle between the wall and the ladder at that instant and how fast is the angle increasing at that instant? (7.7)

20. Use $\tanh y = x$ to express the inverse function $y = \tanh^{-1} x$. (7.9)

$$y = \tanh^{-1} x = \frac{1}{2} \ln\left(\frac{1+x}{1-x}\right) \quad -1 < x < 1$$

21. Given $m, n \in I^+$ show (7.3)

$$\int_0^{2\pi} \sin(mx) \sin(nx)\, dx = \begin{cases} 0 & \text{if } m \neq n \\ \pi & \text{if } m = n \end{cases}$$

22. An airplane at an altitude of 6 miles and a speed of 600 mph is flying directly away from an observer on the ground. What is the rate of change of the angle of elevation when the plane is over a point 4 miles away from the observer? (7.9)

Answers to Chapter 7, Part C Questions

1. (a) $f'(x) = 2\sqrt{16 + 16x^4} = 8\sqrt{1 + x^4} > 0$ for all x.
   $f^{-1}(0) = \frac{1}{2}$ because $f\left(\frac{1}{2}\right) = 0$

   (b) since $f(1/2) = 0$
   $(f^{-1})'(0) = \frac{1}{f'(1/2)} = \frac{1}{2\sqrt{17}} = \frac{\sqrt{17}}{34}$

2. (a) No; if p is a polynomial of even degree, the $\lim_{x \to \infty} p(x) = \infty$ or $-\infty$

   (b) Yes; for instance $p(x) = x^3$ has an inverse.
   $p(x) = x^3 - x$ does not have an inverse.

3. By the mean-value theorem with $f(x) = \ln x$, there exists c between 1 and x such that
   $$\frac{\ln x - \ln 1}{x - 1} = f'(c) = \frac{1}{c} \text{ so } \ln x = \frac{x-1}{c}$$
   If $x > 1$, the $\frac{1}{x} < \frac{1}{c} < 1$ and $x - 1 > 0$
   so $\frac{x-1}{x} < \ln x < x - 1$
   If $0 < x < 1$, the $1 < \frac{1}{c} < \frac{1}{x}$ and $x - 1 < 0$
   so $\frac{x-1}{x} < \ln x < x - 1$

4. (a) because $\sqrt{1 - \sin^2 x} > 0$ for all $x \in [0;$

   (b) $\int_0^\pi \sqrt{1 - \sin^2 x}\, dx = \int_0^\pi \sqrt{\cos^2 x}\, dx$
   $= 2\int_0^{\pi/2} \cos x\, dx$
   $= 2[\sin x]_0^{\pi/2} = 2.$

5. $\frac{d^n}{dx^n}\left(\ln \frac{1}{x}\right) = -\frac{d^n}{dx^n}(\ln x) = (-1)^n \frac{(n-1)!}{x^n}$

6. $\lim_{x \to 1} \frac{e^x - e}{\ln x}$
   $= \lim_{x \to 1}\left(\frac{e^x - e^1}{x - 1}\right)\left(\lim_{x \to 1} \frac{x-1}{\ln x - \ln 1}\right)$
   $= e \cdot 1 = e \cong 2.72$

7. $\int_0^1 \frac{5p^{\sqrt{x+1}}}{\sqrt{x+1}}\, dx = \left[\frac{10p^{\sqrt{x+1}}}{\ln p}\right]_0^1$
   $= \frac{10}{\ln p}(p^{\sqrt{2}} - p)$

8. $\frac{dv}{dt} = -kv \to v = ce^{-kt}$
   $V(0) = ce^0 = c$, so c is velocity when power is shut off

9. $A = \int_{-3}^{3} \frac{3}{9 + x^2}\, dx = 3\left[\frac{1}{3}\tan^{-1}\left(\frac{x}{3}\right)\right]_{-3}^{3}$
   $= \frac{\pi}{4} + \frac{\pi}{4} = \frac{\pi}{2}$

10. $V = \int_{2\sqrt{3}}^{6} 2\pi x \cdot \frac{1}{x^2\sqrt{x^2 - 9}}\, dx$
    $= 2\pi \int_{2\sqrt{3}}^{6} \frac{dx}{x\sqrt{x^2 - 9}}$
    $= 2\pi \left[\frac{1}{3}\sec^{-1}\left(\frac{x}{3}\right)\right]_{2\sqrt{3}}^{6}$
    $= \frac{\pi^2}{9}$

11. $\lim_{x \to \infty} \frac{\sinh x}{e^x} = \lim_{x \to \infty} \frac{e^x - e^{-x}}{2e^x}$
    $= \lim_{x \to \infty}\left(\frac{1}{2} - \frac{e^{-2x}}{2}\right) = \frac{1}{2}$

12. $\lim_{x \to \infty} \frac{\cosh x}{e^{ax}} = \lim_{x \to \infty} \frac{e^x + e^{-x}}{2e^{ax}}$
    $= \lim_{x \to \infty} \frac{1}{2}(e^{x-ax} + e^{-x-ax})$
    For $0 < a < 1$ $\lim = \infty$, for $a > 1$ limit $= 0$

13. $\int \frac{1}{\sqrt{x^2 - a^2}}\, dx = \frac{1}{a}\int \frac{1}{\sqrt{(x/a)^2 - 1}}\, dx$
    $= \int \frac{du}{\sqrt{u^2 - 1}} = \cosh^{-1}(u) + c$
    $= \cosh^{-1}\left(\frac{x}{a}\right) + c$

14. $\log_a(x + \sqrt{x^2 - 1}) = -\log_a\left(\frac{1}{x + \sqrt{x^2 - 1}}\right)$
    $= -\log_a\left(x - \sqrt{x^2 - 1}\right)$

15. Let $f(x) = x + e^x$
    $f'(x) = 1 + e^x > 1$
    hence $f(x)$ is monotonic and
    $f(-1) = -1 + e^{-1} < 0$ and
    $f(1) = 1 + e > 0$
    by the intermediate value theorem there exists at least one such x, $x + e^x = 0$

16. $\lim_{x \to 0} \frac{a^x - 1}{x} = \lim_{x \to 0} \frac{a^x - a^0}{x - 0} = f(x) = a^x = f'(0)$
    $= a^0 \ln a = \ln a$

17. $\lim_{x \to \infty} \dfrac{e^x}{x^n}$

$e^x > x^n$ for sufficiently large x

$e^x > \dfrac{x^{n+1}}{(n+1)!} = \left(\dfrac{x}{(n+1)!}\right) x^n$

$\dfrac{e^x}{x^n} > \dfrac{x}{(n+1)!}$ but as $x \to \infty$ $\dfrac{x}{(n+1)!} \to \infty$

therefore $\lim_{x \to \infty} \dfrac{e^x}{x^n} = \infty$

18. Solution after t hours $1000 \cdot 2^t$.

$10^9 = 1000 \cdot 2^t \Rightarrow t = \dfrac{\log 10^6}{\log 2}$ $t = 19.93$ hour

19. $\sin \theta = y/15$

$\theta = \sin^{-1}\left(\dfrac{x}{15}\right)$

$\dfrac{d\theta}{dt} = \dfrac{1}{\sqrt{225 - x^2}} \dfrac{dx}{dt}$

when $x = 9$

$\theta = \sin^{-1} 9/15 = \sin^{-1}(3/5) \, dt$

and $\dfrac{dx}{dt} = 2$, $\dfrac{d\theta}{dt} = \dfrac{2}{\sqrt{225 - 9^2}} = \dfrac{1}{6} = \dfrac{1}{6}$

θ is increasing at rate of $1/6$ $\dfrac{\text{radians}}{\text{sec}}$

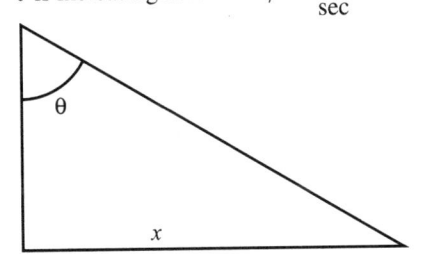

20. $x = \tanh y = \dfrac{e^y - e^{-y}}{e^y + e^{-y}}$

$x(e^y + e^{-y}) = e^y - e^{-y}$

$xe^y + xe^{-y} - e^y + e^{-y} = 0$

$(1-x)e^y - (1+x)e^{-y} = 0$

$\therefore e^{2y} = \dfrac{1+x}{1-x}$ for $|x| < 1$

$\dfrac{1+x}{1-x} > 0$ and $2y = \ln e^{2y} = \ln\left(\dfrac{1+x}{1-x}\right)$

hence $y = \tanh^{-1} x = \dfrac{1}{2} \ln\left(\dfrac{1+x}{1-x}\right) \quad -1 < x < 1$

21. $\displaystyle\int_0^{2\pi} \sin(mv) \sin(nx) \, dx$

$= \dfrac{1}{2} \displaystyle\int_0^{2\pi} ((\cos(m-n)x) - \cos(m+n)x)) \, dx$

$= \dfrac{1}{2} \dfrac{\sin(m-n)x}{m-n}\bigg|_0^{2\pi} - \dfrac{1}{2} \dfrac{\sin(m+n)x}{m+n}\bigg|_0^{2\pi}$

$\therefore$ if $m \neq n \quad \displaystyle\int_0^{2\pi} \sin(mx) \sin nx \, dx = 0$

if $m = n = \dfrac{1}{2} \displaystyle\int_0^{2\pi} (1 - \cos 2mx) \, dx = \pi$

22. $\theta = \tan^{-1} \dfrac{6}{x}$

$\dfrac{d\theta}{dt} = \dfrac{1}{1 + \left(\dfrac{6}{x}\right)^2} \left(\dfrac{-6}{x^2}\right) \dfrac{dx}{dt} = \dfrac{-6}{x^2 + 36} \dfrac{dx}{dt}$

$= \dfrac{-6}{52}(600)$

$\dfrac{d\theta}{dt} = -69.23 \dfrac{\text{radians}}{\text{hr}} = -\dfrac{69.23}{3600} \dfrac{\text{radians}}{\text{sec}}$

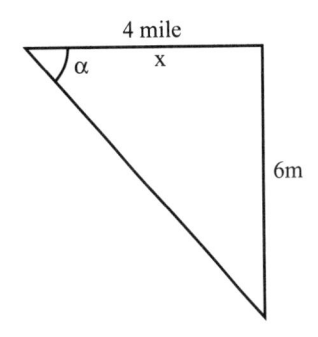

CHAPTER 8 Techniques of Integration
Part A

For Questions 1–3 Determine the appropriate u, du, to integrate the following:

1. $\displaystyle\int \frac{5x}{\sqrt{1-x^2}}\,dx \qquad u = \underline{\hspace{2cm}} \\ du = \underline{\hspace{2cm}}$ (8.1)

2. $\displaystyle\int xe^{-x^2}\,dx \qquad u = \underline{\hspace{2cm}} \\ du = \underline{\hspace{2cm}}$ (8.1)

3. $\displaystyle\int \frac{\sin x}{\sqrt{1+\cos x}}\,dx \qquad u = \underline{\hspace{2cm}} \\ du = \underline{\hspace{2cm}}$ (8.1)

For Questions 4–5 Determine the appropriate u, dv, to integrate the following by parts:

4. $\displaystyle\int x\cos x\,dx \qquad u = \underline{\hspace{2cm}} \qquad dv = \underline{\hspace{2cm}}$ (8.2)

5. $\displaystyle\int \ln x\,dx \qquad u = \underline{\hspace{2cm}} \qquad dv = \underline{\hspace{2cm}}$ (8.2)

6. Fill in:
$\displaystyle\int f(x)g'(x)\,dx = \underline{\hspace{3cm}}$ (8.2)

7. The integral $\sin^m x \cos^n x\,dx$ where m is an odd integer can be done by letting $\sin^m x = (\underline{\hspace{2cm}})(\underline{\hspace{2cm}})$ and using $\sin^2 x = \underline{\hspace{2cm}}$ (8.3)

8. The integral $\sin^m x \cos^n x\,dx$ where both m, n are even integers can be done by using: (8.3)
   (a) $\sin x \cos x = \underline{\hspace{2cm}}$
   (b) $\sin^2 x = \underline{\hspace{2cm}}$
   (c) $\cos^2 x = \underline{\hspace{2cm}}$

9. Verify the reduction formula
$$\int \sin^n x\,dx = -\frac{1}{n}\sin^{n-1} x \cos x + \frac{n-1}{n}\int \sin^{n-2}\,dx$$ (8.3)
for $\displaystyle\int \sin^2 x\,dx$
   (a) substitute $n = 2$ in the formula
   (b) use $\sin^2 x = \dfrac{1}{2}(1 - \cos 2x)$

10. When integrating
    (a) $\sqrt{a^2 - x^2}$ use $x = a\sin u$; label the given triangle appropriately (8.4)

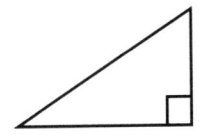

151

(b) $\sqrt{a^2 + x^2}$ use $x = a \tan u$; label the given triangle appropriately (8.4)

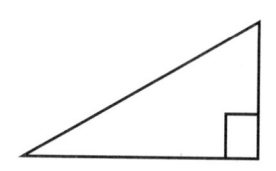

(c) $\sqrt{x^2 - a^2}$ use $x = a \sec u$; label the given triangle appropriately (8.4)

11. Using $A, B, C, Dx + E$ (A, B, C, D, E constants)
fill in
Do Not Solve (8.5)

(a) $\displaystyle\int \frac{1}{(x+2)(x+3)} dx = \int \frac{}{x+2} dx + \int \frac{}{x+3} dx$

(b) $\displaystyle\int \frac{1}{(x^2+1)(x+3)} dx = \int \frac{}{x^2+1} dx + \int \frac{}{x+3} dx$

(c) $\displaystyle\int \frac{1}{(x+3)^3} dx - \int \frac{}{} dx + \int \frac{}{} dx + \int \frac{}{} dx$

12. For $\displaystyle\int \frac{\sqrt{x}}{1+x} dx$ let $u = \sqrt{x}$ $u^2 = x$
$2u\, du = dx$ (8.6)

(a) $2\displaystyle\int \frac{u^2}{1+u^2} du$ is found by _____

(b) $2\displaystyle\int \frac{u^2}{1+u^2} du = ?$

13. $\overline{AB} = y_{k-1}$
$\overline{CD} = y_k$
the area of trapezoid ABCD
= _____ (8.7)

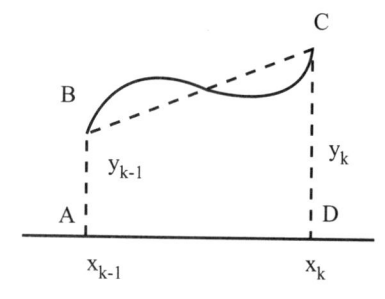

14. Simpson's rule is based on using portions of _____ to approximate $f(x)$ (8.7)

15. Write the equation used to model exponential growth or decay _____ (8.8)

16. A differential equation called a first-order linear differential equation has the form _____ (8.8)

17. A first-order differential equation is separable if f it can be put in the form _____ (8.9)

Answers to Chapter 8, Part A Questions

1. $u = 1 - x^2 \quad du = -2x\,dx$

2. $u = -x^2 \quad du = -2x\,dx$

3. $u = 1 + \cos x \quad du = -\sin x\,dx$

4. $u = x \quad dv = \cos x\,dx$

5. $u = \ln x \quad dv = dx$

6. $= f(x)g(x) - \int f'(x)g(x)\,dx$

7. $\sin^{m-1} x \sin x \quad 1 - \cos^2 x$

8. (a) $\dfrac{1}{2}\sin 2x$

   (b) $\dfrac{1}{2} - \dfrac{1}{2}\cos 2x$

   (c) $\dfrac{1}{2} + \dfrac{1}{2}\cos 2x$

9. (a) $\displaystyle\int \sin^2 x\,dx = -\dfrac{1}{2}\sin x \cos x + \dfrac{1}{2}\int dx$

   $= \dfrac{1}{2}x - \dfrac{1}{2}\sin x \cos x + C$

   (b) $\displaystyle\int \sin^2 x\,dx = \dfrac{1}{2}\int (1 - \cos 2x)\,dx$

   $= \dfrac{1}{2}x - \dfrac{1}{2}\int \cos 2x\,dx$

   $= \dfrac{1}{2}x - \dfrac{1}{4}\sin 2x + C$

   $= \dfrac{1}{2}x - \dfrac{1}{4}(2\sin x \cos x) + C$

   $= \dfrac{1}{2}x - \dfrac{1}{2}\sin x \cos x + C$

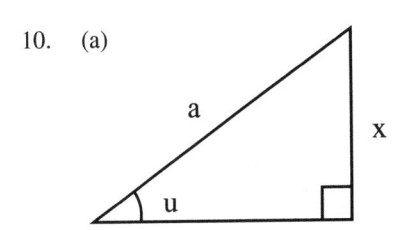

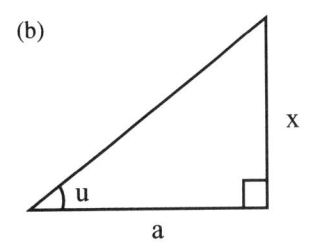

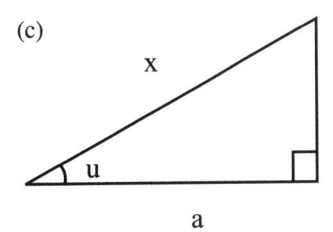

10. (a) [right triangle: hypotenuse a, opposite side x, angle u]

    (b) [right triangle: opposite side x, adjacent side a, angle u]

    (c) [right triangle: hypotenuse x, adjacent side a, angle u]

11. (a) $\displaystyle\int \dfrac{A}{x+2}\,dx + \int \dfrac{B}{x+3}\,dx$

    (b) $\displaystyle\int \dfrac{Dx + E}{x^2 + 1}\,dx + \int \dfrac{A}{x+3}\,dx$

    (c) $\displaystyle\int \dfrac{1}{(x+3)^3}\,dx \int \dfrac{A}{x+3}\,dx + \int \dfrac{B}{(x+3)^2}\,dx$
    $+ \displaystyle\int \dfrac{C}{(x+3)^3}\,dx$

12. $\displaystyle\int \dfrac{\sqrt{x}}{1+x} = 2\int \dfrac{u^2}{1+u^2}\,du$

    (a) Division

    (b) $2\displaystyle\int \left(1 - \dfrac{1}{1+u^2}\right)du$

    $2\displaystyle\int du - 2\int \dfrac{1}{1+u^2}\,du$

    $2u - 2\tan^{-1} u + C$

    $2\sqrt{x} - 2\tan^{-1}\sqrt{x} + C$

13. $\dfrac{1}{2}(y_{k-1} + y_k)(x_k - x_{k-1})$

14. parabolic region

15. $A'(t) = kA(t)$

16. $y' + p(x)y = q(x)$

17. $p(x)\,dx + q(y)y' = 0$

Part B

8.2 Integration by Parts

1. Calculate $\int xe^{-2x}\,dx$.

2. Calculate $\int x\sin 2x\,dx$.

3. Calculate $\int \ln(1+x^2)\,dx$.

4. Calculate $\int x\sqrt{1+x}\,dx$.

5. Calculate $\int e^{-x}\cos 2x\,dx$.

6. Calculate $\int \ln^2 x\,dx$.

7. Calculate $\int \dfrac{x^3}{\sqrt{x^2+1}}\,dx$.

8. Calculate $\int x^2 2^x\,dx$.

9. Calculate $\int \sqrt{x}\ln x^2\,dx$.

10. Calculate $\int e^{5x}\sin 2x\,dx$.

11. Evaluate $\int_1^2 x^3 \ln x\,dx$.

12. Evaluate $\int_0^{\pi/2} x^2 \sin 2x\,dx$.

13. Evaluate $\int_1^2 x\sec^{-1} x\,dx$.

14. Find the centroid of the region under the graph of $f(x) = e^{2x}$, $x \in [0, 1]$.

15. Find the volume generated by revolving the region under the graph of $f(x) = e^{2x}$, $x \in [0, 1]$ about the y-axis.

8.3 Powers and Products of Trigonometric Functions

16. Calculate $\int \cos^3(2x)\sin^2(2x)\,dx$.

17. Calculate $\int \sin^3 x \cos^5 x\,dx$.

18. Calculate $\int \sin^4 \dfrac{\theta}{2} \cos^3 \dfrac{\theta}{2} \, d\theta$.

19. Calculate $\int \sin^3 3\theta \, d\theta$.

20. Evaluate $\int_{\pi/4}^{\pi/3} \dfrac{dx}{\cos^2 x}$.

21. Calculate $\int \cos^4 2x \, dx$.

22. Calculate $\int x \sin^2 x^2 \cos^2 x^2 \, dx$.

23. Evaluate $\int_0^{\pi/2} \dfrac{\cot^2 \theta}{\csc^2 \theta} \, d\theta$.

24. Calculate $\int \cos^2(3x) \sin^2(3x) \, dx$.

25. Calculate $\int \sin^2 \dfrac{x}{2} \cos \dfrac{x}{2} \, dx$.

26. Calculate $\int \sin^2 \dfrac{t}{2} \cos^5 \dfrac{t}{2} \, dt$.

27. Calculate $\int \dfrac{\sin x}{\cos^5 x} \, dx$.

28. Calculate $\int \sin^4 2x \, dx$.

29. Evaluate $\int_0^{\pi/2} \sin^2 2\theta \cos^2 2\theta \, d\theta$.

30. Evaluate $\int_{\pi/3}^{2\pi/3} \sin^4 \theta \cos^3 \theta \, d\theta$.

31. Evaluate $\int_{\pi/6}^{\pi/3} (\tan^2 x - \sec^2 x)^4 \, dx$.

32. Calculate $\int \csc^3(4x) \cot^3(4x) \, dx$.

33. Calculate $\int \tan^3(2x) \sec^6(2x) \, dx$.

34. Calculate $\int \sec^3 \dfrac{x}{2} \tan \dfrac{x}{2} \, dx$.

35. Calculate $\int \tan^5 x \, dx$.

36. Calculate $\int \tan^3 3\theta \, d\theta$.

156 Calculus: One and Several Variables

37. Calculate $\int \sec^6 \dfrac{x}{3} \tan^2 \dfrac{x}{3} \, dx$.

38. Calculate $\int \dfrac{1}{\sec 2x \tan 2x} \, dx$.

39. Calculate $\int \tan^3 \dfrac{x}{2} \sec^4 \dfrac{x}{2} \, dx$.

40. Evaluate $\int_0^{\pi/4} (\tan x + \sec x)^2 \, dx$.

41. Calculate $\int \cot^4 2\theta \, d\theta$.

42. Calculate $\int \tan^5 t \sec^4 t \, dt$.

43. Calculate $\int \cot^2 2x \, dx$.

8.4 Integrals Involving $\sqrt{a^2 - x^2}, \sqrt{a^2 + x^2}, \sqrt{x^2 - a^2}$

44. Calculate $\int \dfrac{x^3}{\sqrt{25 - 4x^2}} \, dx$.

45. Calculate $\int \dfrac{1}{(x^2 + 4)^{3/2}} \, dx$.

46. Calculate $\int \dfrac{1}{(4x^2 - 9)^{3/2}} \, dx$.

47. Calculate $\int \dfrac{1}{x^2 \sqrt{x^2 + 25}} \, dx$.

48. Calculate $\int \dfrac{1}{\sqrt{2 + 4x^2}} \, dx$.

49. Calculate $\int \dfrac{\sqrt{x^2 - 4}}{x} \, dx$.

50. Evaluate $\int_0^4 \sqrt{16 - x^2} \, dx$.

51. Calculate $\int x^3 \sqrt{1 + x^2} \, dx$.

52. Calculate $\int \dfrac{x^2}{\sqrt{x^2 - 3}} \, dx$.

53. Calculate $\int \dfrac{x^2}{\sqrt{3 - x^2}} \, dx$.

54. Calculate $\int \dfrac{1}{x^2 \sqrt{4 - x^2}} \, dx$.

55. Evaluate $\displaystyle\int_2^4 \frac{1}{x^2\sqrt{x^2+4}}\,dx$.

56. Calculate $\displaystyle\int \frac{x^3}{\sqrt{4-x^2}}\,dx$.

57. Calculate $\displaystyle\int \frac{1}{x^2\sqrt{9-x^2}}\,dx$.

58. Calculate $\displaystyle\int \frac{1}{x\sqrt{x^2-4}}\,dx$.

59. Calculate $\displaystyle\int x^3\sqrt{x^2-1}\,dx$.

60. Evaluate $\displaystyle\int_2^3 \frac{dx}{\sqrt{x^2-1}}$.

61. Calculate $\displaystyle\int \frac{1}{x\sqrt{4x^2+9}}\,dx$.

8.5 Rational Functions; Partial Fractions

62. Calculate $\displaystyle\int \frac{x^2-6}{x(x-1)^2}\,dx$.

63. Calculate $\displaystyle\int \frac{x+3}{(x-1)(x^2-4x+4)}\,dx$.

64. Calculate $\displaystyle\int \frac{x+2}{x-x^3}\,dx$.

65. Calculate $\displaystyle\int \frac{x^2}{x^2-2x+1}\,dx$.

66. Calculate $\displaystyle\int \frac{4x^2-3x}{(x-2)(x^2+1)}\,dx$.

67. Calculate $\displaystyle\int \frac{2x-3}{x^3-3x^2+2x}\,dx$.

68. Calculate $\displaystyle\int \frac{2x+1}{x^3+x^2+2x+2}\,dx$.

69. Calculate $\displaystyle\int \frac{x}{(x+1)^2}\,dx$.

70. Calculate $\displaystyle\int \frac{x+1}{x^2+2x-3}\,dx$.

71. Calculate $\displaystyle\int \frac{x+4}{x^3+3x^2-10x}\,dx$.

72. Calculate $\displaystyle\int \frac{x+1}{x^2(x-1)}\,dx$.

73. Calculate $\displaystyle\int \frac{1}{(x+1)(x^2+1)}\,dx$.

74. Calculate $\displaystyle\int \frac{\cos\theta}{\sin^2\theta + 4\sin\theta - 5}\,d\theta$.

75. Calculate $\displaystyle\int \frac{4x}{x^3 - x^2 - x + 1}\,dx$.

76. Calculate $\displaystyle\int \frac{x+4}{x^3 + x}\,dx$.

77. Calculate $\displaystyle\int \frac{x^2 + 3x - 1}{x^3 - 1}\,dx$.

8.6 Some Rationalizing Substitutions

78. Calculate $\displaystyle\int \frac{dx}{2 + \sqrt{3x}}$.

79. Calculate $\displaystyle\int \frac{\sqrt{x}\,dx}{3(1 + \sqrt{x})}$.

80. Calculate $\displaystyle\int \sqrt{3 + e^{2x}}\,dx$.

81. Calculate $\displaystyle\int \frac{dx}{x^2(\sqrt{x} - 1)}$.

82. Calculate $\displaystyle\int (x+2)\sqrt{x-3}\,dx$.

83. Calculate $\displaystyle\int \frac{\sqrt{2x}}{\sqrt{2x}+1}\,dx$.

84. Calculate $\displaystyle\int x^3\sqrt{x+1}\,dx$.

85. Calculate $\displaystyle\int \frac{dx}{\sqrt{2 - e^{2x}}}$.

86. Evaluate $\displaystyle\int_0^2 \frac{x}{(4x+3)^{3/2}}\,dx$.

87. Evaluate $\displaystyle\int_7^{10} \frac{x}{\sqrt{3x-5}}\,dx$.

88. Calculate $\displaystyle\int \frac{dx}{1 - \sin x}$.

89. Calculate $\displaystyle\int \frac{3}{2\cos x + 1}\,dx$.

90. Calculate $\displaystyle\int \frac{dx}{\tan x - \sin x}$.

91. Calculate $\int \dfrac{\cot x}{1+\sin x}\,dx$.

92. Evaluate $\displaystyle\int_0^{\pi/2} \dfrac{\cos x}{2-\cos x}\,dx$.

8.7 Numerical Integration

93. Estimate $\displaystyle\int_0^2 \sqrt{4+x^3}\,dx$ using
    (a) the left-endpoint estimate, $n=4$.
    (b) the right-endpoint estimate, $n=4$.
    (c) the midpoint estimate, $n=4$.
    (d) the trapezoidal rule, $n=4$.

94. Estimate $\displaystyle\int_0^4 \sqrt{4+x^4}\,dx$ using
    (a) the left-endpoint estimate, $n=4$.
    (b) the right-endpoint estimate, $n=4$.

95. Estimate $\displaystyle\int_0^1 \dfrac{1}{1+x^2}\,dx = \dfrac{\pi}{4}$ using
    (a) the trapezoidal rule, $n=8$.
    (b) Simpson's rule, $n=8$.

96. Estimate $\displaystyle\int_0^2 \dfrac{1}{1+x^3}\,dx$ using
    (a) the trapezoidal rule, $n=4$.
    (b) Simpson's rule, $n=4$.

97. Estimate $\displaystyle\int_0^3 (1+x^2)^{3/2}\,dx$ using
    (a) the left-endpoint estimate, $n=6$.
    (b) the right-endpoint estimate, $n=6$.
    (c) the trapezoidal rule, $n=6$.
    (d) Simpson's rule, $n=6$.

98. Determine the values of n for which a theoretical error less than 0.001 can be guaranteed if the integral is estimated using (a) the trapezoidal rule; (b) Simpson's rule.

8.8 Differential Equations; First-Order Linear Equations

99. Find the general solution of $y' - 3y = 6$.

100. Find the general solution of $y' - 2xy = x$.

101. Find the general solution of $y' + \dfrac{4}{x}y = x^4$.

102. Find the general solution of $y' + \dfrac{2}{10+2x}y = 4$.

103. Find the general solution of $y' - y = -e^x$.

104. Find the general solution of $x \ln x y' + y = \ln x$.

105. Find the particular solution of $y' + 10y = 20$ determined by the side condition $y(0) = 2$.

106. Find the particular solution of $y' - y = -e^x$ determined by the side condition $y(0) = 3$.

107. Find the particular solution of $xy' - 2y = x^3 \cos 4x$ determined by the side condition $y(\pi) = 1$.

108. A 100-gallon mixing tank is full of brine containing 0.8 pounds of salt per gallon. Find the amount of salt present t minutes later if pure water is poured into the tank at the rate of 4 gallons per minute and the mixture is drawn off at the same rate.

109. Determine the velocity of time t and the terminal velocity of a 2-kg object dropped with a velocity 3 m/s, if the force due to air resistance is $-50v$ Newtons.

110. Use a suitable transformation to solve the Bernoulli equation $y' + xy = xy^2$.

111. Use a suitable transformation to solve the Bernoulli equation $xy' + y = x^3 y^6$.

8.9 Integral Curves; Separable Equations

112. Find the general solution of $y' = y^2 x^3$.

113. Find the general solution of $y' = \dfrac{x^2 + 7}{y^9 - 3y^4}$.

114. Find the general solution of $y' = y^2 + 1$.

115. Find the general solution of $x(y^2 + 1)y' + y^3 - 2y = 0$.

116. Find the particular solution of $e^x dx - y dy = 0$ determined by the side condition $y(0) = 1$.

117. Find the particular solution of $y' = y(x - 2)$ determined by the side condition $y(2) = 5$.

118. Verify that the equation $y' = \dfrac{y + x}{x}$ is homogeneous, then solve it.

119. Verify that the equation $y' = \dfrac{2y^4 + x^4}{xy^3}$ is homogeneous, then solve it.

120. Verify that the equation $y' + \dfrac{y}{x + \sqrt{xy}}$ is homogeneous, then solve it.

121. Verify that the equation $[2x \sinh(y/x) + 3y \cosh(y/x)]dx - 3x \cosh(y/x) dy = 0$ is homogeneous, then solve it.

122. Find the orthogonal trajectories for the family of curves $x^2 + y^2 = C$.

123. Find the orthogonal trajectories for the family of curves $x^2 + y^2 = Cx$.

Answers to Chapter 8, Part B Questions

1. $\dfrac{-x}{2}e^{-2x} - \dfrac{1}{4}e^{-2x} + C$

2. $\dfrac{-x}{2}\cos 2x + \dfrac{1}{4}\sin 2x + C$

3. $x\ln(1+x^2) - 2x + 2\tan^{-1}x + C$

4. $\dfrac{2x}{3}(1+x)^{3/2} - \dfrac{4}{15}(1+x)^{5/2} + C$

5. $\dfrac{e^{-x}}{5}(2\sin 2x - \cos 2x) + C$

6. $x^2 \ln x - 2x \ln x + 2x + C$

7. $x^2(x^2+1)^{1/2} - \dfrac{2}{3}(2x+1)^{3/2} + C$

8. $\dfrac{2^x}{\ln 2}\left[x^2 - \dfrac{2x}{\ln 2} + \dfrac{2}{(\ln 2)^2}\right] + C$

9. $x^{3/2}\left[\dfrac{2}{3}\ln x^2 - \dfrac{8}{9}\right] + C$

10. $\dfrac{1}{29}e^{5x}(5\sin 2x - 2\cos 2x) + C$

11. $4\ln 2 - 15/16$

12. $\pi^2/8 - 1/2$

13. $\dfrac{2\pi}{3} - \dfrac{\sqrt{3}}{2}$

14. $(\bar{x}, \bar{y}) = \left[\dfrac{e^2+1}{2(e^2-1)}, \dfrac{e^2+1}{4}\right]$

15. $\dfrac{\pi}{2}(e^2 + 1)$

16. $\dfrac{1}{6}\sin^3 2x - \dfrac{1}{10}\sin^5 2x + C$

17. $-\dfrac{1}{6}\cos^6 x + \dfrac{1}{8}\sin^8 x + C$

18. $\dfrac{2}{5}\sin^5\dfrac{\theta}{2} + \dfrac{2}{7}\sin^7\dfrac{\theta}{2} + C$

19. $\dfrac{-1}{3}\cos 3\theta + \dfrac{1}{9}\cos^3 3\theta + C$

20. $\sqrt{3} - 1$

21. $\dfrac{3x}{8} + \dfrac{1}{8}\sin 4x + \dfrac{1}{64}\sin 8x + C$

22. $\dfrac{x^2}{16} - \dfrac{1}{64}\sin 4x^2 + C$

23. $\pi/4$

24. $\dfrac{x}{8} - \dfrac{1}{96}\sin 12x + C$

25. $\dfrac{2}{3}\sin^3\dfrac{x}{2} + C$

26. $\dfrac{2}{3}\sin^2\dfrac{t}{2} - \dfrac{4}{5}\sin^5\dfrac{t}{2} + \dfrac{2}{7}\sin^7\dfrac{t}{2} + C$

27. $\dfrac{1}{4}\sec^4 x + C$

28. $\dfrac{3x}{8} - \dfrac{1}{8}\sin 4x + \dfrac{1}{64}\sin 8x + C$

29. $\pi/16$

30. 0

31. $\pi/6$

32. $\dfrac{1}{20}\csc^5 4x + \dfrac{1}{12}\csc^3 4x + C$

33. $\dfrac{1}{16}\tan^8 2x + \dfrac{1}{6}\tan^6 2x + \dfrac{1}{8}\tan^4 2x + C$

34. $\dfrac{2}{3}\sec^3\dfrac{x}{2} + C$

35. $\dfrac{\sec^4 x}{4} - \sec^2 x + C$

36. $\dfrac{1}{6}\tan^2 3\theta + \dfrac{1}{3}\ln|\cos 3\theta| + C$

37. $\dfrac{3}{7}\tan^7\dfrac{x}{3} + \dfrac{6}{5}\tan^5\dfrac{x}{3} + \tan^3\dfrac{x}{3} + C$

38. $\dfrac{1}{2}\ln|\csc 2x - \cot 2x| + \dfrac{1}{2}\cos 2x + C$

39. $\dfrac{1}{3}\tan^6\dfrac{x}{2} + \dfrac{1}{2}\tan^4\dfrac{x}{2} + C$

40. $2\sqrt{2} - \pi/4$

41. $\dfrac{-1}{6}\cos^3 2\theta + \dfrac{1}{2}\cos 2\theta + \theta + C$

42. $\dfrac{1}{8}\tan^8 t + \dfrac{1}{6}\tan^6 t + C$

43. $-\dfrac{1}{2}\cot 2x - x + C$

44. $\dfrac{-25}{16}(25-4x^2)^{1/2} - \dfrac{1}{48}(25-4x^2)^{3/2} + C$

45. $\dfrac{x}{4\sqrt{x^2+4}} + C$

46. $\dfrac{-x}{9\sqrt{4x^2-9}} + C$

47. $\dfrac{\sqrt{25+x^2}}{25x} + C$

48. $\dfrac{1}{2}\ln\left|\sqrt{1+2x^2} + \sqrt{2}x\right| + C$

49. $\sqrt{x^2-4} - 2\sec^{-1}\dfrac{x}{2} + C$

50. 4π

51. $\dfrac{1}{5}(1+x^2)^{5/2} - \dfrac{1}{3}(1+x^2)^{3/2} + C$

52. $\dfrac{x}{2}\sqrt{x^2-3} + \dfrac{3}{2}\ln\left|x + \sqrt{x^2-3}\right| + C$

53. $\dfrac{3}{2}\sin^{-1}\dfrac{x}{\sqrt{3}} - \dfrac{x}{2}\sqrt{3-x^2} + C$

54. $\dfrac{-\sqrt{4-x^2}}{4x} + C$

55. $\dfrac{2\sqrt{2}-\sqrt{5}}{8}$

56. $-x^2\sqrt{4-x^2} - \dfrac{2}{3}(4-x^2)^{3/2} + C$

57. $\dfrac{-\sqrt{9-x^2}}{9x} + C$

58. $\dfrac{1}{2}\sec^{-1}\dfrac{x}{2} + C$

59. $\dfrac{x^2}{3}(x^2-1)^{3/2} - \dfrac{2}{15}(x^2-1)^{5/2} + C$

60. $\ln\left(\dfrac{3+\sqrt{8}}{2+\sqrt{3}}\right)$

61. $\dfrac{1}{3}\ln\left|\dfrac{\sqrt{4x^2+9}}{2x} - \dfrac{3}{2x}\right| + C$

62. $-6\ln|x| + 7\ln|x-1| + \dfrac{5}{x-1} + C$

63. $4\ln\left|\dfrac{x-1}{x-2}\right| - \dfrac{5}{(x-2)} + C$

64. $2\ln|x| - \dfrac{3}{2}\ln|1-x| - \dfrac{1}{2}\ln|1+x| + C$

65. $x + 2\ln|x-1| - \dfrac{1}{x-1} + C$

66. $2\ln|x-2| + \ln|x^2+1| + \tan^{-1}x + C$

67. $\dfrac{-3}{2}\ln|x| + \ln|x-1| + \dfrac{1}{2}\ln|x-2| + C$

68. $-\dfrac{1}{3}\ln|x+1| + \dfrac{1}{6}\ln(x^2+2)$
$+\dfrac{5}{3\sqrt{2}}\tan^{-1}\dfrac{x}{\sqrt{2}} + C$

69. $\ln|x+1| + \dfrac{1}{x+1} + C$

70. $\dfrac{1}{2}\ln|x^2+2x-3| + C$

71. $-\dfrac{2}{5}\ln|x| + \dfrac{3}{7}\ln|x-2| - \dfrac{1}{35}\ln|x+5| + C$

72. $2\ln\dfrac{x-1}{x} + \dfrac{1}{x} + C$

73. $\dfrac{1}{2}\ln|x+1| - \dfrac{1}{4}\ln(x^2+1) + \dfrac{1}{2}\tan^{-1}x + C$

74. $\dfrac{1}{6}\ln\left|\dfrac{\sin\theta-1}{\sin\theta+5}\right| + C$

75. $\ln|x-1| - \dfrac{2}{x-1} - \ln|x+1| + C$

76. $4\ln|x| - 2\ln(x^2+1) + \tan^{-1}x + C$

77. $\ln|x-1| + \dfrac{4}{\sqrt{3}}\tan^{-1}\dfrac{2x+1}{\sqrt{3}} + C$

78. $\dfrac{2\sqrt{x}}{\sqrt{3}} - \dfrac{4}{3}\ln\left|2+\sqrt{3x}\right| + C$

79. $\dfrac{x}{3} - \dfrac{2}{3}\sqrt{x} + \dfrac{2}{3}\ln\left|\sqrt{x}+1\right| + C$

Answers

80. $\sqrt{3+e^{2x}} + \dfrac{\sqrt{3}}{2}\ln\left|\dfrac{\sqrt{3+e^{2x}} - \sqrt{3}}{\sqrt{3+e^{2x}} + \sqrt{3}}\right| + C$

81. $\dfrac{1}{x} + 2\left(\dfrac{1}{\sqrt{x}} + \ln\left|\dfrac{\sqrt{x}-1}{\sqrt{x}}\right|\right) + C$

82. $\dfrac{2}{5}(x-3)^{5/2} + \dfrac{10}{3}(x-3)^{3/2} + C$

83. $x - \sqrt{2x} + \ln\left|\sqrt{2x}+1\right| + C$

84. $\dfrac{2}{9}(x+1)^{9/2} - \dfrac{6}{7}(x+1)^{7/2} + \dfrac{6}{5}(x+1)^{5/2}$
    $- \dfrac{2}{3}(x+1)^{3/2} + C$

85. $\dfrac{1}{2\sqrt{2}}\ln\left|\dfrac{\sqrt{2-e^{2x}} - \sqrt{2}}{\sqrt{2-e^{2x}} + \sqrt{2}}\right| + C$

86. $\dfrac{1}{4}\left(\dfrac{7}{\sqrt{11}} - \sqrt{3}\right)$

87. $152/27$

88. $\dfrac{2}{1-\tan(x/2)} + C$

89. $\sqrt{3}\ln\left|\dfrac{\tan(x/2)+\sqrt{3}}{\tan(x/2)-\sqrt{3}}\right| + C$

90. $-\dfrac{1}{4}\left(\tan\dfrac{x}{2}\right)^{-2} - \dfrac{1}{2}\ln\left|\tan\dfrac{x}{2}\right| + C$

91. $\ln\left|\dfrac{\tan(x/2)}{[1+\tan(x/2)]^2}\right| + C$

92. $\pi\left(\dfrac{4}{3\sqrt{3}} - \dfrac{1}{2}\right)$

93. (a) 4.4914   (c) 4.8030
    (b) 5.2234   (d) 4.8574

94. (a) 17.9277   (b) 32.0253

95. (a) 0.7847   (b) 0.7854

96. (a) 1.0865   (b) 1.0968

97. (a) 20.8933   (c) 28.0490
    (b) 36.2047   (d) 27.9586

98. (a) $n \geq 19$ (b) $n \geq 2$

99. $y = Ce^{3x} - 2$

100. $y = Ce^{x^2} - \dfrac{1}{2}$

101. $y = \dfrac{C}{x^4} + \dfrac{1}{9}x^5$

102. $y = \dfrac{40x + 4x^2 + C}{10 + 2x}$

103. $y = (C - x)e^x$

104. $y = \dfrac{\ln^2 x + C}{2\ln x}$

105. $y = 2$ (identically)

106. $y = (3 - x)e^x$

107. $y = \dfrac{1}{4}x^2 \sin 4x + \left(\dfrac{x}{\pi}\right)^2$

108. $80e^{-0.04t}$ pounds

109. $v = 0.392 + 2.608e^{-25t}$;
     terminal velocity 0.392 m/s

110. $y = \dfrac{1}{1 + Ce^{x^2/2}}$

111. $y = \left(\dfrac{5}{2}x^3 + Cx^5\right)^{-1/5}$

112. $y = \dfrac{-4}{x^4 + C}$

113. $\dfrac{1}{10}y^{10} - \dfrac{3}{5}y^5 - \dfrac{1}{3}x^3 - 7x = C$

114. $y = \tan(x + C)$

115. $(y^2 - 2)^3 x^4 = Cy^2$

116. $y = \sqrt{2e^x - 1}, \; x > \ln\dfrac{1}{2}$

117. $y = 5e^{(x-2)^2/2}$

118. $y = x\ln|Cx|$

119. $x^8 = C(y^4 + x^4)$

120. $-2\sqrt{x/y} + \ln|y| = C$

121. $x^2 = C\sinh^3(y/x)$

122. $y = Kx$

123. $x^2 + y^2 = Ky$

Part C

1. Evaluate $\displaystyle\int_{-1}^{1} \frac{x^2}{x^e + 1} dx$ (8.1)

2. Let $f(x) = \dfrac{\ln x}{x}$ on $[1, 2e]$ (8.2)
   (a) Find the area bounded by $f(x)$ and the x-axis
   (b) Find the volume of the solid generated by revolving the region in part (a) around the x-axis

3. Find the volume generated by revolving the region under the graph about the y-axis
   $f(x) = x \cos x \quad x \in [0, \pi/2]$

4. Consider the identity
   $$f(b) - f(a) = \int_a^b f'(x)\,dx$$
   assume that f has a continuous second derivative;
   use integration by parts to derive the identity (8.2)
   $$f(b) - f(a) = f'(a)(b - a) - \int_a^b f''(x)(x - b)\,dx$$

5. Given $n \in I^+$ use integration by parts to derive the reduction formula (8.3)
   $$\int (\ln x)^n dx = x(\ln x)^n - n \int (\ln x)^{n-1} dx$$
   Let $u = (\ln x)^n$

6. Given $\displaystyle\int \sin^n x\, dx = -\frac{1}{n}\sin^{n-1} x \cos x + \frac{n-1}{n}\int \sin^{n-2} x\, dx$

   Show that (a) $\displaystyle\int_0^{\pi/2} \sin^n x\, dx = \frac{n-1}{n} \int_0^{\pi/2} \sin^{n-2} dx$ (8.3)

   Show that (b) $\displaystyle\int_0^{\pi/2} \sin^n x\, dx = \begin{cases} \dfrac{(n-1)\cdots 5.3.1}{n\cdots 6.4.2} \cdot \dfrac{\pi}{2} & n \text{ even}, n \geq 2 \\ \dfrac{(n-1)\cdots 4.2}{n\cdots 5.3} & n \text{ odd}, n \geq 3 \end{cases}$

7. Find the area of the region bounded on the left and right by the two branches of the hyperbola
   $\dfrac{x^2}{a^2} - \dfrac{y^2}{b^2} = 1$ and above and below by the lines $y = \pm b$ (8.4)

8. Derive a formula for $\displaystyle\int \frac{du}{u(a + bu)} \quad a, b$ constants $\neq 0$ (8.5)

9. Use $u = \tanh\left(\dfrac{x}{2}\right)$ to evaluate $\displaystyle\int \frac{1 - e^x}{1 + e^x} dx$ (8.6)

10. Show that the function $g(x) = Ax^2 + Bx + C$ satisfies the condition
    $$\int_a^b g(x)\,dx = \frac{b-a}{b}\left[g(a) + 4g\left(\frac{a+b}{2}\right) + g(b)\right] \text{ for every interval } [a, b]$$ (8.7)

11. Given a certain population P has known birth rate $\dfrac{dB}{dt}$ and known death rate $\dfrac{dD}{dt}$. The rate of change
    of P is given by $\dfrac{dP}{dt} = \dfrac{dB}{dt} - \dfrac{dD}{dt}$ (8.8)

11. (a) Assume $\dfrac{dB}{dt} = aP$ and $\dfrac{dD}{dt} = bP$, a, b constants find $P(t)$ if $P(0) = P > 0$
    (b) Compare P when (i) $a > b$
    (ii) $a = b$
    (iii) $a < b$
    (c) Find the $\lim\limits_{t \to \infty} P(t)$ in each of the above cases (8.9)

12. Find the orthogonal trajectory for the following family of curves (8.9)
    $y = Cx$

13. Given a circle radius r and a chord in the circle h units from the center. Find a formula for the shaded area (8.4)

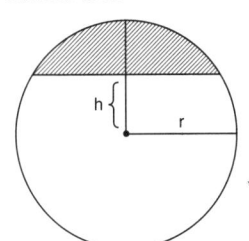

14. Use partial fractions to show (8.5)
    $$\int \frac{dx}{a^2 - x^2} = \frac{1}{2a} \ln \frac{(a+x)}{(a-x)}$$

15. Use trigonometric substitution to show (8.4)
    $$\int \frac{dy}{a^2 - x^2} = \frac{1}{2a} \ln \frac{(a+x)}{(a-x)}$$

16. Find (8.4)
    $$\int \frac{-x^2}{\sqrt{1-x^2}} dx$$

17. Use integration by parts to show (8.3)
    $$\int \sqrt{a^2 - x^2}\, dx = x\sqrt{a^2 - x^2} + \int \frac{x^2}{\sqrt{a^2 - x^2}} dx$$
    Then let $x^2 = -(-x^2) = -(a^2 - x^2 - a^2)$ in the numerator of the integral on the right side of the equation to show
    $$\int \sqrt{a^2 - x^2}\, dx = \frac{1}{2} x\sqrt{a^2 - x^2} + \frac{1}{2} a^2 \sin^{-1} \frac{x}{a} + C$$

18. Derive $\int x^n e^{ax}\, dx = \frac{1}{a} x^n e^{ax} - \frac{n}{a} \int x^{n-1} e^{ax}\, dx$ (8.2)

19. Find $\int x^2 e^{-2x}\, dx$ using number 18 (8.2)

20. Derive $\int x^m (\ln x)^n\, dx = \frac{x^{m+1}(\ln x)^n}{m+1} - \frac{n}{m+1} \int x^m (\ln x)^{n-1}\, dx$ (8.2)

21. Find $\int x^3 (\ln x)^2\, dx$ using number 20 (8.2)

22. Derive $\int \sec^n x\, dx = \frac{1}{n-1} \sec^{n-2} \tan x + \frac{n-2}{n+1} \int \sec^{n-2} x\, dx$ (8.3)

23. Find $\int \sec^3 x\, dx$ using number 22 (8.3)

Answers to Chapter 8, Part C Questions

1. $\int_{-1}^{1} \frac{x^2}{x^2+1} dx = \int_{-1}^{1} \frac{x^2+1-1}{x^2+1} dx$

   $= \int_{-1}^{1} \left(1 - \frac{1}{x^2+1}\right) dx$

   $= [x - \tan^{-1} x]_{-1}^{1} = 2 - \frac{\pi}{2}$

2. (a) $A = \int_{1}^{2e} \frac{\ln x}{x} dx = \left[\frac{1}{2}(\ln x)^2\right]_{1}^{2e} = \frac{(\ln 2e)^2}{2}$

   $= \frac{(1 + \ln 2)^2}{2}$

   (b) $\boxed{\begin{array}{ll} u = (\ln x)^2 & dv = \frac{1}{x^2} dx \\ du = \frac{2 \ln x}{x} dx & v = -\frac{1}{x} \end{array}}$

   $V = \int_{1}^{2e} \pi \left(\frac{\ln x}{x}\right)^2 dx$

   $= \pi \left[-\frac{(\ln x)^2}{x}\right]_{1}^{2e} - \pi \int_{1}^{2e} -\frac{2 \ln x}{x^2} dx$

   $= \pi \left[-\frac{(\ln 2e)^2}{2e}\right] + 2\pi \left[-\frac{\ln x}{x}\right]_{1}^{2e}$

   $- 2\pi \int_{1}^{2e} -\frac{1}{x^2} dx$

   $= \pi \left[-\frac{(\ln 2e)^2}{2e} - \frac{\ln 2e}{e} - \frac{1}{2e} + 2\right]$

3. $V_y = \int_{0}^{\pi/2} 2\pi x^2 \cos x \, dx$

   $= 2\pi [x^2 \sin x + 2x \cos x + 2 \sin x]_{0}^{\pi/2}$

   $= \frac{\pi}{2}(\pi^2 - 8)$

4. $f(b) - f(a) = \int_{a}^{b} f'(x) dx = [f'(x)(x-b)]_{a}^{b}$

   $\quad - \int_{a}^{b} f''(x)(x-b) dx$

   $= f'(a)(b-a) - \int_{a}^{b} f''(x)(x-b) dx$

   $u = f'(x) \quad dv = dx$

   $du = f''(x) dx \quad v = (x-b)$

5. $\int (\ln x)^n dx = x(\ln x)^n - n \int \frac{x(\ln x)^{n-1}}{x} dx$

   $= x(\ln x)^n - n \int (\ln x)^{n-1} dx$

   $u = (\ln x)^n \quad dv = dy$

   $du = \frac{n(\ln x)^{n-1}}{x} dx \quad v = x$

6. (a) $\int_{0}^{\pi/2} \sin^n x \, dx = \left[-\frac{1}{n} \sin^{n-1} x \cos x\right]_{0}^{\pi/2}$

   $+ \frac{n-1}{n} \int_{0}^{\pi/2} \sin^{n-2} x \, dx$

   $= \frac{n-1}{n} \int_{0}^{\pi/2} \sin^{n-2} x \, dx$

   (b) For $n = 2$, $\int_{0}^{\pi/2} \sin^2 x \, dx$

   $= \left[\frac{x}{2} - \frac{\sin 2x}{4}\right]_{0}^{\pi/2} = \frac{\pi}{4} = \frac{2-1}{2} \cdot \frac{\pi}{2}$

   For $n = 3$, $\int_{0}^{\pi/2} \sin^3 x \, dx$

   $= \frac{2}{3} \int_{0}^{\pi/2} \sin x \, dx = \frac{2}{3}[-\cos x]_{0}^{\pi/2} = \frac{2}{3}$

   The result then follows from (a) by induction.

7. $A = 4 \int_{0}^{b} \frac{a}{b} \sqrt{b^2 + y^2} \, dy = \frac{4a}{b} \left[\frac{y}{2} \sqrt{b^2 + y^2}\right.$

   $\left. + \frac{b^2}{2} \ln(y + \sqrt{b^2 + y^2})\right]_{0}^{b}$

   $= 2ab[\sqrt{2} + \ln(1 + \sqrt{2})]$

8. $\int \frac{1}{u(a+bu)} du = \frac{1}{a} \int \frac{1}{u} du - \frac{1}{a} \int \frac{b}{a+bu} du$

   $= \frac{1}{a}(\ln|u| - \ln|a+bu|) + C$

   $= \frac{1}{a} \ln \left|\frac{u}{a+bu}\right| + C$

9. $\int \frac{1-e^x}{1+e^x} dx = \int \frac{1 - \cosh x - \sinh x}{1 + \cosh x + \sinh x} dx$

   $= \int \frac{1 - \frac{1+u^2}{1-u^2} - \frac{2u}{1-u^2}}{1 + \frac{1+u^2}{1-u^2} + \frac{2u}{1-u^2}} \cdot \frac{2}{1-u^2} du$

   $= \int -\frac{2u}{1-u^2} du = \ln|1-u^2| + C$

   $= \ln \left|1 - \tanh^2 \left(\frac{x}{2}\right)\right| + C$

10. $\frac{b-a}{6} \left[g(a) + 4g\left(\frac{a+b}{2}\right) + g(b)\right]$

    $= \frac{b-a}{6} \left\{(Aa^2 + Ba + C) + 4\left[A\left(\frac{a+b}{2}\right)^2\right.\right.$

$$+ B\left(\frac{a+b}{2}\right) + C\right] + \left(Ab^2 + Bb + C\right)\}$$

$$= \frac{b-a}{6}\{A(b^2+a^2) + B(b+a) + 2C$$

$$+ A(a^2 + 2ab + b^2) + 2B(a+b) + 4C\}$$

$$= \frac{b-a}{6}\{2A(b^2 + ab + a^2) + 3B(b+a) + 6C\}$$

$$= \frac{1}{3}A(b^3 - a^3) + \frac{1}{2}B(b^2 - a^2) + C(b-a)$$

$$= \int_a^b Ax^2\,dx + \int_a^b Bx\,dx + \int_a^b C\,dx$$

$$= \int_a^b g(x)\,dx$$

11. (a) $\dfrac{dP}{dt} + (b-a)P = 0$, integrating factor $e^{(b-a)t}$

$$H(t) = \int (b-a)dt$$

$$e^{(b-a)t}\frac{dP}{dt} + (b-a)e^{(b-a)t}P = 0$$

$$\frac{d}{dt}[e^{(b-a)t}P] = 0$$

$$e^{(b-a)t}P = C$$

$$P = Ce^{(a-b)t}$$

$$P(0) = P_0 \Rightarrow P(t) = P_0 e^{(a-b)t}$$

(b) (i) $a > b \Rightarrow P_0 e^{(a-b)t}$ is increasing
(ii) $a = b \Rightarrow P(t) = P_0$ is constant
(iii) $a < b \Rightarrow P_0 e^{(a-b)t}$ is decreasing

(c) (i) $p(t) = \infty$ as $t \to \infty$
(ii) $p(t) = P_0$ as $t \to \infty$
(iii) $p(t) \to 0$ as $t \to \infty$

12. Curves: $y = Cx$, $y' = C = \dfrac{y}{x}$

orthogonal trajectories $y' = -\dfrac{x}{y}$

$$\int y\,dy + \int x\,dx = k_1;\ x^2 + y^2 = k(=2k_1)$$

13. $A = 2\displaystyle\int_0^{\sqrt{r^2-h^2}}\left(\sqrt{r^2-x^2} - h\right)dx$

$$= 2\left[\frac{1}{2} \times \sqrt{r^2-x^2}\right.$$

$$\left. + \frac{1}{2}r^2 \sin^{-1}\left(\frac{x}{r}\right) - hx\right]_0^{\sqrt{r^2-h^2}}$$

$$= r^2 \sin^{-1}\left(\frac{\sqrt{r^2-h^2}}{r}\right) - h\sqrt{r^2-h^2}$$

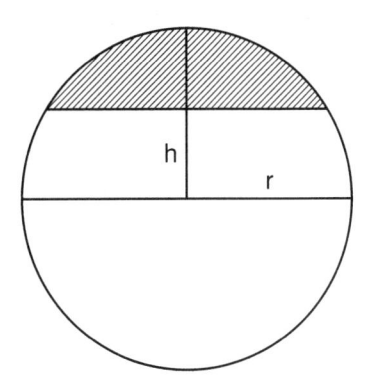

14. $\dfrac{1}{a^2 - x^2} = \dfrac{A}{a-x} + \dfrac{B}{a+x}$

so
$$1 = A(a+x) + B(a-x)$$

for $x = a$
$$1 = A(2a) \Rightarrow A = \frac{1}{2a}$$

for $x = -a \Rightarrow B = \dfrac{1}{2a}$

$$\int \frac{dx}{a^2 - x^2} = \frac{1}{2a}\int \frac{dx}{a-x} + \frac{1}{2a}\int \frac{dx}{a-x}$$

$$= -\frac{1}{2a}\ln|a-x| + \frac{1}{2a}\ln|a+x| + C$$

$$= \frac{1}{2a}\ln\left|\frac{a+x}{a-x}\right| + C$$

15. $\displaystyle\int \frac{dx}{a^2 - x^2}$

Let $x = a\sin u$
$dx = a\cos u\,du$

$$\int \frac{dx}{a^2 - x^2} = \int \frac{a\cos u\,du}{a^2(\cos u)^2} = \frac{1}{a}\int \frac{1}{\cos u}\,du$$

$$= \frac{1}{a}\int \sec u\,du = \frac{1}{a}\ln|\sec u + \tan u| + C$$

$$= \frac{1}{2a}\ln\left|\frac{a+x}{a-x}\right| + C$$

16. $x = \sin\theta$
$dx = \cos\theta\,d\theta$

$$\int \frac{-x^2}{\sqrt{1-x^2}}\,dx = -\int \frac{\sin^2\theta}{\cos\theta}\cos\theta\,d\theta$$

$$= -\int \sin^2\theta\,d\theta$$

$$= \frac{1}{2}(\sin\theta\cos\theta - \theta) + C$$

$$= \frac{1}{2}\left(\frac{x}{1} \cdot \frac{\sqrt{1-x^2}}{1} - \sin^{-1}\left(\frac{x}{1}\right)\right)$$
$$+C$$
$$= \frac{1}{2} \times \sqrt{1-x^2} - \frac{1}{2}\sin^{-1}(x) + C$$

[Figure: right triangle with hypotenuse 1, vertical side x, horizontal side $\sqrt{1-x^2}$, and angle θ]

17. $\int \sqrt{a^2 - x^2}\, dx = x\sqrt{a^2 - x^2} + \int \frac{x^2}{\sqrt{a^2 - x^2}}\, dx$

$u = \sqrt{a^2 - x^2} \quad dv = dx$
$du = \frac{-x\, dx}{\sqrt{a^2 - x^2}} \quad v = x$

$\int \frac{x^2}{\sqrt{a^2 - x^2}}$
$= \int \frac{-(a^2 - x^2 - a^2)}{\sqrt{a^2 - x^2}}\, dx$
$= -\int \frac{a^2 - x^2}{\sqrt{a^2 - x^2}}\, dx$
$\quad + \int \frac{a^2\, dx}{\sqrt{a^2 - x^2}}$
$= -\int \sqrt{a^2 - x^2}\, dx$
$\quad + \int \frac{a^2\, dx}{\sqrt{a^2 - x^2}}$

$\therefore 2\int \sqrt{a^2 - x^2}\, dx = x\sqrt{a^2 - x^2} + a^2 \sin^{-1}\frac{x}{a} + C$

$\therefore \int \sqrt{a^2 - x^2}\, dx = \frac{1}{2}x\sqrt{a^2 - x^2}$
$\quad + \frac{1}{2}a^2 \sin^{-1}\frac{x}{a} + C$

18. Let $u = (\ln x)^n \quad dv =$
$\int x^n e^{ax}\, dx = \frac{1}{a}x^n e^{ax} - \frac{n}{a}\int x^{n-1}e^{ax}\, dx$

$u = x^n \quad v = \frac{1}{a}e^{ax}$
$du = nx^{n-1} \quad dv = e^{ax}$

19. $\int x^2 e^{-2x}\, dx = -\frac{1}{2}x^2 e^{-2x} + \int xe^{-2x}\, dx$
$= -\frac{1}{2}x^2 e^{-2x} - \frac{1}{2}xe^{-2x} + \frac{1}{2}\int e^{-2x}\, dx$
$= -\frac{1}{2}x^2 e^{-2x} - \frac{1}{2}xe^{-2x} - \frac{1}{4}e^{-2x} + C$

20. $\int x^m (\ln x)^n\, dx$
$= \frac{x^{m+1}}{m+1}(\ln x)^n - \frac{n}{m+1}\int x^m (\ln x)^{n-1}\, dx$

let $u = (\ln x)^n \quad v = \frac{x^{m+1}}{m+1}$.
$du = n(\ln x)^{n-1}\left(\frac{1}{x}\right) dx, \quad dv = x^m\, dx$

21. $\int x^3 (\ln x)^2\, dx = \frac{x^4}{4}(\ln x)^2 - \frac{2}{4}\int x^3 (\ln x)\, dv$
$= \frac{x^4}{4}(\ln x)^2$
$\quad -\frac{1}{2}\left(\frac{x^4}{4}(\ln x) - \frac{1}{4}\int x^3\, dx\right)$
$= \frac{x^4}{4}(\ln x)^2 - \frac{x^4}{8}\ln x + \frac{1}{32}x^4 + C$

22. $\int \sec^n x\, dx = \sec^{n-2} x \tan x - (n-2)$
$\quad \int \sec^{n-2} x \tan^2 x\, dx$

let $u = \sec^{n-2} x$
$du = (n-2)\sec^{n-3} x \sec x \tan x\, dx$
$du = (n-2)\sec^{n-2} \tan x\, dx$
$dv = \sec^2 x\, dx$
$v = \tan x$

$\int \sec^n x\, dx = \sec^{n-2} x \tan x - (n-2)$
$\quad \int \sec^{n-2} x \tan^2 x\, dx$

$= \sec^{n-2} x \tan x - (n-2)$
$\quad \int \sec^{n-2} x(\sec^2 x - 1)\, dx$
$= \sec^{n-2} x \tan x - (n-2)$
$\quad \int \sec^n x\, dx + (n-2)$
$\quad \int \sec^{n-2} x\, dx$

$\therefore \int \sec^n x\, dx = \frac{1}{n-1}\sec^{n-2} x \tan x$
$\quad + \frac{n-2}{n-1}\int \sec^{n-2} x\, dx$

23. $\int \sec^3 x\, dx = \frac{1}{2}\sec x \tan x + \frac{1}{2}\int \sec x\, dx$
$= \frac{1}{2}\sec x \tan x + \frac{1}{2}\ln|\sec x + \tan x|$
$\quad + C$

CHAPTER 9 Conic Sections; Polar Coordinates; Parametric Equations

Part A

1. True or false: The distance d from the point $(-1, -4)$ to this line $2x - 3y - 7 = 0$ is
$$d = \frac{2(-1) - 3(-4) - 4}{\sqrt{13}}$$ (9.1)

2. True or false: For the parabola $x^2 = 16y$ the vertex is $(0, 0)$, the focus is $(0, 4)$, and the directrix is $y = 4$. (9.1)

3. Write the equation of the ellipse in standard position if the horizontal axis is the major axis and is 16 units and the vertical axis is 4 units. (9.2)

4. Write the equation of the hyperbola with foci $(+6, 0)$, $(-6, 0)$ and the lines $y = \frac{2}{5}\sqrt{5}x$, $y = -\frac{2}{5}\sqrt{5}x$ as asymptotes. (9.2)

5. Write the equation of the hyperbola determined by vertices $(\pm 3, 0)$ focus $(5, 0)$. (9.2)

6. Identify (9.2)
   (a) The set of all points P for which $d(P, F_1) + d(P, F_2) = k$ is ——— F_1, F_2 are points
   (b) The set of all points equidistant from a point F and a line ℓ is ———

7. Find the rectangular coordinates of the point with polar coordinates $(3, \pi)$. (9.3)

8. Find 2 pairs of polar coordinates, with r's having opposite signs for the rectangular coordinates $(1, 2)$. (9.3)

9. Identify the graphs of the following: (9.4)
   (a) $r = 1$ (b) $r = 2 \cos \theta$ (c) $r = 4 \sec \theta$
   (a) ——— (b) ——— (c) ———

10. Identify the graphs of the following: (9.4)
    (a) $r = \theta, \theta \geq 0$ (b) $r = 1 + \cos \theta$ (c) $r^2 = 2 \sin 2\theta$
    (a) ——— (b) ——— (c) ———

11. Find the points of intersection and express your answer in rectangular coordinates $r = 1 - \cos \theta$
    $r = \sin \theta$ (9.4)

12. (a) The area bounded by $\theta = \alpha$, $\theta = \beta$, and $r = p(\theta)$ is given by A =

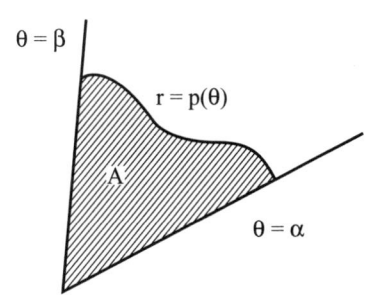

(b) The area bounded by $\theta = \alpha$, $\theta = \beta$ $r = P_1(\theta)$, and $r = P_2(\theta)$ A =

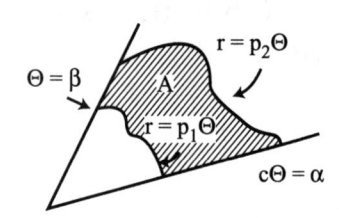

13. State the functions that parametrize the line that passes through (x_0, y_0) and (x_1, y_1) $(x_0, y_0) \pm (x_1, y_1)$ (9.6)
    $x(t) =$ $\qquad$ $y(t) =$

14. Parametrize the circle $x^2 + y^2 = 4$. (9,6)

15. Parametrize the ellipse $\dfrac{x^2}{9} + \dfrac{y^2}{4} = 1$. (9.6)

16. Find the equation in x and y for the tangent line to the curve $x(t) = t^2$, $y(t) = t + 5$ at $t = 2$. (9.7)

17. Given curve c parametrized by a pair of continuously differentiable functions $x = x(t)$ $\quad y = y(t)$ $t \in [a, b]$ the length of arc of C from a to b is (9.8)
    $L(c) =$

18. Given c an arc length of the continuous differentiable function $f(x) = y$ $\quad x \in [a, b]$ then the length of c is given by $L(c) =$ (9.8)

Answers to Chapter 9, Part A Questions

1. false

2. false

3. $\dfrac{x^2}{4} + \dfrac{y^2}{2} = 1$

4. $\dfrac{x^2}{20} - \dfrac{y^2}{16} = 1$

5. $\dfrac{x^2}{9} - \dfrac{y^2}{16} = 1$

6. (a) ellipse
   (b) parabola

7. $(-3, 0)$

8. $r = \sqrt{1^2 + 2^2} = \sqrt{5}$ $\tan\theta = 2$ first quadrant
   $\theta = \tan^{-1} 2$ so $(\sqrt{5}, \tan^{-1} 2)\,(-\sqrt{5}, \tan^{-1} 2)$

9. (a) circle with radius 1 centered at $(0, 0)$
   (b) circle with radius 1 centered at $(1, 0)$
   (c) line-vertical $x = 4$

10. (a) spiral
    (b) cardioids
    (c) lemniscate

11. $(0, 0)\,(0, 1)$

12. (a) $A = \displaystyle\int_\alpha^\beta \dfrac{1}{2}[p(\theta)]^2 d\theta$

    (b) $A = \displaystyle\int_\alpha^\beta \dfrac{1}{2}([p_2(\theta)]^2 - [p_1(\theta)]^2)\, d\theta$

13. $x(t) = x_0 + t(x_1 - x_0)$
    $y(t) = y_0 + t(y_1 - y_0)$

14. $x(t) = 2\cos t$
    $y(t) = 2\sin t$

15. $x(t) = 3\cos t$
    $y(t) = 2\sin t$

16. $x' = 2t\quad y' = 1$
    $ = 4$
    $\dfrac{dy}{dx} = \dfrac{1}{4}$ at $t = 2$ tangent $y - 7 = \dfrac{1}{4}(x - 4)$

17. $\displaystyle\int_a^b \sqrt{[x'(t)]^2 + [y'(t)]^2}\, dt$

18. $\displaystyle\int_a^b \sqrt{1 + [f'(x)]}\, dx$

Part B

9.1 Translations; The Parabola

1. Sketch and give an equation for the parabola with vertex $(0, 0)$ and directrix $x = 5/2$.

2. Sketch and give an equation for the parabola with vertex $(1, 2)$ and focus $(1, 4)$.

3. Sketch and give an equation for the parabola with focus $(6, -2)$ and directrix $y = 2$.

4. Sketch and give an equation for the parabola with vertex $(-1, 2)$ and focus $(2, 2)$.

5. Find the vertex, focus, axis, and directrix for the parabola $y^2 + 6y + 6x = 0$.

6. Find the vertex, focus, axis, and directrix for the parabola $x^2 - 4x - 2y - 8 = 0$.

7. Find the vertex, focus, axis, and directrix for the parabola $2x^2 - 10x + 5y = 0$.

8. Find an equation for the parabola with directrix $x = -2$ and vertex $(1, 3)$. Where is the focus?

9. Find an equation for the parabola with directrix $y = 3$ and vertex $(-2, 2)$. Where is the focus?

10. Find an equation for the parabola with directrix $x = 5$ and focus $(-1, 0)$. Where is the vertex?

11. Find an equation for the parabola that has vertex $(2, 1)$, passes through $(5, -2)$, and has axis of symmetry parallel to the x-axis.

12. Find the length of the latus rectum for the parabola $(y - 2) = 3(x + 1)^2$.

9.2 The Ellipse and Hyperbola

13. For the ellipse $\dfrac{x^2}{16} + \dfrac{y^2}{9} = 1$, find
    (a) the center
    (b) the foci
    (c) the length of the major axis
    (d) the length of the minor axis
    Then sketch the figure.

14. For the ellipse $5x^2 + 3y^2 = 15$, find
    (a) the center
    (b) the foci
    (c) the length of the major axis
    (d) the length of the minor axis
    Then sketch the figure.

15. For the ellipse $36(x - 1)^2 + 4y^2 = 144$, find
    (a) the center
    (b) the foci
    (c) the length of the major axis
    (d) the length of the minor axis
    Then sketch the figure.

16. For the ellipse $9x^2 + 16y^2 - 36x + 96y + 36 = 0$, find
    (a) the center
    (b) the foci
    (c) the length of the major axis
    (d) the length of the minor axis
    Then sketch the figure.

17. For the ellipse $9x^2 + 5y^2 + 36x - 30y + 36 = 0$, find
    (a) the center
    (b) the foci
    (c) the length of the major axis
    (d) the length of the minor axis
    Then sketch the figure.

18. Find an equation for the ellipse with foci at $(0, 3)$, $(0, -3)$ and major axis 10.

19. Find an equation for the ellipse with foci at $(1, 0)$, vertices at $(-1, 0)$ $(3, 0)$, and foci at $(1 - \sqrt{3}, 0)$ and $(1 + \sqrt{3}, 0)$.

20. Find an equation for the ellipse with focus $(2, 2)$, center at $(2, 1)$, and major axis 10.

21. Find an equation for the ellipse with foci at $(1, -1)$, $(7, -1)$, and minor axis 6.

22. Find the equation of the parabola that has vertex at the origin and passes through the ends of the minor axis of the ellipse $y^2 - 10y + 25x^2 = 0$.

23. Determine the eccentricity of the ellipse $\dfrac{x^2}{25} + \dfrac{(y-1)^2}{9} = 1$.

24. Write an equation for the ellipse with major axis from $(-2, 0)$ to $(2, 0)$, eccentricity $\frac{1}{2}$.

25. Find an equation for the hyperbola with foci at $(3, 0)$, $(-3, 0)$, and transverse axis 4.

26. Find an equation for the hyperbola with asymptotes $y = \pm \dfrac{4}{3}x$ and foci at $(10, 0)$, $(-10, 0)$.

27. Find an equation for the hyperbola with the center at $(2, 2)$, a vertex at $(2, 10)$, and a focus at $(2, 11)$.

28. Find an equation for the hyperbola with vertices at $(7, -1)$, $(-5, -1)$, and a focus at $(9, -1)$.

29. For the hyperbola $x^2 - \dfrac{y^2}{4} = 1$, find the center, the vertices, the foci, the asymptotes, and the length of the transverse axis. Then sketch the figure.

30. For the hyperbola $\dfrac{(x-1)^2}{4} - \dfrac{y^2}{16} = 1$, find the center, the vertices, the foci, the asymptotes, and the length of the transverse axis. Then sketch the figure.

31. For the hyperbola $9x^2 - 16y^2 = 144$, find the center, the vertices, the foci, the asymptotes, and the length of the transverse axis. Then sketch the figure.

32. For the hyperbola $16x^2 - 9y^2 - 160x - 72y + 112 = 0$, find the center, the vertices, the foci, the asymptotes, and the length of the transverse axis. Then sketch the figure.

33. For the hyperbola $\dfrac{(x-3)^2}{9} + \dfrac{(y+4)^2}{16} = 1$, find the center, the vertices, the foci, the asymptotes, and the length of the transverse axis. Then sketch the figure.

34. For the hyperbola $4y^2 - 9x^2 - 36x - 8y - 68 = 0$, find the center, the vertices, the foci, the asymptotes, and the length of the transverse axis. Then sketch the figure.

35. Determine the eccentricity of the hyperbola $\dfrac{x^2}{49} - y^2 = 1$.

36. Determine the eccentricity of the hyperbola $\dfrac{y^2}{16} - \dfrac{x^2}{12} = 1$.

9.3 Polar Coordinates

37. Find the rectangular coordinates of the point with polar coordinates $[4, 2\pi/3]$.

38. Find the rectangular coordinates of the point with polar coordinates $[3, -\pi/4]$.

39. Find the rectangular coordinates of the point with polar coordinates $[-2, -\pi/3]$.

40. Find all possible polar coordinates for the point with rectangular coordinates $(-4, 4\sqrt{3})$.

41. Find all possible polar coordinates for the point with rectangular coordinates $(2, -2)$.

42. Find the point $[r, \theta]$ symmetric to the point $[2, \pi/6]$ about
    (a) the x-axis
    (b) the y-axis
    (c) the origin
    Express your answer with $r > 0$ and $\theta \in [0, 2\pi]$.

43. Find the point $[r, \theta]$ symmetric to the point $[2/3, 5\pi/4]$ about
    (a) the x-axis
    (b) the y-axis
    (c) the origin
    Express your answer with $r > 0$ and $\theta \in [0, 2\pi]$.

44. Test the curve $r = 3 + 2\cos\theta$ for symmetry about the coordinate axes and the origin.

45. Test the curve $r \sin 2\theta = 1$ for symmetry about the coordinate axes and the origin.

46. Write the equation $x + y^2 = x - y$ in polar coordinates.

47. Write the equation $x^2 + y^2 - 6y = 0$ in polar coordinates.

48. Write the equation $x^2 + y^2 + 8y = 0$ in polar coordinates.

49. Write the equation $x^4 + x^2 y^2 = y^2$ in polar coordinates.

50. Write the equation $y^3 + x^2 y = x$ in polar coordinates.

51. Write the equation $y^6 + x^2 y^4 = x^4$ in polar coordinates.

52. Write the equation $(x^2 + y^2)^2 = 4xy$ in polar coordinates.

53. Write the equation $x(x^2 + y^2) = 2(3x^2 - y^2)$ in polar coordinates.

54. Write the equation $(x^2 + y^2)^{3/2} = x^2 - y^2 - 2xy$ in polar coordinates.

Conic Sections; Polar Coordinates; Parametric Equations 175

55. Identify the curve given by $r \sin \theta = 2$ and write the equation in rectangular coordinates.

56. Identify the curve given by $r = 4 \cos \theta$ and write the equation in rectangular coordinates.

57. Identify the curve given by $\theta^2 = \frac{4}{9}\pi^2$ and write the equation in rectangular coordinates.

58. Identify the curve given by $r = 4 \sin \theta - 6 \cos \theta$ and write the equation in rectangular coordinates.

59. Identify the curve given by $r = 3 \cos \theta - \sin \theta$ and write the equation in rectangular coordinates.

60. Identify the curve given by $r = \dfrac{1}{\cos \theta - 1}$ and write the equation in rectangular coordinates.

61. Identify the curve given by $r = \dfrac{10}{2 + \cos \theta}$ and write the equation in rectangular coordinates.

62. Identify the curve given by $r = \dfrac{1}{1 - \cos \theta}$ and write the equation in rectangular coordinates.

63. The parabola $r = \dfrac{1}{1 + \cos \theta}$ has focus at the pole and directrix $x = 2$. Without resorting to xy-coordinates,
    (a) locate the vertex of the parabola
    (b) find the width of the latus rectum
    (c) sketch the parabola

64. The ellipse $r = \dfrac{10}{8 + 5 \cos \theta}$ has right focus at the pole, major axis horizontal. Without resorting to xy-coordinates,
    (a) find the eccentricity of the ellipse
    (b) locate the ends of the major axis
    (c) locate the center of the ellipse
    (d) locate the second focus
    (e) determine the length of the minor axis
    (f) determine the width of the ellipse at the foci
    (g) sketch the ellipse

65. The hyperbola $r = \dfrac{9}{2 + 6 \cos \theta}$ has left focus at the pole, transverse axis horizontal. Without resorting to xy-coordinates,
    (a) find the eccentricity of the hyperbola
    (b) locate the ends of the transverse axes
    (c) locate the center of the hyperbola
    (d) locate the second focus
    (e) determine the width of the hyperbola at the foci, and sketch the hyperbola

66. Find the points at which the curves $r = 2 \cos \theta$ and $r = -1$ intersect. Express your answers in rectangular coordinates.

67. Find the points at which the curves $r = 1 + \cos \theta$ and $r = 2 \cos \theta$ intersect. Express your answers in rectangular coordinates.

68. Find the points at which the curves $r = \sin 3\theta$ and $r = 2 \sin \theta$ intersect. Express your answers in rectangular coordinates.

69. Find the points at which the curves $r = \frac{1}{2} + \cos \theta$ and $\theta = \pi/4$ intersect. Express your answers in rectangular coordinates.

176 Calculus: One and Several Variables

70. Find the points at which the curves $r = \dfrac{1}{1 + \cos\theta}$ and $r \sin\theta = 2$ intersect. Express your answers in rectangular coordinates.

9.4 Graphing in Polar Coordinates

71. Sketch and identify the polar curve $r^2 = 9 \sin 2\theta$.

72. Sketch and identify the polar curve $r = 1 + \cos\theta$.

73. Sketch and identify the polar curve $r = 2\cos\theta$.

74. Sketch and identify the polar curve $r = \sin 3\theta$.

75. Sketch and identify the polar curve $r = 4 + 4\cos\theta$.

76. Sketch and identify the polar curve $r = \sqrt{2}$.

77. Sketch and identify the polar curve $r^2 = 4\cos 2\theta$.

78. Sketch and identify the polar curve $r = 2 - 4\sin\theta$.

79. Sketch and identify the polar curve $r = \cos 3\theta$.

80. Sketch and identify the polar curve $r = 2\sin 2\theta$.

81. Sketch and identify the polar curve $r = 2 + 4\sin\theta$.

82. Sketch and identify the polar curve $r = 4 + 2\sin\theta$.

83. Sketch and identify the polar curve $r = 3\sin\theta$.

84. Sketch and identify the polar curve $r = 1 - 2\cos\theta$.

85. Sketch and identify the polar curve $r = 2 + 4\cos\theta$.

86. Sketch and identify the polar curve $r = 3 + 2\cos\theta$.

87. Sketch and identify the polar curve $r = 4(1 - \cos\theta)$.

88. Sketch and identify the polar curve $r = 4(1 - \sin\theta)$.

9.5 Area in Polar Coordinates

89. Find the area of the region enclosed by $r = 4\sin 3\theta$.

90. Find the area of the region enclosed by $r = 2 + \sin\theta$.

91. Find the area of the region that is inside $r = 5\sin\theta$ but outside $r = 2 + \sin\theta$.

92. Find the area of the region that is outside $r = 1 + \sin\theta$ but inside $r = 3 + \sin\theta$.

93. Find the area of the region that is common to $r = 3\cos\theta$ and $r = 1 + \cos\theta$.

94. Find the area of the region that is common to $r = 1 + \sin\theta$ and $r = 1$.

95. Find the area of the region that is inside $r = 1$ but outside $r = 1 - \cos\theta$.

96. Find the area of the region enclosed by $r = 2 + \cos\theta$.

97. Find the area of the region that is inside $r = 2\cos\theta$ but outside $r = \sin\theta$.

98. Find the area of the region enclosed by $r = 1 - \sin\theta$.

99. Find the area of the region that is common to $r = 3a\cos\theta$ and $r = a(1 + \cos\theta)$.

100. Find the area of the region that is inside $r = 2$ but outside $r = 1 + \cos\theta$.

101. Find the area of the region enclosed by $r = 2\cos 3\theta$.

102. Find the area of the region that is inside $r = 3(1 + \sin\theta)$ but outside $r = 3\sin\theta$.

103. Find the area of the region that is common to $r = a\cos 3\theta$ and $r = a/2$. Take $a > 0$.

104. Find the area of the region enclosed by $r^2 = \cos 2\theta$.

105. Find the area of the region enclosed by the inner loop of $r = 1 - 2\sin\theta$.

106. Find the centroid of the region enclosed by $r = 2\cos\theta$.

9.6 Curves Given Parametrically

107. Express the curve by an equation in x and y: $x(t) = 2 + \sin t$, $y(t) = 3 - \cos t$.

108. Express the curve by an equation in x and y: $x(t) = e^t - 1$, $y(t) = 3 + e^{2t}$.

109. Express the curve by an equation in x and y: $x(t) = 2\cos t$, $y(t) = 3\sin t$.

110. Express the curve by an equation in x and y: $x(t) = 3 + \cosh t$, $y(t) = 2 + \sinh t$.

111. Express the curve by an equation in x and y: $x(t) = 3 + \cos t$, $y(t) = 3 - 2\sin t$.

112. Express the curve by an equation in x and y: $x(t) = 2t^2 + t - 3$, $y(t) = t - 1$.

113. Express the curve by an equation in x and y: $x(t) = \cos 2t$, $y(t) = \sin t$.

114. Express the curve by an equation in x and y: $x(t) = -1 + 3\cos t$, $y(t) = \sin t$.

115. Find the parametrization $x = x(t)$, $y = y(t)$, $t \in [0, 1]$, for the line segment from $(2, 5)$ to $(5, 8)$.

116. Find the parametrization $x = x(t)$, $y = y(t)$, $t \in [0, 1]$, for the curve $y = x^2$ from $(1, 1)$ to $(3, 9)$.

9.7 Tangents to Curves Given Parametrically

117. Find an equation in x and y for the line tangent to the curve $x(t) = 3t$, $y(t) = t^2 - 1$ at $t = 1$.

118. Find an equation in x and y for the line tangent to the curve $x(t) = 2t^2$, $y(t) = (1 - t)^2$ at $t = 1$.

119. Find an equation in x and y for the line tangent to the curve $x(t) = 2e^t$, $y(t) = \frac{1}{2}e^{-t}$ at $t = 0$.

120. Find an equation in x and y for the line tangent to the curve $x(t) = 2/t$, $y(t) = 2t^2 + 3$ at $t = 1$.

121. Find an equation in x and y for the line tangent to the polar curve $r = 3 + 2\sin\theta$ at $\theta = \pi/2$.

122. Find an equation in x and y for the line tangent to the polar curve $r = 3\sin 3\theta$ at $\theta = \pi/6$.

123. Find an equation in x and y for the line tangent to the polar curve $r = \dfrac{3}{2 + \cos\theta}$ at $\theta = \pi/2$.

124. Find the points (x, y) at which the curve $x(t) = 2t - t^3$, $y(t) = t - 1$ has (a) a horizontal tangent; (b) a vertical tangent.

125. Find the points (x, y) at which the curve $x(t) = 4 + 3\sin t$, $y(t) = 3 + 4\cos t$ has (a) a horizontal tangent; (b) a vertical tangent.

126. Find the tangent(s) to the curve $x(t) = t^2 - 2t$, $y(t) = 1 - t$ at the point $(-1, 0)$.

127. Calculate $\dfrac{d^2y}{dx^2}$ at the point $t = 1$ without eliminating the parameter if $x(t) = e^t - 1$ and $y(t) = 3 + e^{2t}$.

128. Calculate $\dfrac{d^2y}{dx^2}$ at the point $t = 3\pi/4$ without eliminating the parameter if $x(t) = 5 - 2\cos t$ and $y(t) = 3 + \sin t$.

9.8 Arc Length and Speed

129. Find the arc length of the curve $f(x) = 2x^{3/2}$, $x \in [0, 8/9]$ and compare it to the straight-line distance between the endpoints.

130. Find the arc length of the curve $f(x) = \dfrac{x^3}{6} + \dfrac{1}{2x}$, $x \in [1, 3]$ and compare it to the straight-line distance between the endpoints.

131. Find the arc length of the curve $f(x) = \dfrac{2}{3}(x + 1)^{3/2}$, $x \in [1, 2]$ and compare it to the straight-line distance between the endpoints.

132. Find the arc length of the curve $f(x) = x^{2/3}$, $x \in [0, 8]$ and compare it to the straight-line distance between the endpoints.

133. Find the arc length of the curve $f(x) = \dfrac{2}{3}x^{3/2} - \dfrac{1}{2}x^{1/2}$, $x \in [1, 4]$ and compare it to the straight-line distance between the endpoints.

134. Find the arc length of the curve $f(y) = \dfrac{3}{5}y^{5/3} - \dfrac{3}{4}x^{1/3}$, $y \in [1, 8]$ and compare it to the straight-line distance between the endpoints.

135. Find the arc length of the curve $f(x) = 2x^{3/2}$, $x \in [0, 3]$ and compare it to the straight-line distance between the endpoints.

136. Find the arc length of the curve $f(x) = \dfrac{1}{3}(x^2 + 2)^{3/2}$, $x \in [0, 3]$ and compare it to the straight-line distance between the endpoints.

137. The equations $x(t) = 2 + \sin t$, $y(t) = 3 - \cos t$ give the position of a particle at time t from $t = 0$ to $t = \pi/2$. Find the initial speed of the particle, the terminal speed, and the distance traveled.

138. The equations $x(t) = e^t \sin t$, $y(t) = e^t \cos t$ give the position of a particle at time t from $t = 0$ to $t = \pi$. Find the initial speed of the particle, the terminal speed, and the distance traveled.

139. The equations $x(t) = t^2 + 2$, $y(t) = t^3 - 3$ give the position of a particle at time t from $t = 0$ to $t = 1$. Find the initial speed of the particle, the terminal speed, and the distance traveled.

140. The equations $x(t) = 3(t-1)^2$, $y(t) = 8t^{3/2}$ give the position of a particle at time t from $t = 0$ to $t = 1$. Find the initial speed of the particle, the terminal speed, and the distance traveled.

141. Find the length of the polar curve $r = 2e^{3\theta}$ from $\theta = 0$ to $\theta = \pi$.

142. Find the length of the polar curve $r = 2\cos 2\theta$ from $\theta = 0$ to $\theta = 2\pi$.

9.9 The Area of a Surface of Revolution; The Centroid of a Curve; Pappus's Theorem on Surface Area

143. Find the length of the curve, locate its centroid, and determine the area of the surface generated by revolving $f(x) = 3x$, $x \in [0, 1]$ about the x-axis.

144. Find the length of the curve, locate its centroid, and determine the area of the surface generated by revolving $y = \frac{1}{2}x$, $x \in [0, 2]$ about the x-axis.

145. Find the area of the surface generated by revolving $f(x) = 2\sqrt{x+1}$, $x \in [-1, 1]$ about the x-axis.

146. Find the area of the surface generated by revolving $y = \sin x$, $x \in [0, \pi/2]$ about the x-axis.

Answers to Chapter 9, Part B Questions

1. $y^2 = -10x$

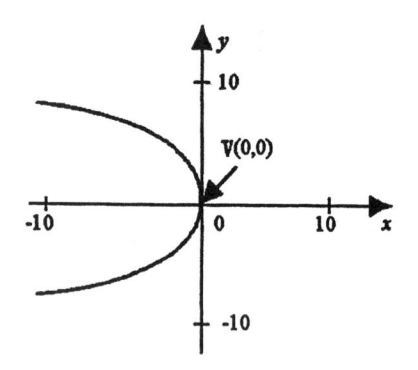

2. $8(y-2) = (x-1)^2$

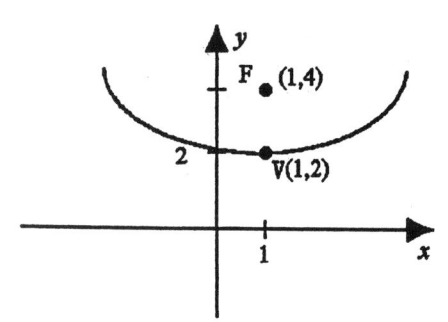

3. $8y = -(x-6)^2$

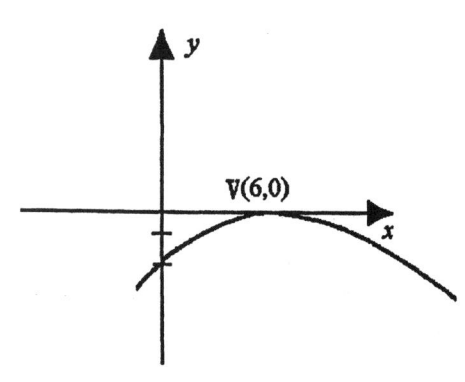

4. $12(x+1) = (y-2)^2$

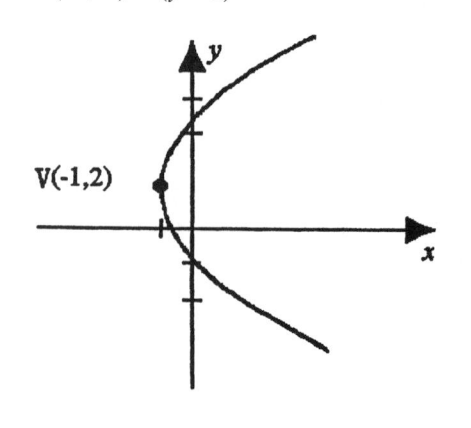

5. $V: (3/2, -3)$ $\quad F: (-3/2, -3)$
   axis: $y = -3$ $\quad$ directrix: $x = 9/2$

6. $V: (2, -6)$ $\quad F: (2, -11/2)$
   axis: $x = 2$ $\quad$ directrix: $y = -13/2$

7. $V: (5/2, 5/2)$ $\quad F: (5/2, 15/8)$
   axis: $x = 5/2$ $\quad$ directrix: $y = 25/8$

8. $(y-3)^2 = 12(x-1)$; $\quad F: (4, 3)$

9. $(x+2)^2 = -4(y-2)$; $\quad F: (-2, 1)$

10. $y^2 = -12(x-2)$; focus: $(0, 2)$

11. $(y-1)^2 = 3(x-2)$

12. 3

13. (a) $(0, 0)$ $\qquad$ (c) 8
    (b) $(\sqrt{7}, 0), (-\sqrt{7}, 0)$ $\quad$ (d) 6

14. (a) $(0, 0)$ $\qquad$ (c) $2\sqrt{5}$
    (b) $(0, \sqrt{2}), (0, -\sqrt{2})$ $\quad$ (d) $2\sqrt{3}$

15. (a) $(1, 0)$ $\qquad$ (c) 12
    (b) $(1, 4\sqrt{2})$, $\qquad$ (d) 4
    $(1, -4\sqrt{2})$

16. (a) $(2, -3)$ $\qquad$ (c) 8
    (b) $(2 + \sqrt{7}, -3)$, $\quad$ (d) 6
    $(2 - \sqrt{7}, -3)$

17. (a) $(-2, 3)$ $\qquad$ (c) 6
    (b) $(-2, 1), (-2, 5)$ $\quad$ (d) $2\sqrt{5}$

18. $\dfrac{x^2}{25} + \dfrac{y^2}{10} = 1$

19. $\dfrac{(x-1)^2}{4} + y^2 = 1$

20. $\dfrac{(x-2)^2}{25} + (y-1)^2 = 1$

21. $\dfrac{(x-4)^2}{18} + \dfrac{(y+1)^2}{9} = 1$

22. $x^2 = \dfrac{1}{5}y$

23. $4/5$

24. $\dfrac{x^2}{4} + \dfrac{y^2}{3} = 1$

Answers

25. $\dfrac{x^2}{4} - \dfrac{y^2}{5} = 1$

26. $\dfrac{x^2}{36} - \dfrac{y^2}{64} = 1$

27. $\dfrac{(y-2)^2}{64} - \dfrac{(x-2)^2}{17} = 1$

28. $\dfrac{(x-1)^2}{36} - \dfrac{(y+1)^2}{28} = 1$

29. center: (0, 0)
    vertices: (−1, 0), (1, 0)
    foci: $(-\sqrt{5}, 0), (\sqrt{5}, 0)$
    asymptotes: $y = \pm 2x$
    length of transverse axis: 2

30. center: (1, 0)
    vertices: (−1, 0), (3, 0)
    foci: $(1 - 2\sqrt{5}, 0), (1 + 2\sqrt{5}, 0)$
    asymptotes: $y = \pm 2(x - 1)$
    length of transverse axis: 4

31. center: (0, 0)
    vertices: (−4, 0), (4, 0)
    foci: (−5, 0), (5, 0)
    asymptotes: $y = \pm \dfrac{3}{4}x$
    length of transverse axis: 8

32. center: (5, −4)
    vertices: (2, −4), (8, −4)
    foci: (0, −4), (10, −4)
    asymptotes: $y = -4 \pm \dfrac{4}{3}(x - 5)$
    length of transverse axis: 6

33. center: (3, −4)
    vertices: (0, −4), (6, −4)
    foci: (−2, −4), (8, −4)
    asymptotes: $y = -4 \pm \dfrac{4}{3}(x - 3)$
    length of transverse axis: 6

34. center: (−2, 1)
    vertices: (−2, −2), (−2, 4)
    foci: $(-2, 1 - \sqrt{13}), (-2, 1 + \sqrt{13})$
    asymptotes: $y = 1 \pm \dfrac{3}{2}(x + 2)$
    length of transverse axis: 6

35. $\dfrac{5\sqrt{2}}{7}$

36. $\dfrac{\sqrt{7}}{2}$

37. $(-2, 2\sqrt{3})$

38. $(\sqrt{2}, -\sqrt{2})$

39. $(-1, \sqrt{3})$

40. $\left[8, \dfrac{2\pi}{3} + 2n\pi\right], \left[-8, \dfrac{5\pi}{3} + 2n\pi\right]$

41. $\left[2\sqrt{2}, \dfrac{7\pi}{4} + 2n\pi\right], \left[-2\sqrt{2}, \dfrac{3\pi}{4} + 2n\pi\right]$

42. (a) $\left[2, \dfrac{11\pi}{6}\right]$ (b) $\left[2, \dfrac{5\pi}{6}\right]$ (c) $\left[2, \dfrac{7\pi}{6}\right]$

43. (a) $\left[\dfrac{2}{3}, \dfrac{3\pi}{4}\right]$ (b) $\left[\dfrac{2}{3}, \dfrac{7\pi}{4}\right]$ (c) $\left[\dfrac{2}{3}, \dfrac{\pi}{4}\right]$

44. x-axis only

45. symmetric with respect to x-axis, y-axis, origin

46. $r = \dfrac{\cos\theta - \sin\theta}{1 + 2\sin\theta\cos\theta}$

47. $r = 6\cos\theta$

48. $r = -8\sin\theta$

49. $r = \tan\theta$

50. $r^2 = \cot\theta$

51. $r = \cot^2\theta$

52. $r^2 = 2\sin 2\theta$

53. $r = 2(3\cos\theta - \sin\theta\tan\theta)$

54. $r = \cos 2\theta - \sin 2\theta$

55. $y = 2$; line

56. circle; center (2, 0)
    $x^2 + y^2 = 4x$ or $(x - 2)^2 + y^2 = 4$

57. half lines; $\theta = \pm\dfrac{2\pi}{3}$
    $y = -3\sqrt{x}\ (x \le 0)$
    $y = 3\sqrt{x}\ (x \le 0)$

58. $(x + 3)^2 + (y - 2)^2 = 13$; circle

59. $(x - 3/2)^2 + (y + 1/2)^2 = 5/2$; circle

60. $y^2 = -2(x - \frac{1}{2})$; parabola

61. $\dfrac{(x+10/3)^2}{400/9} + \dfrac{y^2}{400/12} = 1$; ellipse

62. $y^2 = 2(x + 1/2)$; parabola

63. (a) $(1, 0)$ (b) 4

64. (a) 5/8   (d) $(-100/39, 0)$
    (b) $(-10/3, 0)$,   (e) $20/\sqrt{39}$
         $(10/13, 0)$ (f) 5/2
    (c) $(-50/39, 0)$

65. (a) 3   (d) $(27/8, 0)$
    (b) $(9/8, 0), (9/4, 0)$ (e) 9
    (c) $(27/16, 0)$

66. $\left(\dfrac{1}{2}, \dfrac{\sqrt{3}}{2}\right), \left(\dfrac{1}{2}, \dfrac{-\sqrt{3}}{2}\right)$

67. $(2, 0), (0, 0)$

68. $\left(\dfrac{-\sqrt{2}}{2}, \dfrac{1}{2}\right), \left(\dfrac{\sqrt{3}}{2}, \dfrac{1}{2}\right), (0, 0)$

69. $(0, 0), \left(\dfrac{2+\sqrt{2}}{4}, \dfrac{2+\sqrt{2}}{4}\right), \left(\dfrac{2-\sqrt{2}}{4}, \dfrac{2-\sqrt{2}}{4}\right)$

70. $(-3/2, 2)$

71. Lemniscate

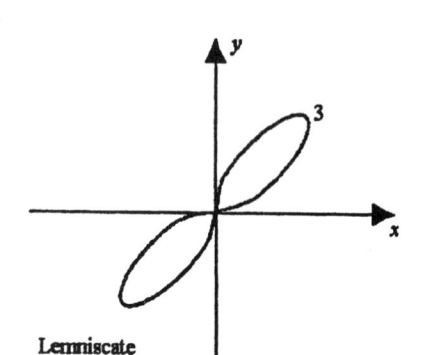

72. Cardioid

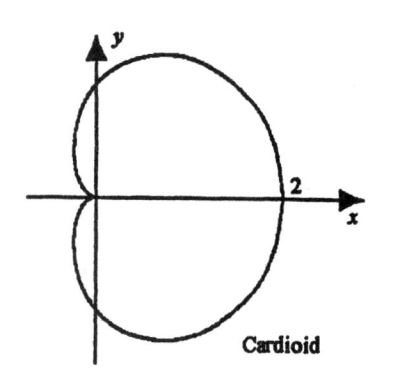

73. Circle

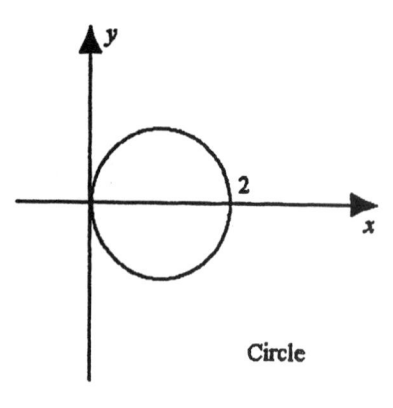

74. Three-petal rose

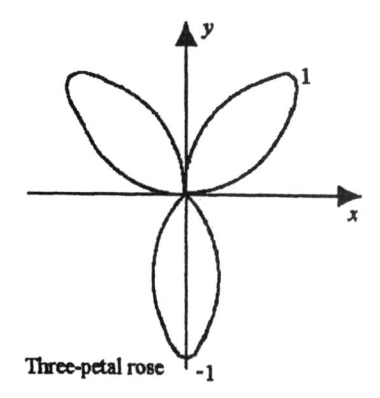

75. Cardioid

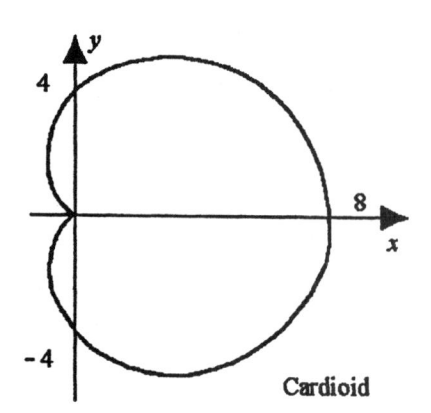

76. Circle

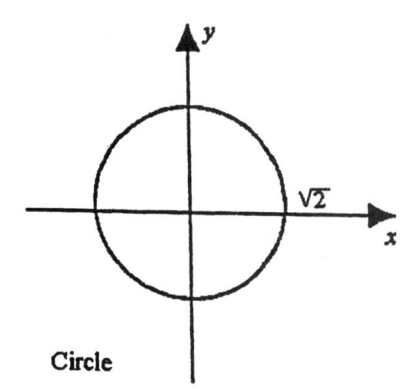

77. Lemniscate

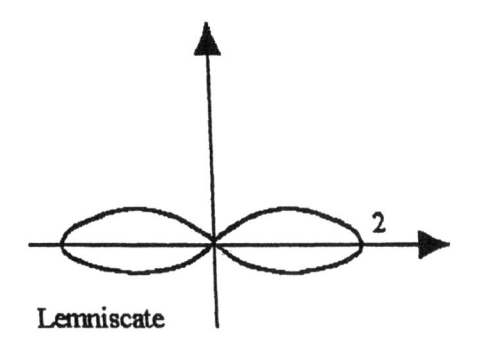

78. Limaçon

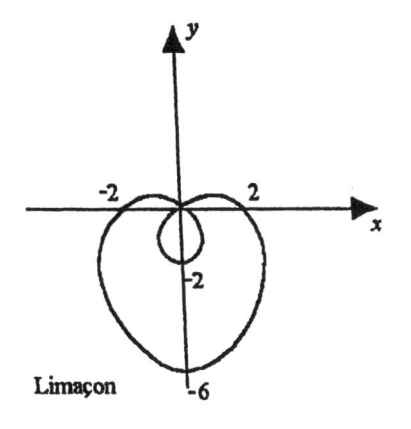

79. Three-petal rose

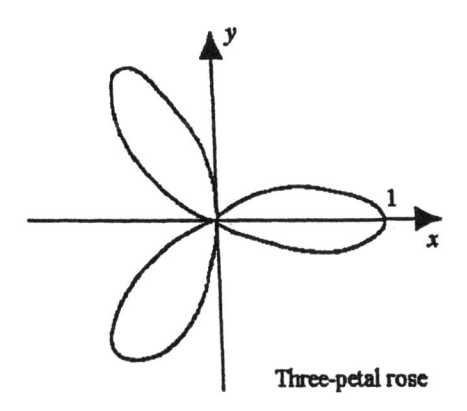

80. Four-petal rose

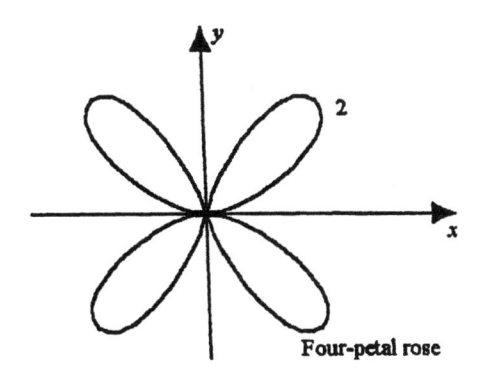

81. Limaçon

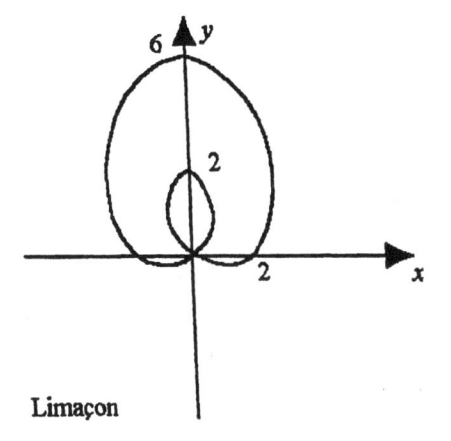

82. Circle

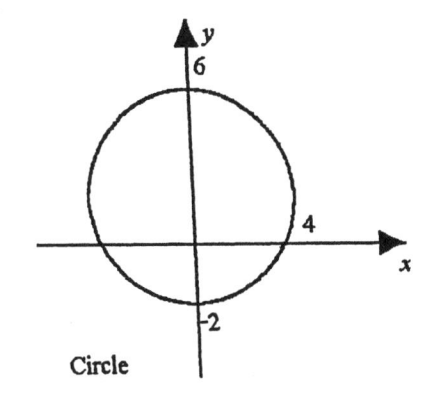

83. Circle

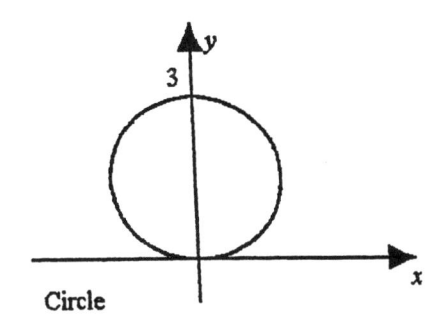

84. Limaçon

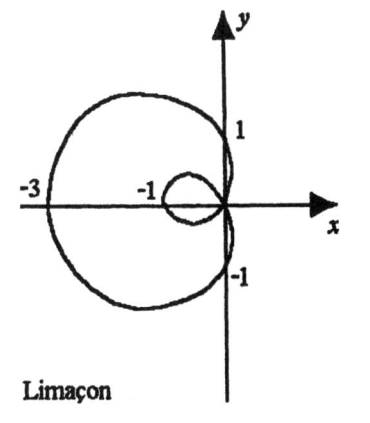

184   Calculus: One and Several Variables

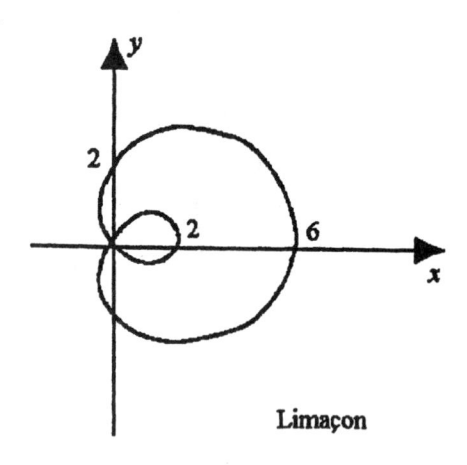

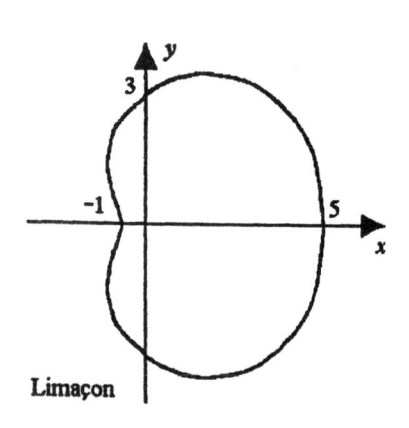

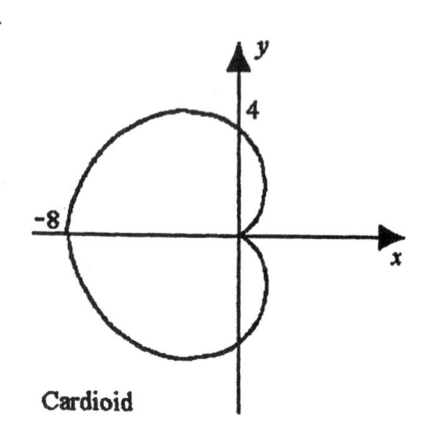

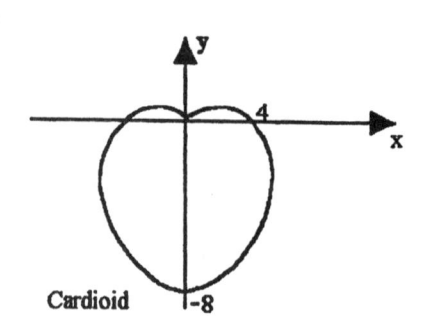

85. Limaçon

86. Limaçon

87. Cardioid

88. Cardioid

89. 4π

90. $9\pi/2$

91. $8\pi/2 + \sqrt{3}$

92. 8π

93. $5\pi/4$

94. $5\pi/4 - 2$

95. $2 - \pi/4$

96. $9\pi/2$

97. $\dfrac{\pi}{2} + \dfrac{3}{4}\tan^{-1} 2 + \dfrac{1}{2}$

98. $3\pi/2$

99. $5\pi a^2/4$

100. $5\pi/2$

101. π

102. $45\pi/4$

103. $a^2\left(\dfrac{\pi}{6} - \dfrac{\sqrt{3}}{8}\right)$

104. 1

105. $\pi - \dfrac{3\sqrt{3}}{2}$

106. $(1, 0)$

107. $(x - 2)^2 + (y - 3)^2 = 1$; circle

108. $y = x^2 + 2x + 4 = (x + 1)^2 + 3$; parabola

109. $x^2/4 + y^2/9 = 1$; ellipse

110. $(x - 3)^2 - (y - 2)^2 = 1$; hyperbola

111. $(x - 3)^2 + \dfrac{(y - 3)^2}{4} = 1$; ellipse

112. $x = 2y^2 + 5y$; parabola

113. $x = 1 - 2y^2$; parabola

114. $\dfrac{(x + 1)^2}{9} + \dfrac{y^2}{1} = 1$; ellipse

115. $x = 2 + 3t$; $y = 5 + 3t$

Answers

116. $x = 2 + 3t; y = 1 + 4t + 4t^2$

117. $2x - 3y = 6$

118. $y = 0$

119. $x + 4y = 4$

120. $2x + y = 9$

121. $y = 5$

122. $\sqrt{3}x + y = 6$

123. $x + 2y = 3$

124. (a) None
     (b) $\left(\dfrac{4\sqrt{2}}{3\sqrt{3}}, \dfrac{\sqrt{2}}{\sqrt{3}} - 1\right); \left(\dfrac{-4\sqrt{2}}{3\sqrt{3}}, \dfrac{-\sqrt{2}}{\sqrt{3}} - 1\right)$

125. (a) $(4, 7), (4, -11)$
     (b) $(7, 3), (1, 3)$

126. vertical tangent

127. 2

128. $\dfrac{-\sqrt{2}}{2}$

129. 52/27

130. 14/3

131. $\dfrac{2}{3}(8 - 3\sqrt{3})$

132. $\dfrac{8}{27}(10\sqrt{10} - 1)$

133. 31/6

134. 387/20

135. $\dfrac{2}{27}(56\sqrt{7} - 1)$

136. 12

137. initial speed $= 1$; terminal speed $= 1$
     distance $= \pi/2$

138. initial speed $= \sqrt{2}$; terminal speed $= e^\pi \sqrt{2}$
     distance $= \sqrt{2}(e^\pi - 1)$

139. initial speed $= 0$; terminal speed $= \sqrt{13}$
     distance $= \dfrac{1}{27}(13\sqrt{13} - 8)$

140. initial speed $= 0$; terminal speed $= 12$
     distance $= 9$

141. $\dfrac{2\sqrt{10}}{3}(e^3\pi - 1)$

142. 4π

143. length: $\sqrt{10}$; centroid: $\left(\dfrac{1}{2}, \dfrac{3}{2}\right)$ area: $3\pi\sqrt{10}$

144. length: $\sqrt{5}$; centroid: $\left(1, \dfrac{1}{2}\right)$ area: $\pi\sqrt{5}$

145. $8\pi\left(\sqrt{3} - \dfrac{1}{3}\right)$

146. $\pi[\sqrt{2} + \ln(1 + \sqrt{2})]$

Part C

1. For $x^2 = 4y$ What is the slope of the parabola at the endpoints of the latus rectum? (9.1)

2. Show that the set of all points $(a \cos t, b \sin t)$ with t real lie on an ellipse. (9.2)

3. Find functions $x = x(t)$, $y = y(t)$ such that, as t ranges over the set of real numbers, the points $(x(t), y(t))$ transverse
   (a) the right branch of the hyperbola
   (b) the left branch of the hyperbola (9.2)

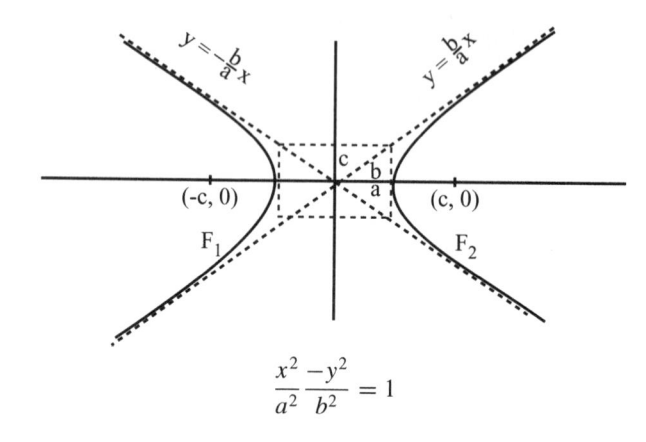

$$\frac{x^2}{a^2} - \frac{y^2}{b^2} = 1$$

4. Write an equation for the ellipse major axis from $(-3, 0)$ to $(3, 0)$ eccentricity $1/3$. (9.2)

5. What happens to a hyperbola if e tends to 1? (9.2)

6. Write the equation in rectangular coordinates and identify it. $r = \dfrac{6}{1 + 2 \sin \theta}$ (9.3)

7. Find a polar equation for the set of points $P[r, \theta]$ such that the distance from P to the pole is twice the distance from P to the line $x = -d$. (9.3)

8. Show that the point $[2, \frac{1}{2}\pi]$ lies on both $r^2 \sin \theta = 4$ and $r = 2 \cos 2\theta$. (9.4)

9. Sketch the curves and find the points at which they intersect. Express your answers in rectangular coordinates $r = 1 - \cos \theta$ and $r = \sin \theta$ (9.4)

10. Fix $a > 0$ and let n be a positive integer. Prove that the petal curves $r = a \cos([2n + 1]\theta)$ and $r = a \sin([2n + 1]\theta)$, all enclose exactly the same area. Find the area. (9.5)

11. Find a parametrization $x(t) = \sin f(t)$ $y(t) = \cos f(t)$ $t \in (0, 1)$ which traces out the unit circle infinitely often. (9.6)

12. Suppose that the curve $C : x = x(t)$, $y = y(t)$, $t \in [c, d]$ is the graph of a nonnegative function $y = f(x)$ over an interval $[a, b]$. Suppose that $x'(t)$ and $y'(t)$ are continuous $x(c) = a$ and $x(d) = b$. Show that the area below $C = \int_c^d y(t) x'(t) dt$.
    Hint: Since C is the graph of f, we know that $y(t) = f(x(t))$. (9.6)

13. Suppose that $x = x(t)$, $y = y(t)$ are twice differentiable functions that parametrize a curve. Take a point on the curve at which $x'(t) \neq 0$ and $\dfrac{d^2 y}{dx^2}$ exists.

Show that $\dfrac{d^2y}{dx^2} = \dfrac{x'(t)y''(t) - y'(t)x''(t)}{[x'(t)]^3}$. (9.7)

14. The curve defined parametrically by $x(\theta) = \theta \cos\theta$, $y(\theta) = \theta \sin\theta$ is called an Archimedean Spiral. Find the length of this spiral for $0 \leq \theta \leq 2\pi$. (9.8)

15. (a) Show that the circle $(x-a)^2 + y^2 = b^2$ can be parametrized by $x(t) = a + b\cos t$, $y(t) = b\sin t$ $0 \leq t \leq 2\pi$.
    (b) Suppose that $0 < b < a$. The solid generated by revolving the circle around the y-axis is a torus. Find the volume of the torus. (9.9)
    (c) Find the surface area of the torus. (9.9)

16. Sketch the graph of parametric equations: $x = t + \dfrac{1}{t}$ and $y = t - \dfrac{1}{t}$. Find its rectangular equation. (9.9)

17. Given $x^3 + y^3 = 3axy$ using $t = y/x$ as parameter (9.9)
    (a) find the parametric equations
    (b) show that $x + y + a = 0$ is an asymptote
    (c) show the graph is symmetric to the line $y = x$
    (d) sketch the graph*
        * the graph is known as the "folium of Descartes"

18. The parametric equations: $x = a(\theta - \sin\theta)$, $y = a(1 - \cos\theta)$ graph is the cycloid. Show that the area under one arch of the cycloid is 3 times the area of the rolling circle (Torricelli's theorem). (9.9)

19. Find the area inside the circle $r = a$ and outside the cardioid $r = a(1 - \cos\theta)$. (9.9)

20. Given $r^2 = 2a^2 \cos 2\theta$ (lemniscate) consider the part in the first quadrant. Show that at any point on this curve the angle between the radial direction and the outward normal is 2θ. (9.9)

21. Show that the spirals $r = \theta$ and $r = \dfrac{1}{\theta}$ intersect orthogonally at $\theta = 1$. (9.7)

22. Find the surface area generated by revolving $r = 2a\cos\theta$ about $\theta = \pi/2$. (9.9)

23. Show that $y = Ax^2$ intersects every ellipse $x^2 + 2y^2 = B$ at right angles. (9.7)

24. Show that the distance from either focus of $\dfrac{x^2}{a^2} - \dfrac{y^2}{b^2} = 1$ to either asymptote is b. (9.7)

Answers to Chapter 9, Part C Questions

1. $y = \frac{1}{4}x^2 \Rightarrow \frac{dy}{dx} = \frac{x}{2}$ at $x = \pm 2$ $\frac{dy}{dx} = \pm 1$

2. The points lie on an ellipse: $b^2x^2 + a^2y^2 = a^2b^2$
   $b^2a^2\cos^2 t + a^2b^2\sin^2 t = a^2b^2$

3. (a) take $x(t) = a\cos ht$ $y(t) = b\sin ht$
   then $\frac{x^2}{a^2} - \frac{y^2}{b^2} = \cos h^2 t - \sin h^2 t = 1$
   range $(x) = [a, \infty]$
   range $(y) = (-\infty, \infty)$

   (b) take $x(t) = -a\cos ht$ $y(t) = b\sin ht$
   then $\frac{x^2}{a^2} - \frac{y^2}{b^2} = 1$
   range $(x) = [-\infty, -a]$
   range $(y) = (-\infty, \infty)$

4. $a = 3$ $c = ea = \frac{1}{3} \cdot 3 = 1$ $b = \sqrt{a^2 - c^2} = \sqrt{8}$
   center $(0, 0)$ so $\frac{x^2}{9} + \frac{y^2}{8} = 1$

5. It tends to the union of two oppositely-directed half lines that begin at the ends of the transverse axis.

6. $r = \dfrac{6}{1 + 2\sin\theta}$ $\qquad 6 = r + 2r\sin\theta$
   $6 = \sqrt{x^2 + y^2} + 2y$, $x^2 + y^2 = 4y^2 - 24y + 36$
   $x^2 - 3y^2 + 24y = 36$ a hyperbola

7. $r = 2(d + r\cos\theta)$
   $= 2d + 2r\cos\theta$

8. $[2, \pi/2] = [-2, \pi/2]$ the coordinates $[2, \pi/2]$ satisfy $r^2\sin\theta = 4$ and the coordinates $[-2, -\pi/2]$ satisfy $r = 2\cos 2\theta$.

9.

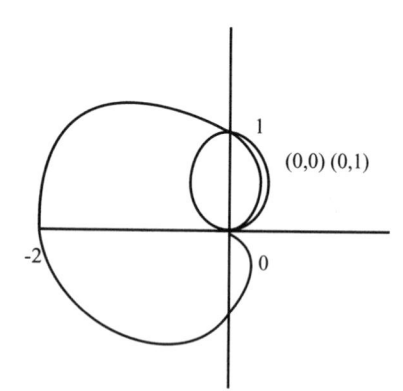

10. Since there are $2n + 1$ petals,
    total area $= (2n + 1)$(area of one petal)
    $\therefore A = 2\displaystyle\int_0^{\pi/4n} \frac{1}{2}(a\cos 2n\theta)^2 d\theta$

$\quad = a^2 \displaystyle\int_0^{\pi/4n} \cos^2 2n\theta\, d\theta$

$\quad = a^2 \displaystyle\int_0^{\pi/4n} \left(\frac{1}{2} + \frac{\cos 4n\theta}{2}\right) d\theta$

$\quad = a^2 \left[\frac{1}{2}\theta + \frac{\sin 4n\theta}{8n}\right]_0^{\pi/4n}$

$\quad = \dfrac{\pi a^2}{8n}$

$\rightarrow$ total area enclosed by $r = a\cos 2n\theta = \dfrac{\pi a^2}{2}$

the area of one petal of $r = a\sin 2n\theta$
$A = 2\displaystyle\int_0^{\pi/4n} \frac{1}{2}(a\sin 2n\theta)^2 d\theta$

$\quad = a^2 \displaystyle\int_0^{\pi/4n} \left(\frac{1}{2} + \frac{\cos 4n\theta}{2}\right) d\theta$

$\quad = \dfrac{\pi a^2}{8n} \rightarrow$ total area $= \dfrac{\pi a^2}{2}$

11. Any continuous function unbounded on $(0, 1)$ will do; for example $f(t) = \dfrac{1}{t}$

12. $\displaystyle\int_c^d y(t)x'(t)\,dt = \displaystyle\int_c^d f(x(t))x'(t)\,dt$
    $\quad = \displaystyle\int_a^b f(x)\,dx =$ area below C

13. $\dfrac{d^2y}{dx^2} = \dfrac{d}{dt}\left(\dfrac{dy}{dx}\right)\dfrac{dt}{dx} = \dfrac{d}{dt}\left[\dfrac{y'(t)}{x'(t)}\right] \cdot \dfrac{1}{x'(t)}$
    $\quad = \dfrac{x'(t)y''(t) - y'(t)x''(t)}{x'(t)^3}$

14. $\sqrt{[x'(\theta)]^2 + [y'(\theta)]^2}$
    $= \sqrt{(\cos\theta - \theta\sin\theta)^2 + (\sin\theta + \theta\cos\theta)^2}$
    $= \sqrt{1 + \theta^2}$
    $\therefore L = \displaystyle\int_0^{2\pi} \sqrt{1 + \theta^2}\,d\theta$
    $= \left[\dfrac{1}{2}\theta\sqrt{1 + \theta^2} + \dfrac{1}{2}\ln/\theta + \sqrt{1 + \theta^2}\right]_0^{2\pi}$
    $= \pi\sqrt{1 + 4\pi^2} + \dfrac{1}{2}\ln(2\pi + \sqrt{1 + 4\pi^2})$

15. (a) Clear since this is a circle $x^2 + y^2 = b^2$ translated horizontally by a.
    $x(t) = b\cos t$, $y(t) = b\sin t$
    (b) Area of the disc is πb^2, centroid is $(2, 0)$.
    So $V_y = 2\pi\bar{x}A = 2\pi 2\pi b^2 = 4\pi^2 b^2$
    (c) Length of circle is $2\pi b$, centroid is $(2, 0)$. So
    $A_y = 2\pi\bar{x}L = 2\pi 2(2\pi b) = 8\pi^2 b$

16. $x = t + \dfrac{1}{t}$ $y = t - \dfrac{1}{t}$ $x^2 - y^2 = 4$

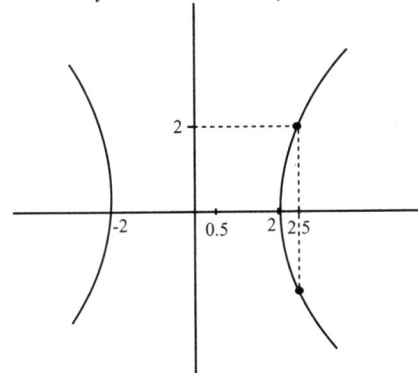

17. (a) $t = \dfrac{y}{x} \Rightarrow y = xt$

$x^3 + x^3 t^3 = 3ax^2 t \Leftrightarrow x = \dfrac{3at}{1+t^3}$

$\Rightarrow y = \dfrac{3at^2}{1+t^3}$

(b) $\lim\limits_{t \to -1}(x+y) = \lim \dfrac{3at + 3at^2}{1+t^3}$

$= 3a \lim\limits_{t \to -1} \dfrac{t(t+1)}{(t+1)(t^2 - t + 1)} = -a$

notice that $x \to \infty$ when $t \to -1$

$x + y + a = 0 \quad t = y/x \quad \therefore x = -\dfrac{a}{t+1}$

$y = -\dfrac{at}{t+1}$

$\therefore -\dfrac{at}{t+1} - \dfrac{3at^2}{1+t^3} = -\dfrac{at + 2at^2 + at^3}{1+t^3}$

$= -\dfrac{at(1+t)^2}{1+t^3}$

$= -\dfrac{at(1+t)}{1-t+t^3} \to 0$ when $t \to -1$

so $x + y + a = 0$ is an asymptote

(c) if (π, y) is on $x^3 + y^3 = 3axy$, then (y, x) is on $x^3 + y^3 = 3axy$ so it's symmetric to $y = x$

(d)

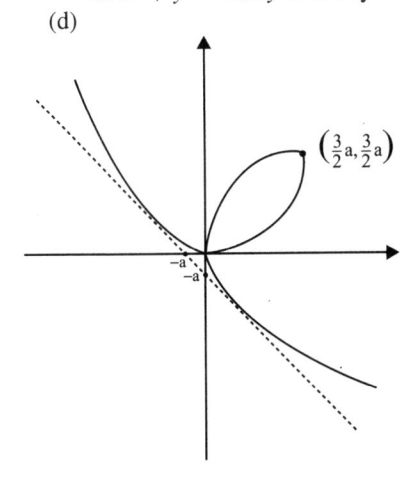

18. $A = \left|\dfrac{1}{2}\int_0^{2\pi}(xdy - ydx)\right| = \dfrac{1}{2}\left|\int_0^{2\pi}(a(\theta - \sin\theta)\right.$

$\left. \cdot a\sin\theta) - a(1 - \cos\theta) \cdot a(1 - \cos\theta))d\theta\right|$

$= \left|\dfrac{a^2}{2}\int_0^{2\pi}(\theta\sin\theta - \sin^2\theta - 1 + 2\cos\theta\right.$

$\left. - \cos^2\theta)d\theta\right|$

$= \left|\dfrac{a^2}{2}\int_0^{2\pi}(\theta\sin\theta - 2 + 2\cos\theta)d\theta\right|$

$= \dfrac{a^2}{2}[-\theta\cos\theta + \sin\theta - 2\theta + 2\sin\theta]_0^{2\pi}$

$= \dfrac{a^2}{2} \cdot 6\pi = 3\pi a^2$

19. $A = \dfrac{1}{2}\int_{-\frac{\pi}{2}}^{\frac{\pi}{2}}(a^2 - a^2(1-\cos\theta)^2)d\theta$

$= \int_0^{\frac{\pi}{2}}(a^2 - a^2 + 2a^2\cos\theta - a^2\cos^2\theta)d\theta$

$= \int_0^{\frac{\pi}{2}}(2a^2\cos\theta - a^2\cos^2\theta)d\theta$

$= 2a^2[\sin\theta]_0^{\frac{\pi}{2}} - \dfrac{a^2}{2}\int_0^{\frac{\pi}{2}}(1+\cos 2\theta)d\theta$

$= 2a^2 - \dfrac{a^2}{2}[\theta]_0^{\frac{\pi}{2}} - \dfrac{a^2}{4}[\sin 2\theta]_0^{\frac{\pi}{2}}$

$= 2a^2 - \dfrac{\pi a^2}{4}$

20.

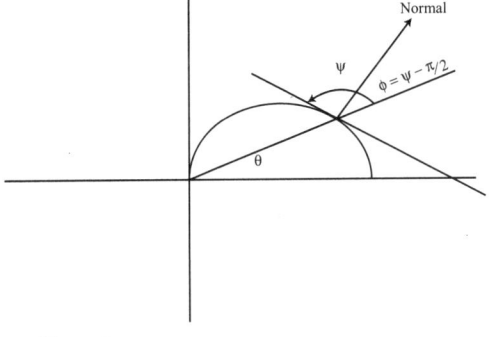

$\varphi - \pi/2 = \phi$
angle $= -\pi/2$
$\tan\phi = r\dfrac{d\theta}{dr}$
$r^2 = 2a^2\cos 2\theta$
$2r\dfrac{dr}{d\theta} = -4a^2\sin 2\theta$
so $\dfrac{dr}{d\theta} = -\dfrac{2a^2\sin 2\theta}{r}$
and $\tan\phi = -\dfrac{r^2}{2a^2\sin 2\theta} = -\cot 2\theta$
$\tan(\phi - \pi/2) = -\cot\phi = -\dfrac{1}{\tan\phi} = \tan 2\theta$
Since both $\phi - \pi/2, 2\theta \in (-\pi/2, \pi/2)$
$\phi - \pi/2 = 2\theta$

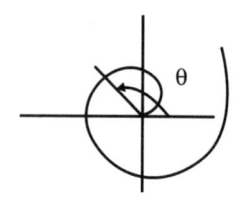

21.

Write
$$x(\theta) = r\cos\theta = \theta\cos\theta$$
$$y(\theta) = r\sin\theta = \theta\sin\theta$$
$$(*)\frac{y'}{x'} = \frac{\sin\theta + \theta\cos\theta}{\cos\theta - \theta\sin\theta}$$
$$x'(\theta) = \cos\theta - \theta\sin\theta \quad y'(\theta) = \sin\theta + \theta\cos\theta$$
Similarly $x(\theta) = r\cos\theta = \dfrac{\cos\theta}{\theta}$
$$y(\theta) = r\sin\theta = \frac{\sin\theta}{\theta}$$
$$(**)\frac{y'}{x'} = -\frac{\theta\cos\theta - \sin\theta}{\theta\sin\theta + \cos\theta}$$
$$x'(\theta) = \frac{-\theta\sin\theta - \cos\theta}{\theta^2}$$
$$y'(\theta) = \frac{\theta\cos\theta - \sin\theta}{\theta^2}$$
$$r = \theta \quad \frac{dr}{d\theta} = 1 \text{ so } \tan\psi_1 = r\frac{d\theta}{dr} = \theta$$
$$r = \frac{1}{\theta} \quad \frac{dr}{d\theta} = -\frac{1}{\theta^2} \tan\psi_2 = -\theta$$
at $\theta = 1 \tan\psi_2 \tan\psi_1 = (-1)(1) = -1$
when $\theta = 1$

$(*)$ becomes $\dfrac{\sin 1 + \cos 1}{\cos 1 - \sin 1}$

$(**)$ becomes $-\dfrac{\cos 1 - \sin 1}{\sin 1 + \cos 1}$

hence $\dfrac{\sin 1 + \cos 1}{\cos 1 - \sin 1} = -\dfrac{1}{\frac{\cos 1 - \sin 1}{\sin 1 + \cos 1}}$

hence orthogonal

22. Consider
$$r = 2a\cos\theta$$
line $\theta = \pi/2$ about the y-axis

So $dA = 2\pi \times ds$
$$x = r\cos\theta = 2a\cos^2\theta$$
$$ds = 2a\,d\theta$$

So $\displaystyle\int_{-\pi/2}^{\pi/2} 2\pi(2a\cos^2\theta)2a\,d\theta$

$= 4\pi a^2 \displaystyle\int_{-\pi/2}^{\pi/2}(1 + \cos 2\theta)d\theta$

$= 4\pi a^2\left[\theta + \dfrac{1}{2}\sin 2\theta\right]_{-\pi/2}^{\pi/2}$

$= 4\pi^2 a^2$

23. $y = Ax^2 \longrightarrow$ slope(1) $\dfrac{dy}{dx} = 2Ax$ at any point x

$x^2 + 2y^2 = B \longrightarrow \dfrac{dy}{dx} = \dfrac{x}{-2y}$ slope (2)

at the point of intersection $y = Ax^2$

so $\dfrac{dy}{dx} = \dfrac{x}{-2y} = \dfrac{x}{-2Ax^2} = -\dfrac{1}{2Ax}$

[slope(1)] [slope (2)] $2Ax\left(-\dfrac{1}{2Ax}\right) = -1$

so they intersect at right angles

24. the focus is $(\pm\sqrt{a^2 + b^2}, 0)$ and the asymptotes are $y = \pm\dfrac{b}{a}x$

$\therefore$ the distance from focus to $bx + ay = 0$ is

$D = \dfrac{a\cdot 0 \pm b(\pm\sqrt{a^2 + b^2})}{\sqrt{a^2 + b^2}} = \dfrac{bc}{c} = b$

by symmetry the other distances equal "b"

CHAPTER 10 Sequences; Intermediate Forms; Improper Integrals

Part A

1. Complete: M is the least upper bound of the set S a nonempty set of real numbers iff (10.1)
   (i) _____
   (ii) _____

2. Given a nonempty set of real numbers S that has an upper bound then S must have _____ (10.1)

3. Define a sequence of real numbers. (10.2)

4. Given a sequence $\{a_n\}$ (10.2)
   (a) $\{a_n\}$ is nondecreasing iff _____
   (b) $\{a_n\}$ is decreasing iff _____

5. True or false
   Given $\{a_n\}$ (10.2)
   (a) $\{a_n\}$ defined by $a_n = (1)^n$, $\{a_n\}$ is monotonic
   (b) $\{a_n\}$ defined by $a_n = (-1)^n$, $\{a_n\}$ is monotonic
   (c) $\{a_n\}$ defined by $a_n = n^n$, $\{a_n\}$ is monotonic
   (d) $\{a_n\}$ defined by $a_n = \dfrac{n+1}{n}$, $\{a_n\}$ is monotonic

6. Complete: $\lim\limits_{n \to \infty} a_n = L$ (10.3)
   iff for each $\epsilon > 0$ there exists a positive number k such that _____.

7. If $\lim\limits_{n \to \infty} a_n = L_1$ and $\lim\limits_{n \to \infty} a_n = L_2$ and $\lim\limits_{n \to \infty} a_n = L_3$
   then _____. (10.3)

8. (a) Define a sequence $\{a_n\}$ that is said to be convergent. (10.3)
   (b) If a sequence $\{a_n\}$ is divergent then $\lim\limits_{n \to \infty} a_n =$ _____.

9. If $\{a_n\}$ is a nonincreasing sequence then a_n will converge to its _____. (10.3)

10. Given $a_n \to 5$ and $b_n \to 3$ (10.3)
    then (a) $a_n + b_n \to$
    (b) $a_n \cdot b_n \to$
    (c) $\dfrac{1}{a_n} \to$
    (d) $\dfrac{b_n}{a_n} \to$

11. $\lim\limits_{n \to \infty} \left(1 + \dfrac{1}{n}\right)^n =$ _____. (10.3)

12. If $x > 0$ then (10.4)
    (a) $x^{1/n} \to$ _____ as $n \to \infty$
    (b) if $-1 < x < 1$ then
    $x^n \to$ _____ as $n \to \infty$

(c) $\dfrac{1}{n^2} \to$ _____ as $n \to \infty$

(d) $\dfrac{x^n}{n!} \to$ _____ as $n \to \infty$

(e) $\dfrac{\ln n}{n} \to$ _____ as $n \to \infty$

(f) $n^{1/n} \to$ _____ as $n \to \infty$

(g) $\left(1 + \dfrac{x}{n}\right)^n \to$ _____ as $n \to \infty$

13. $\displaystyle\lim_{x \to 1} \dfrac{x^2 - 1}{x^3 - 1} =$ _____ verify using L' Hospital's rule (10.5)

14. Does L' Hospital's rule apply to:
$\displaystyle\lim_{x \to 0} \dfrac{\sin x}{x^2 - 1}$?

15. Complete: The Cauchy mean-value theorem (10.5)
Given f, g differentiable on (a, b) and continuous on $[a, b]$. If g' is never 0 in (a, b) then there exists a number r for which _____

16. (a) $\displaystyle\lim_{x \to -\infty} \dfrac{x^2}{e^{-x}}$ is an example of what indeterminate form? _____
    (b) $\displaystyle\lim_{x \to \infty} e^{-x} \sqrt{x}$ is an example of what indeterminate form? _____
    (c) find the limit to part (a) if it exists
    (d) find the limit to part (b) if it exists

17. Complete: When confronted with the indeterminate forms: (10.6)
$0 \cdot \infty, \infty - \infty, 0°, 1^\infty, \infty°$
we handle these situations by writing the given problem as _____
So that
$\displaystyle\lim_{x \to 1^+} \left(\dfrac{1}{\ln x} - \dfrac{1}{x-1}\right), \infty - \infty$ becomes
$\displaystyle\lim_{x \to 1^+} (\quad)$
and
$\displaystyle\lim_{x \to \infty} e^{-x} \sqrt{x}, 0 \cdot \infty$ becomes
$\displaystyle\lim_{x \to \infty} (\quad)$

18. Give an example of an improper integral. (10.7)

19. Complete: $\displaystyle\int_1^\infty \dfrac{dx}{x^p}$ converges if _____
    diverges if _____

20. which of the following converge?
    (a) $\displaystyle\int_1^\infty \dfrac{dx}{x^{1/2}}$
    (b) $\displaystyle\int_1^\infty \dfrac{dy}{x^{3/2}}$
    (c) $\displaystyle\int_1^\infty x^{-e} dx$
    (d) $\displaystyle\int_1^\infty \dfrac{dy}{x}$

21. Complete: Given f, g are continuous and $0 \leq f(x) \leq g(x)$ for all $x \in [a, \infty)$ (10.7)
    (a) if $\displaystyle\int_a^\infty g(x) dx$ converges then _____
    (b) if $\displaystyle\int_a^\infty f(x) dx$ diverges then _____

For Questions 22–23 Set up the following problems for solution (do not solve)

22. $\displaystyle\int_{-\infty}^{\infty} \frac{e^x}{1+e^{2x}}\,dx = \int + \int$ (10.7)
$= \lim + \lim$

23. $\displaystyle\int_{-3}^{2} \frac{dx}{x^3} = \int + \int$ (10.7)
$= \lim + \lim$

For Questions 24–30 True or False

24. $\displaystyle\lim_{x\to\infty} \frac{x}{\ln x} = 0$ (10.6)

25. $\displaystyle\lim_{x\to 0} \frac{\sin 3x}{\sin 5x} = \frac{3}{5}$ (10.6)

26. $\displaystyle\lim_{x\to\infty} \frac{\sin x}{x} = \lim_{x\to\infty} \cos x$ (10.6)

27. $\displaystyle\int_{-\infty}^{0} e^x\,dx = 1$ (10.7)

28. $\displaystyle\int_{-\infty}^{\infty} f(x)\,dx = \lim_{b\to\infty} \int_{-b}^{b} f(x)\,dx$ (10.7)

29. $\displaystyle\int_{0}^{\infty} x^2 e^{-x}\,dx$ converges (10.7)

30. $\displaystyle\int_{0}^{\pi/2} \cot x\,dx = 0$ (10.7)

Answers to Chapter 10, Part A Questions

1. (i) M is an upper bound for S
   (ii) $M \leq k$ where k is any upper bound of S

2. lub

3. A real-valued function defined on the set of positive integers.

4. (a) $a_n \leq a_{n+1}$ for each positive integer n
   (b) $a_n > a_{n+1}$ for $\cdots n$

5. (a) T
   (b) F
   (c) T
   (d) T

6. if $n > k$, the $|a_n - L| < \epsilon$

7. $L_1 = L_2 = L_3$

8. (a) has limit, $a_n \to L$
   (b) $\infty, a_n \to \infty$

9. glb

10. (a) 8
    (b) 15
    (c) $\dfrac{1}{5}$
    (d) $\dfrac{3}{5}$

11. e

12. (a) 1
    (b) 0
    (c) 0
    (d) 0
    (e) 0
    (f) 1
    (g) e^x

13. $^2/_3 = {}^2/_3$

14. No

15. $\dfrac{f'(r)}{g'(r)} = \dfrac{f(b) - f(a)}{g(b) - g(a)}$

16. (a) ∞/∞
    (b) $0 \cdot \infty$
    (c) 0
    (d) 0

17. quotient
    $$\lim_{x \to 1^+} \frac{x - 1 - \ln x}{(\ln x)(x - 1)}$$
    $$\lim_{x \to \infty} \frac{\sqrt{x}}{e^x}$$

18. $\displaystyle\int_0^\infty e^{2x}\,dx$ ans vary

19. $p > 1$
    $p \leq 1$

20. (a) diverge
    (b) converge
    (c) converge
    (d) diverge

21. $\displaystyle\int_a^\infty f(x)\,dx$ converges
    $\displaystyle\int_a^\infty g(x)\,dx$ diverges

22. $\displaystyle\lim_{a \to -\infty}\int_a^0 \frac{e^x}{1 + e^{2x}}\,dx + \lim_{b \to \infty}\int_0^b \frac{e^x}{1 + e^{2x}}\,dx$

23. $\displaystyle\lim_{a \to 0^-}\int_{-3}^a \frac{dx}{x^3} + \lim_{b \to 0^+}\int_b^2 \frac{dx}{x^3}$

24. F (false)

25. T

26. F

27. T

28. F (false)

29. T

30. F (false)

Part B

10.1 The Least Upper Bound Axiom

1. Find the least upper bound (if it exists) and the greatest lower bound (if it exists) for the interval (0, 4).

2. Find the least upper bound (if it exists) and the greatest lower bound (if it exists) for the set $\{x : x^2 < 5\}$.

3. Find the least upper bound (if it exists) and the greatest lower bound (if it exists) for the set $\{x : x^2 > 9\}$.

4. Find the least upper bound (if it exists) and the greatest lower bound (if it exists) for the set $\{x : |x - 2| > 1\}$.

5. Find the least upper bound (if it exists) and the greatest lower bound (if it exists) for the set $\{x : |x - 3| < 1\}$.

6. Find the least upper bound (if it exists) and the greatest lower bound (if it exists) for the set $\{x : x^2 + 3x + 2 \geq 0\}$.

10.2 Sequences of Real Numbers

7. Give an explicit formula for the nth term of the sequence $\{1, 4/5, 6/8, 8/11, 10/14, 12/17, \ldots\}$.

8. Give an explicit formula for the nth term of the sequence $\{1/2, -1/4, 1/8, -1/16, \ldots\}$.

9. Give an explicit formula for the nth term of the sequence $\{1, 4, 1/9, 16, 1/25, 36, \ldots\}$.

10. Determine the boundedness and monotonicity of the sequence $\dfrac{2n}{n+1}$.

11. Determine the boundedness and monotonicity of the sequence $\dfrac{2n}{2n-1}$.

12. Determine the boundedness and monotonicity of the sequence $\dfrac{2n-5}{3n+2}$.

13. Determine the boundedness and monotonicity of the sequence $1 - \dfrac{2}{n}$.

14. Determine the boundedness and monotonicity of the sequence $\dfrac{3^n}{2^n + 3}$.

15. Determine the boundedness and monotonicity of the sequence $\dfrac{1}{3n} - \dfrac{1}{3n+1}$.

16. Determine the boundedness and monotonicity of the sequence $\dfrac{4^n}{(n+2)^2}$.

17. Determine the boundedness and monotonicity of the sequence $\dfrac{3^n}{e^n}$.

18. Determine the boundedness and monotonicity of the sequence $\dfrac{3^n}{(n+1)!}$.

19. Determine the boundedness and monotonicity of the sequence $\dfrac{n+2}{e^n}$.

196   Calculus: One and Several Variables

20. Determine the boundedness and monotonicity of the sequence $\left(\dfrac{9}{10}\right)^n$.

21. Determine the boundedness and monotonicity of the sequence $\sin\dfrac{n\pi}{3}$.

22. Write the first six terms of the sequence given by $a_1 = 1; a_{n+1} = \dfrac{n+2}{n+1}a_n$, and then give the general term.

23. Write the first six terms of the sequence given by $a_1 = 1; a_{n+1} = 2a_n - 1$, and then give the general term.

24. Write the first six terms of the sequence given by $a_1 = 1; a_{n+1} = 2a_n - n(n+1)$, and then give the general term.

25. Write the first six terms of the sequence given by $a_1 = 1; a_{n+1} = 1 - 2a_n$, and then give the general term.

26. Write the first six terms of the sequence given by $a_1 = 1; a_2 = 4; a_{n+1} = 2a_n - a_{n-1}$, and then give the general term.

27. Write the first six terms of the sequence given by $a_1 = 2; a_2 = 2; a_{n+1} = 2a_n - a_{n-1}$, and then give the general term.

28. Write the first six terms of the sequence given by $a_1 = 3; a_2 = 5; a_{n+1} = 3a_n - 2a_{n-1} - 2$, and then give the general term.

10.3 The Limit of a Sequence

29. State whether or not the sequence $\dfrac{1}{n}\sin\dfrac{\pi}{n}$ converges as $n \to \infty$, and, if it does, find the limit.

30. State whether or not the sequence $\dfrac{\ln n}{n}$ converges as $n \to \infty$, and, if it does, find the limit.

31. State whether or not the sequence $(-1)^{n+1}\dfrac{n}{n+2}$ converges as $n \to \infty$, and, if it does, find the limit.

32. State whether or not the sequence $(1+n)^{1/n}$ converges as $n \to \infty$, and, if it does, find the limit.

33. State whether or not the sequence $\dfrac{n}{n+2}$ converges as $n \to \infty$, and, if it does, find the limit.

34. State whether or not the sequence $1 + (-1)^n$ converges as $n \to \infty$, and, if it does, find the limit.

35. State whether or not the sequence $\dfrac{n^3 + 6n^2 + 11n + 6}{2n^3 + 3n^2 + 1}$ converges as $n \to \infty$, and, if it does, find the limit.

36. State whether or not the sequence $\dfrac{\sqrt{n}}{\ln n}$ converges as $n \to \infty$, and, if it does, find the limit.

37. State whether or not the sequence $\dfrac{1-n^2}{2+3n^2}$ converges as $n \to \infty$, and, if it does, find the limit.

38. State whether or not the sequence $n\sin\dfrac{1}{n}$ converges as $n \to \infty$, and, if it does, find the limit.

39. State whether or not the sequence $\dfrac{2n}{\sqrt{n^2-1}}$ converges as $n \to \infty$, and, if it does, find the limit.

40. State whether or not the sequence $\dfrac{n}{2n+1}$ converges as $n \to \infty$, and, if it does, find the limit.

41. State whether or not the sequence $\dfrac{\sin n}{n}$ converges as $n \to \infty$, and, if it does, find the limit.

42. State whether or not the sequence $\dfrac{n+6}{2n+3}$ converges as $n \to \infty$, and, if it does, find the limit.

43. State whether or not the sequence $\dfrac{(n+1)^2}{4^n}$ converges as $n \to \infty$, and, if it does, find the limit.

44. State whether or not the sequence $\dfrac{3^n}{n!}$ converges as $n \to \infty$, and, if it does, find the limit.

45. State whether or not the sequence $\dfrac{2n-5}{3n+2}$ converges as $n \to \infty$, and, if it does, find the limit.

46. State whether or not the sequence $\dfrac{e^n}{\sqrt{n}}$ converges as $n \to \infty$, and, if it does, find the limit.

47. State whether or not the sequence $\dfrac{1-n^2}{2+3n^2}$ converges as $n \to \infty$, and, if it does, find the limit.

48. State whether or not the sequence $\ln\left(\dfrac{2n}{n+1}\right)$ converges as $n \to \infty$, and, if it does, find the limit.

49. State whether or not the sequence $\ln\left(\dfrac{2n}{4n+1}\right)$ converges as $n \to \infty$, and, if it does, find the limit.

50. State whether or not the sequence $\ln\left(\dfrac{n^2}{n+1}\right)$ converges as $n \to \infty$, and, if it does, find the limit.

51. State whether or not the sequence defined recursively by $a_1 = 1$, $a_{n+1} = -\dfrac{1}{3}a_n$ converges as $n \to \infty$, and, if it does, find the limit.

52. State whether or not the sequence defined recursively by $a_1 = 1$, $a_{n+1} = (-1)^n a_n$ converges as $n \to \infty$, and, if it does, find the limit.

53. State whether or not the sequence defined recursively by $a_1 = 1$, $a_{n+1} = 1 - \dfrac{1}{2}a_n$ converges as $n \to \infty$, and, if it does, find the limit.

10.5 The Indeterminate Form (0/0)

54. Find $\lim\limits_{x \to e} \left[\dfrac{\ln(\ln x)}{\ln x - 1}\right]$.

55. Find $\lim\limits_{x \to \infty} \dfrac{x^{-3/2}}{\sin 1/x}$.

56. Find $\lim\limits_{x \to 0} \dfrac{x}{1 + \sin x}$.

57. Find $\lim\limits_{x \to 0} \dfrac{\sin x - x}{\tan x - x}$.

198 Calculus: One and Several Variables

58. Find $\lim\limits_{x\to 0} \dfrac{\tan x}{x}$.

59. Find $\lim\limits_{x\to 0} \dfrac{x - \sin x}{2 + 2x + x^2 - 2e^x}$.

60. Find $\lim\limits_{x\to 0} \dfrac{5^x - 3^x}{x}$.

61. Find $\lim\limits_{x\to 0} \left[\dfrac{x - \ln(1+x)}{x^2}\right]$.

62. Find $\lim\limits_{x\to \pi/4} \dfrac{1 - \tan x}{\cos 2x}$.

63. Find $\lim\limits_{x\to 0} \dfrac{e^{2x} - 1}{x^2 - \sin x}$.

64. Find $\lim\limits_{x\to 0} \dfrac{xe^{3x}}{1 - e^{3x}}$.

65. Find $\lim\limits_{x\to 4} \dfrac{x^2 - 16}{x^2 + x - 20}$.

66. Find $\lim\limits_{x\to 1} \dfrac{\ln x}{x - 1}$.

67. Find $\lim\limits_{x\to 1} \dfrac{x^2 + 2x - 3}{x^2 + 3x - 4}$.

68. Find $\lim\limits_{x\to 0} \dfrac{\sinh x}{\cosh x - 1}$.

69. Find $\lim\limits_{x\to 0} \dfrac{xe^x}{1 - e^x}$.

70. Find $\lim\limits_{x\to 0} \dfrac{x - \tan x}{1 - \cos x}$.

71. Find $\lim\limits_{x\to a} \dfrac{\frac{1}{x} - \frac{1}{a}}{x - a}$.

10.6 The Indeterminate Form (∞/∞); Other Indeterminate Forms

72. Find $\lim\limits_{x\to \infty} \dfrac{4x^3 - 2x + 1}{4x^3 + 2}$.

73. Find $\lim\limits_{x\to \infty} \dfrac{e^x}{x^3}$.

74. Find $\lim\limits_{x\to \pi/2} (\sec x - \tan x)$.

75. Find $\lim\limits_{x\to 0} (\csc x - 1/x)$.

Sequences; Intermediate Forms; Improper Integrals   **199**

76. Find $\lim_{x \to 1^-} (x - 1) \tan \dfrac{\pi x}{2}$.

77. Find $\lim_{x \to 0^+} \sin x \ln x$.

78. Find $\lim_{x \to 0^+} x \ln x$.

79. Find $\lim_{x \to \infty} (x + e^x)^{2/x}$.

80. Find $\lim_{x \to \frac{\pi}{2}^+} (\tan x)^{\cos x}$.

81. Find $\lim_{x \to 0} \left(\dfrac{1}{2} - \dfrac{1}{x} \right)$.

82. Find $\lim_{x \to 0} (\cos 3x)^{1/x}$.

83. Find $\lim_{x \to 0^+} x \ln \sin x$.

84. Find $\lim_{x \to 0} (e^x + 3x)^{1/x}$.

85. Find $\lim_{x \to \infty} \left(1 + \dfrac{1}{x^2} \right)^{x^2}$.

86. Find $\lim_{x \to 0} (\sin x)^x$.

87. Find $\lim_{x \to 0} (\sin 2x + 1)^{1/x}$.

88. Find $\lim_{x \to \infty} (2e^x + x^2)^{3/x}$.

89. Find $\lim_{x \to 0} (\cosh x)^{4/x}$.

10.7   Improper Integrals

90. Evaluate $\displaystyle\int_0^1 \dfrac{x}{\sqrt{1 - x^2}} dx$.

91. Evaluate $\displaystyle\int_0^{\pi/4} \dfrac{\sec^2 x}{\sqrt{\tan x}} dx$.

92. Evaluate $\displaystyle\int_1^{\infty} \dfrac{dx}{x^3}$.

93. Evaluate $\displaystyle\int_0^1 \dfrac{1}{\sqrt{x}} dx$.

94. Evaluate $\displaystyle\int_1^4 \dfrac{1}{\sqrt[3]{x - 3}} dx$.

95. Evaluate $\displaystyle\int_0^3 \dfrac{x}{(x^2 - 1)^{2/3}} dx$.

96. Evaluate $\int_{3}^{4} \frac{1}{(x-4)^3} dx$.

97. Evaluate $\int_{0}^{8} \frac{1}{x^{1/3}} dx$.

98. Evaluate $\int_{0}^{\infty} xe^{-x^2} dx$.

99. Evaluate $\int_{0}^{1} \frac{1}{\sqrt{1-x^2}} dx$.

100. Evaluate $\int_{-1}^{2} \frac{1}{x^2} dx$.

101. Evaluate $\int_{-2}^{0} \frac{1}{x+2} dx$.

102. Evaluate $\int_{0}^{4} \frac{1}{(x-1)^3} dx$.

103. Evaluate $\int_{1}^{\infty} \frac{1}{\sqrt{x}} dx$.

104. Evaluate $\int_{1}^{2} \frac{1}{x \ln x} dx$.

105. Evaluate $\int_{0}^{\infty} \frac{1}{x^{1/3}} dx$.

106. Evaluate $\int_{2}^{\infty} \frac{1}{(x-1)^3} dx$.

107. Evaluate $\int_{\infty}^{1} e^{(x-e^x)} dx$.

Answers to Chapter 10, Part B Questions

1. lub: 4; glb: 0
2. lub: $\sqrt{5}$; glb: $-\sqrt{5}$
3. lub: none; glb: none
4. lub: none; glb: none
5. lub: 4; glb: 2
6. lub: none; glb: none
7. $\dfrac{2n}{3n-1}$
8. $\dfrac{(-1)^{n+1}}{2^n}$
9. $n^{2(-1)^n}$
10. Bounded: below by 1, above by 2; monotone increasing
11. Bounded: below by 1, above by 2; monotone decreasing
12. Bounded: below by $-3/5$, above by $2/3$; monotone increasing
13. Bounded: below by -1, above by 1; monotone increasing
14. Bounded: below by $3/5$, no upper bound; monotone increasing
15. Bounded: below by 0, above by $1/12$; monotone decreasing
16. Bounded: below by $4/9$, no upper bound; monotone increasing
17. Bounded: below by $3/e$, no upper bound; monotone increasing
18. Bounded: below by 0, above by $3/2$; monotone decreasing
19. Bounded: below by 0, above by $3/e$; monotone decreasing
20. Bounded: below by 0, above by $9/10$; monotone decreasing
21. Bounded: below by $\dfrac{-\sqrt{3}}{2}$, above by $\dfrac{\sqrt{3}}{2}$; not monotone
22. 1, 3/2, 2, 5/2, 3, 7/2; $a_n = \dfrac{n+1}{2}$
23. 1, 1, 1, 1, 1, 1; $a_n = 1$
24. 1, 0, -6, -24, -68, -166; $a_n = -\dfrac{7}{2}(2)^n + n^2 + 3n + 4$
25. 1, -1, 3, -5, 11, -21; $a_n = \dfrac{1}{3}[1 - (-2)^n]$
26. 1, 4, 7, 10, 13, 16; $a_n = 3n - 2$
27. 2, 2, 2, 2, 2, 2; $a_n = 2$
28. 3, 5, 7, 9, 11, 13; $a_n = 2n + 1$
29. converges; 0
30. converges; 0
31. diverges
32. converges; 1
33. converges; 1
34. diverges
35. converges; ½
36. diverges
37. converges; $-1/3$
38. converges; 1
39. converges; 2
40. converges; ½
41. converges; 0
42. converges; ½
43. converges; 0
44. converges; 0
45. converges; 2/3
46. diverges
47. converges; $-1/3$
48. converges; ln 2

49. converges; 0
50. diverges
51. converges; 0
52. diverges
53. converges; 2/3
54. 1
55. 0
56. 0
57. $-\frac{1}{2}$
58. 1
59. $-\frac{1}{2}$
60. $\ln 5/3$
61. $\frac{1}{2}$
62. 1
63. -2
64. $-1/3$
65. 8/9
66. 1
67. 4/5
68. $+\infty$
69. -1
70. 0
71. $\dfrac{-1}{a^2}$
72. 1
73. $+\infty$
74. 0
75. 0
76. $-2/\pi$
77. 0

78. 0
79. e^2
80. 1
81. $+\infty$
82. 1
83. 0
84. e^4
85. e
86. 1
87. e^2
88. e^3
89. 1
90. 1
91. 2
92. $\frac{1}{2}$
93. 2
94. $\dfrac{3}{2}(1 - \sqrt[3]{4})$
95. 9/2
96. divergent
97. 6
98. $\frac{1}{2}$
99. $\pi/2$
100. divergent
101. divergent
102. divergent
103. divergent
104. divergent
105. divergent
106. $\frac{1}{2}$
107. $1 - \dfrac{1}{e^e}$

Part C

1. Let S and T be nonempty sets of real numbers such that $x \leq y$ for all $x \in S$ and $y \in T$. (10.1)
   (a) Prove that lub $S \leq y$ for all $y \in T$
   (b) Prove that lub $S \leq$ glb T

2. Let S be a bounded set of real numbers and suppose that lub $S =$ glb S. What can you conclude about S? (10.1)

For Questions 3–4 Determine whether the sequence with the given nth term is monotonic

3. $a_n = (-1)^n \left(\dfrac{1}{n}\right)$

4. $a_n = \sin \dfrac{n\pi}{6}$ (10.2)

5. If $a_1 = 3$ and $a_{n+1} = a_n + 5$ then $a_n = 5n - 2$. Prove using mathematical induction. (10.2)

6. Let $\{a_n\}$ and $\{b_n\}$ be sequences such that $a_n \to 0$ and $\{b_n\}$ is bounded. Prove that $a_n b_n \to 0$.

7. Find the indicated limit $\lim\limits_{n \to \infty} \dfrac{1^2 + 2^2 + \cdots + n^2}{(1+n)(2+n)}$ Hint: $1^2 + 2^2 + \cdots + n^2 = \dfrac{n(n+1)(2n+1)}{6}$

8. A sequence $\{a_n\}$ is said to be a Cauchy sequence iff for each $\epsilon > 0$ there exists a positive integer k such that $|a_n - a_m| < \epsilon$ for all $m, n \geq k$. Show that every convergent sequence is a Cauchy sequence. (10.4)

9. Show that, if $a > 0$ then $\lim\limits_{n \to \infty} n(a^{1/n} - 1) = \ln a$. (10.5)

10. Let f be a twice differentiable function and fix a value of x. (10.5)
    (a) Prove that $\lim\limits_{h \to 0} \dfrac{f(x+h) - (f(x-h))}{2h} = f'(x)$
    (b) Prove that $\lim\limits_{h \to 0} \dfrac{f(x+h) - 2f(x) + f(x-h)}{h^2} = f''(x)$

11. Show that the graphs of $y = \cosh x$ and $y = \sinh x$ are asymptotic. (10.6)

12. Let α be a positive number. Show that $\lim\limits_{x \to \infty} \dfrac{x^\alpha}{e^x} = 0$. (10.6)

13. (a) For what values of r is $\int_0^\infty x^r e^{-x}\, dx$ convergent? (10.7)
    (b) Use mathematical induction to show that
    $$\int_0^\infty x^n e^{-x}\, dx = n! \quad n = 1, 2, 3, \ldots$$

14. Use the comparison test to determine if the integral converges $\int_1^\infty 2^{-x^2}\, dx$

15. Let f be a continuous, positive, decreasing function on $[1, \infty)$ show that $\int_1^\infty f(x)\, dx$ converges iff $\left\{\int_1^n f(x)\, dx\right\}$ converges

16. The gamma function $\Gamma(n)$ is defined by
$$\Gamma(n) = \int_0^\infty x^{n-1} e^{-x} dx \quad n > 0$$
   (a) Find $\Gamma(1)$
   (b) Find $\Gamma(2)$
   (c) Show that $\Gamma(n+1) = n\Gamma(n)$
      Hint: Use integration by parts
   (d) Express $\Gamma(n)$ in factorial notation

17. A nonnegative function f is called a probability density function, f
$$\int_{-\infty}^\infty f(t)\,dt = 1$$
   show that $f(t) = \begin{cases} \frac{1}{\eta} e^{-\frac{1}{\eta}t} & t \geq 0 \\ 0 & t < 0 \end{cases}$ is a probability density function.

18. The expected value of a probability density function is $E(x) = \int_{-\infty}^\infty t f(t)\,dt$.
    Find the expected value for #17

Answers to Chapter 10, Part C Questions

1. (a) Any $y \in T$ is an upper bound for S. So by definition lub $S \leq y$.
   (b) By (a), lub S is the lower bound for T. So by definition lub $S \leq$ glb T.

2. S consists of a single element, equal to lub S.

3. no

4. no

5. True for $n = 1$. Assume true for n
   then $a_{n+1} = a_n + 5 = 5n - 2 + 5 = 5(n+1) - 2$

6. Let M be a bound for $\{b_n\}$ then $|a_n b_n| \leq |a_n| M$.
   Given $\epsilon > 0$ choose k such that $|a_n| < \epsilon/M$ for $n \geq k$, then $|a_n b_n| < \epsilon$ for $n \geq k$.

7. $\dfrac{1^2 + 2^2 + \cdots n^2}{(1+n)(2+n)}$
   $= \dfrac{n(2n+1)\cancel{(n+1)}}{\cancel{(n+1)}(n+2)} \to \infty$ diverge

8. Let $\epsilon > 0$, if $a_n \to L$ there exists a positive integer $k > 0$ such that
   $|a_p - L| < \dfrac{\epsilon}{2}$ for all $p \geq k$
   with $m, n \geq k$
   $|a_m - a_n| \leq |a_m - L| + |a_n - L| < \dfrac{\epsilon}{2} + \dfrac{\epsilon}{2} = \epsilon$
   Cauchy.

9. $\lim_{n \to \infty} n(a^{1/n} - 1) = \lim_{n \to \infty} \dfrac{a^{1/n} - 1}{1/n}$
   $\doteq \lim_{x \to \infty} \dfrac{a^{1/x} \ln a \left(-\frac{1}{x^2}\right)}{\left(-\frac{1}{x^2}\right)} = \ln a$

10. (a) $\lim_{h \to 0} \dfrac{f(x+h) - f(x-h)}{2h}$
    $\doteq \lim_{h \to 0} \dfrac{f'(x+h) - f'(x-h)(-1)}{2} = f'(x)$
    (b) $\lim_{h \to 0} \dfrac{f(x+h) - 2f(x) + f(x-h)}{h^2}$
    $\doteq \lim_{h \to 0} \dfrac{f'(x+h) - f'(x-h)}{2h}$
    $= \lim_{h \to 0} \dfrac{f''(x+h) - f''(x-h)}{2} = f''(x)$

11. $\cosh x - \sinh x = \dfrac{1}{2}(e^x + e^{-x}) - \dfrac{1}{2}(e^x - e^{-x})$
    $= e^{-x} \to 0$ as $x \to \infty$

12. $\lim_{x \to \infty} \dfrac{x^d}{e^x} \doteq \lim_{x \to \infty} \dfrac{\alpha x^{\alpha-1}}{e^x}$
    $\doteq \cdots \doteq \lim_{x \to \infty} \dfrac{\alpha(\alpha-1) \cdots (\alpha - k) x^{\alpha-(k+1)}}{e^x} = 0$
    let $k < \alpha \leq k + 1$, k integer

13. (a) For any r, we can find k such that $x^r < e^{x/2}$ for $x \geq k$
    then $\int_0^\infty x^r e^{-x} dx < \int_0^k x^r e^{-x} dx + \int_k^\infty e^{-x/2} dx$
    converges thus $\int_0^\infty x^r e^{-x} dx$ converges for all r.
    (b) For $n = 1$: $\int_0^\infty x e^{-x} dx$
    $= \lim_{b \to \infty} [-xe^{-x} - e^{-x}]_0^b = 1$
    Assume true for n.
    $\int_0^\infty x^{n+1} e^{-x} dx = \lim_{b \to \infty} \left([-x^{n+1} e^{-x}]_0^b + (n+1) \int_0^b x^n e^{-x} dx \right)$
    $= \lim_{b \to \infty} (-b^{n+1} e^{-n})$
    $+ (n+1) \int_0^\infty x^n e^{-x} dx$
    $= 0 + (n+1)n! = (n+1)!$

14. Converges by comparison with $\int_1^\infty e^{-x} dx$

15. Observe that $F(t) = \int_1^t f(x) dx$ is continuous and increasing that $a_n = \int_1^n f(x) dx$ is increasing and that $(*)$ $a_n \leq \int_1^t f(x) dx \leq a_{n+1}$ for $t \in [n, n+1]$
    If $\int_1^\infty f(x) dx$ converges then F being continuous, is bounded and, by $(*)$, $\{a_n\}$ is bounded and therefore convergent. If $\{a_n\}$ converges, then $\{a_n\}$ is bounded and, by $(*)$, F is bounded. Being increasing, F is also convergent; that is, $\int_1^\infty f(x) dx$ converges.

16. (a) $\Gamma(1) = \int_0^\infty e^{-x} dx = [-e^{-x}]_0^\infty = 1$
    (b) $\Gamma(2) = \int_0^\infty x e^{-x} dx$
    $= [-xe^{-x}]_0^\infty + \int_0^\infty e^{-x} dx = 1$
    (c) $\Gamma(n+1) = \int_0^\infty x^n e^{-x} dx$
    $= [-x^n e^{-x}]_0^\infty + n \int_0^\infty x^{n-1} e^{-x} dx$
    $= n \int_0^\infty x^{n-1} e^{-x} dx = n\Gamma(n)$
    (d) $\Gamma(n) = (n-1)!$

17. $\displaystyle\int_{-\infty}^{\infty} f(t)\,dt = \int_0^{\infty} \frac{1}{7}e^{-\frac{1}{7}t} = \lim_{b\to\infty}\left[(-e^{\frac{-1}{7}t})\right]_0^b$

$\qquad\qquad\qquad = \lim_{b\to\infty}(-e^{-\frac{1}{7}b} + 1)$

$\qquad\qquad\qquad = 1$

18. $\displaystyle E(x) = \int_{-\infty}^{\infty} tf(t)\,dt = \frac{1}{7}\int_0^{\infty} t\cdot e^{-\frac{1}{7}t}$

$\qquad\qquad = [-te^{-\frac{1}{7}t}]_0^{\infty} + \int_0^{\infty} e^{\frac{-1}{7}t}\,dt$

$\qquad\qquad = 7.$

CHAPTER 11 Infinite Series
Part A

1. Expand $\sum_{k=1}^{5} a_k$. (11.1)

2. Given $\sum_{k=0}^{\infty} a_k$ (11.1)
   find (a) $S_1 =$
   (b) $S_3 =$
   (c) $S_4 =$

3. Fill in: If the partial sums of an infinite series diverges then _____. (11.1)

4. If $-1 < x < 1$ then $\sum_{k=0}^{\infty} x^k$ will _____ the series is called _____ series. (11.1)

5. If $\sum_{k=0}^{\infty} a_k = 7$ and $\sum_{k=0}^{\infty} b_k = 2$ (11.1)
   (a) then $\sum_{k=0}^{\infty} (a_k + b_k) =$
   (b) $2 \sum_{k=0}^{\infty} a_k =$

6. If $\sum_{k=0}^{\infty} a_k = 7$ and $a_0 + a_1 + a_2 = 3$ (11.1)
   then $\sum_{k=3}^{\infty} a_k =$

7. If $\sum_{k=0}^{\infty} a_k$ converges then $a_k \to$ _____ (11.1)
   as $k \to$ _____.

8. $\sum_{k=1}^{\infty} \frac{k+1}{12}$ does not converge because (11.1)
   a_k _____.

9. Complete: A series with nonnegative terms converges iff _____. (11.2)

10. Complete: Given $f(k) = \frac{1}{k^2}$ on $[1, \infty)$
    then $\sum_{k=1}^{\infty} \frac{1}{k^2}$ converges iff $\int$ (11.2)

11. Given $\int_{1}^{\infty} \frac{1}{x^3} = \frac{1}{2}$ (11.2)
    (a) verify this
    (b) then $\sum_{k=1}^{\infty} \frac{1}{k^3}$ _____

207

12. Complete:

$\sum_{k=1}^{\infty} \frac{1}{k}$ is called the _____ series and _____ converge. (11.2)

13. Complete:

$\sum_{k=1}^{\infty} \frac{1}{k^3}$ converges then $\sum_{k=1}^{\infty} \frac{1}{3k^3+1}$ will _____ by the _____ test. (11.2)

14. Complete:

$\sum_{k=1}^{\infty} \frac{1}{k}$ diverges then $\sum_{k=1}^{\infty} \frac{1}{k+5}$ will _____ by the _____ test.

15. Given $\dfrac{\sum_{k=1}^{\infty} \frac{k+2}{2k^3-3}}{\sum_{k=1}^{\infty} \frac{1}{2k^2}} \to 1$ as $k \to \infty$

therefore since $\sum_{k=1}^{\infty} \frac{1}{2k^2}$ _____

so $\sum_{k=1}^{\infty} \frac{k+2}{2k^3-3}$ _____ by the _____ test.

16. Let $\sum a_k$ be a series with nonnegative terms and suppose $(a_k)^{1/2} \to P$ as $k \to \infty$ (11.3)
    then if (a) $P < 1$ then $\sum a_k$ _____
    if (b) $P > 1$ then $\sum a_k$ _____
    if (c) $P = 1$ then $\sum a_k$ _____

17. $\sum_{k=2}^{\infty} \frac{1}{(\ln k)^k}$ has limit = _____ (11.3)

    then the series _____.

18. Let $\sum a_k$ be a series of positive terms and suppose that (11.3)
    $\dfrac{a_{k+1}}{a_k} \to \lambda$ as $k \to \infty$
    if (a) $\lambda < 1$ $\sum a_k$ _____
    (b) $\lambda > 1$ $\sum a_k$ _____
    (c) $\lambda = 1$ $\sum a_k$

19. For $\sum_{k=0}^{\infty} \frac{3^k}{k!}$ $a_k =$ $a_{k+1} =$

    $\lim_{k \to \infty} \dfrac{a_{k+1}}{a_k} =$

    therefore $\sum_{k=0}^{\infty} \frac{3^k}{k!}$ will _____ because λ is _____.

20. True or false
    (a) If $\sum |a_k|$ converges, then $\sum a_k$ converges (11.4)
    (b) If $\sum a_k$ converges, then $\sum |a_k|$ converges

21. If $\sum_{k=1}^{\infty}(-1)\dfrac{k+1}{2k-1}$ converges to $\pi/4$ does $\sum \left|(-1)^{k+1}\dfrac{1}{2k-1}\right|$ converge? (11.4)

22. Explain:

    $\sum_{k=1}^{\infty}(-1)^{k+1}\dfrac{1}{2k-1}$ is called _____ convergent. (11.4)

23. $\sum_{k=1}^{\infty}(-1)^{k+1}\dfrac{1}{2k-1}$ has the following properties (11.4)

    $|a_{k+1}|$ _____ $|a_k|$

    $|a_k| \to$ _____ as $k \to$ _____

    hence by the _____ test

    the series _____.

24. Suppose that f has $n+1$ continuous derivatives on (a,b) containing 0. Then for $x \in (a,b)$. (11.5)
    Complete: $f(x) = f(0) +$ ___ $+$ ___ $+$ ___ $+ \cdots +$ ___ $+$ ___
    where the remainder $R_{k+1}(x)$ is
    $R_{k+1}(x) =$

25. In Taylor's theorem if $R_{n+1}(x) = 0$ then the Taylor series converges to _____.

26. Complete: Suppose that f has $n+1$ continuous derivatives on an open interval I containing O. Then for each $x \in I$

    $f(x) = f(0) + f'(0)x + \dfrac{f''(0)}{2!}x^2 + \cdots + \dfrac{f^{(n)}(0)}{n!} + R_{n+1}(x)$ (11.5)

    $R_{n+1}(x) =$ _____

27. For $f(x) = e^x \Rightarrow f(0) =$ _____ (11.5)
    $f'(x) = e^x \Rightarrow f'(0) =$ _____
    $f''(x) = e^x \Rightarrow f''(0) =$ _____
    etc.
    Write the Taylor polynomial for
    $e^x =$

    $R_{n+1}(x) = \left|\dfrac{f^{(n+1)}(c)}{(n+1)!}x^{n+1}\right| = |\quad| \Rightarrow$ _____.

28. Write the Maclaurin series for: (11.5)
    (a) $\sin x =$ (c) $\ln(1+x) =$
    (b) $\cos x =$ for $-1 < x \le 1$

29. Expand $g(x) = x^2$ in powers of $x - 2$. (11.6)

30. Expand $g(x) = \cos x$ in powers of $x - \dfrac{1}{2}\pi$.

31. A power series $\sum a_k x^k$ is said to converge at c iff _____. (11.7)

32. A power series $\sum a_k x^k$ is said to converge on the set S iff _____. (11.7)

33. True or false every power series $\sum a_k x^k$ has a radius of convergence r, where $0 \le r < \infty$ with the property that the series converges absolutely if $|x| < r$ and diverges if $|x| > r$. (11.7)

Calculus: One and Several Variables

34. Find the interval of convergence (11.7)

   for $\sum_{k=0}^{\infty} \dfrac{k+2}{3^k} x^k = 2 + \dfrac{3}{3}x + \dfrac{4}{3^2}x^2 + \cdots$

   $\left| \dfrac{a_k}{a_{k+1}} \right| = $ _____

   So $r = $ _____.

   Interval of convergence is

35. If $\sum_{k=0}^{\infty} a_k x^k = a_0 + a_1 x + a_2 x^2 + \cdots + a_n x^4 + \cdots$ (11.8)

    converges on $(-c, c)$ then

    $\sum_{k=0}^{\infty} \dfrac{d}{dx}(a_k x^k) = $ _____ and it will _____ on $(-c, c)$.

36. True or false

    If $\sum_{k=0}^{\infty} a_k x^k$ has radius of convergence r

    then $\sum \dfrac{d^{10}}{dx^{10}}(a_k x^k)$ has radius of convergence r. (11.8)

37. Given $\sin x = x - \dfrac{x^3}{3!} + \dfrac{x^5}{5!} \cdots$ (11.8)

    confirm $\dfrac{d^2}{dx^2}(\sin x) = -\sin x$

38. True or false

    $\int \left(\sum_{k=0}^{\infty} a_k x^k \right) dx = \left(\sum_{k=0}^{\infty} \dfrac{a_k}{k+1} x^k \right) + C$

39. If $f(x) = \sum_{k=0}^{\infty} a_k x^k$ converges on $(-c, c)$

    then $g(x) = \sum_{k=0}^{\infty} \dfrac{a_k}{k+1} x^{k+1}$ converges on _____ and $\int f(x) dx = $ _____ (11.8)

40. $\int \dfrac{1}{1+x^2} dx = \tan^{-1} x$

    find a power series expansion for $\tan^{-1} x$ (11.8)

    $\dfrac{1}{1+x^2} = $ _____

    if $|x| < 1$

    $\tan^{-1} x = \int$

41. $(1+x)^n = $ (11.9)

42. Evaluate (a) $\binom{10}{9}$ (b) $\binom{10}{1}$ (c) $\binom{10}{10}$ (d) $\binom{10}{0}$ (11.9)

Answers to Chapter 11, Part A Questions

1. $a_1 + a_2 + a_3 + a_4 + a_5$

2. $S_1 = a_1 + a_0$
   $S_3 = a_0 + a_1 + a_2 + a_3$
   $S_4 = a_0 + a_1 + a_2 + a_3 + a_4$

3. the series diverges

4. converge geometric series

5. (a) 9 (b) 14

6. 4

7. $a_k \to 0$ as $k \to \infty$

8. $a_k \to \infty$

9. the sequence of partial sums is bounded

10. $\int_1^\infty \frac{1}{k^2} dk$ converge

11. (a) $\int_1^\infty \frac{1}{x^3} dx = \lim_{a \to \infty} \left[-\frac{1}{2} x^{-2} \right]_1^a = \frac{1}{2}$
    (b) converges

12. harmonic, fails to

13. converge, list comparison

14. diverge, list comparison

15. converges, converges, list comparison

16. converges
    diverges
    inconclusive

17. 0
    converges

18. converges
    diverges
    inconclusive

19. $a_k = \frac{3^k}{k!}$ $a_{k+1} = \frac{3^{k+1}}{(k+1)!}$ $\frac{a_{k+1}}{a_k} = \frac{3}{k+1}$
    $\lim_{n \to \infty} {a_{k+1}}/{a_k} = 0$
    converges
    < 1

20. (a) T (b) F

21. no
    $\sum \left| (-1)^{k-1} \frac{1}{2k-1} \right| = \sum \frac{1}{2k-1}$ diverge
    compare with $\sum \frac{1}{k}$

22. conditionally

23. $<$
    0   ∞ alternating series test
    converges

24. $f(0) + f'(0)x + \frac{f''(0)}{2!} x^2 + \cdots + \frac{f^{(n)}(0)}{n!} x^n$
    $+ R_{n+1}(x)$
    $R_{n+1}(x) = \frac{1}{n!} \int_0^x f^{(n+1)}(t)(x-t)^n \, dt$

25. a Taylor polynominal
    $f(0) + f'(0)x + \frac{f''(0)}{2!} x^2 \cdots \frac{f^{(n)}(0)}{n!} x^n$

26. $\frac{f^{(n+1)}(c)}{(n+1)!} x^{n+1}$ where $C \in (0, x)$

27. 1 1 1
    $e^x = 1 + x + \frac{1}{2!} x^2 + \cdots$
    $\left| \frac{e^c}{(n+1)!} x^{n+1} \right| = \frac{e^c}{(n+1)!} (x)^{n+1}$

28. (a) $\sin x = \sum_{n=1}^\infty (-1)^{n+1} \frac{x^{2n-1}}{(2n-1)!}$
    (b) $\cos x = \sum_{n=0}^\infty (-1)^n \frac{x^{2n}}{(2n)!}$
    (c) $\ln(1+x) = \sum_{n=1}^\infty (-1)^{n+1} \frac{x^n}{n}$

29. $g(x) = 4 + 4(x-2) + (x-2)^2$

30. $\sum_{k=0}^\infty \frac{(-1)^{k+1}}{(2k+1)!} \left(x - \frac{\pi}{2} \right)^{2k+1}$

31. $\sum a_k c^k$ converges

32. iff $\sum a_k x^k$ converges at each $x \in S$

33. F

34. $\left|\dfrac{3(k+2)}{(k+3)x}\right| \to 3 \quad (-3, 3)$

35. $a_1 + 2a_2 x + \cdots n a_n x^{n-1} \cdots$
    converge

36. T

37. $\dfrac{d}{dx}(\sin x) = 1 - \dfrac{x^2}{2!} + \dfrac{x^4}{4!} \cdots$

    $\dfrac{d^2}{dx^2}(\sin x) = -x + \dfrac{x^3}{3!} - \dfrac{x^5}{5!} \cdots$
    $= -\sin x$

38. False

39. converges on $(-c, c)$
    $\int f(x)\,dx = g(x) + C$

40. $x - \dfrac{x^3}{3} + \dfrac{x^5}{5} - \dfrac{x^7}{7} \cdots$

    $\dfrac{1}{1+x^2} = \displaystyle\sum_{k=0}^{\infty} (-1)^k x^{2k}$

41. $1 + nx + \dfrac{n(n-1)}{2!}x^2 + \cdots n x^{n-1} + x^n$

42. (a) 10   (c) 1
    (b) 10   (d) 1

Part B

11.1 Infinite Series

1. Evaluate $\sum_{k=0}^{3}(4k+2)$. (11.1)

2. Evaluate $\sum_{k=1}^{4}(5k-2)$.

3. Evaluate $\sum_{k=0}^{3}(3^k+1)$.

4. Evaluate $\sum_{k=0}^{3}(-1)^k 3^k$.

5. Evaluate $\sum_{k=1}^{4}(-1)^{k+1}3^{k+1}$.

6. Evaluate $\sum_{k=3}^{5}\left(\frac{1}{3}\right)^{2k}$.

7. Evaluate $\sum_{k=0}^{4}\frac{2}{e^k}$.

8. Evaluate $\sum_{k=0}^{3}\frac{3^k}{2^{k+1}}$.

9. Express $2^0 + 2^1 + \cdots + 2^{10}$ in sigma notation.

10. Express $2x^2 - 2^2 x^3 + \cdots - 2^6 x^7$ in sigma notation.

11. Express $\frac{3^3}{e^2} + \frac{3^2}{e^3} + \cdots + \frac{1}{3^2 e^7}$ in sigma notation.

12. Express $\frac{1}{3(4)} + \frac{1}{4(5)} + \cdots + \frac{1}{9(10)}$ in sigma notation.

13. Determine whether $\sum_{k=1}^{\infty}\frac{1}{2k(k+1)}$ converges or diverges. If it converges, find the sum.

14. Determine whether $\sum_{k=2}^{\infty}\frac{(-1)^k}{5^k}$ converges or diverges. If it converges, find the sum.

15. Determine whether $\sum_{k=0}^{\infty}\frac{3^{k+2}}{4^{k+1}}$ converges or diverges. If it converges, find the sum.

16. Determine whether $\sum_{k=0}^{\infty} \dfrac{3}{10^k}$ converges or diverges. If it converges, find the sum.

17. Determine whether $\sum_{k=1}^{\infty} \dfrac{3}{e^k}$ converges or diverges. If it converges, find the sum.

18. Determine whether $\sum_{k=1}^{\infty} \dfrac{1}{(k+3)(k+4)}$ converges or diverges. If it converges, find the sum.

19. Determine whether $\sum_{k=1}^{\infty} \dfrac{2^k}{5}$ converges or diverges. If it converges, find the sum.

20. Determine whether the series given by $\sum_{k=0}^{\infty} u_k = 1 - \dfrac{2}{5} + \dfrac{4}{25} - \dfrac{8}{125} + \cdots$ converges or diverges. If it converges, find the sum.

21. Determine whether $\sum_{k=1}^{\infty} \dfrac{(-1)^{k+1}}{4^k}$ converges or diverges. If it converges, find the sum.

22. Determine whether $\sum_{k=3}^{\infty} \left(\dfrac{1}{2}\right)^k$ converges or diverges. If it converges, find the sum.

23. Determine whether $\sum_{k=1}^{\infty} \dfrac{2}{(2k-1)(2k+1)}$ converges or diverges. If it converges, find the sum.

24. Determine whether $\sum_{k=1}^{\infty} \left(-\dfrac{3}{7}\right)^{k+1}$ converges or diverges. If it converges, find the sum.

25. Determine whether $\sum_{k=1}^{\infty} 4^{k-1}$ converges or diverges. If it converges, find the sum.

26. Determine whether $\sum_{k=1}^{\infty} \left(-\dfrac{2}{3}\right)^{k+1}$ converges or diverges. If it converges, find the sum.

27. Determine whether $\sum_{k=1}^{\infty} \dfrac{1}{(k+1)(k+2)}$ converges or diverges. If it converges, find the sum.

28. Write the decimal fraction 0.21212121... as an infinite series and express the sum as the quotient of two integers.

29. Write the decimal fraction 0.251251251251... as an infinite series and express the sum as the quotient of two integers.

30. Write the decimal fraction 0.315315315315... as an infinite series and express the sum as the quotient of two integers.

31. Find a series expansion for $\dfrac{x}{1-x^3}$, $|x| < 1$.

32. Find a series expansion for $\dfrac{2}{3+x}$, $|x| < 3$.

11.2 The Integral Test; Comparison Tests

33. Determine whether $\sum_{k=1}^{\infty} \dfrac{3}{3k+2}$ converges or diverges. Justify your answer.

34. Determine whether $\sum_{k=1}^{\infty} \dfrac{1}{k\sqrt{k}}$ converges or diverges. Justify your answer.

35. Determine whether $\sum_{k=1}^{\infty} \dfrac{1}{3k+4}$ converges or diverges. Justify your answer.

36. Determine whether $\sum_{k=1}^{\infty} \dfrac{k^2}{2k+1}$ converges or diverges. Justify your answer.

37. Determine whether $\sum_{k=1}^{\infty} \dfrac{k}{\sqrt{2k^2+1}}$ converges or diverges. Justify your answer.

38. Determine whether $\sum_{k=1}^{\infty} \dfrac{3}{e^k}$ converges or diverges. Justify your answer.

39. Determine whether $\sum_{k=1}^{\infty} \dfrac{1}{4k+1}$ converges or diverges. Justify your answer.

40. Determine whether $\sum_{k=2}^{\infty} \dfrac{1}{k(\ln k)^3}$ converges or diverges. Justify your answer.

41. Determine whether $\sum_{k=1}^{\infty} \dfrac{1}{(2k+3)^3}$ converges or diverges. Justify your answer.

42. Determine whether $\sum_{k=1}^{\infty} \dfrac{k+1}{k(k+2)}$ converges or diverges. Justify your answer.

43. Find the sum of $\sum_{k=0}^{\infty} \left(\dfrac{5}{10^k} - \dfrac{6}{100^k} \right)$.

44. Determine whether $\sum_{k=1}^{\infty} \dfrac{1}{\sqrt{(k+1)^3}}$ converges or diverges. Justify your answer.

45. Determine whether $\sum_{k=1}^{\infty} \dfrac{2k}{1+k^4}$ converges or diverges. Justify your answer.

46. Determine whether $\sum_{k=1}^{\infty} \dfrac{1}{k\sqrt{k^2-1}}$ converges or diverges. Justify your answer.

47. Determine whether $\sum_{k=1}^{\infty} \dfrac{k}{e^k}$ converges or diverges. Justify your answer.

48. Determine whether $\sum_{k=1}^{\infty} \dfrac{1}{\cosh^2 k}$ converges or diverges. Justify your answer.

49. Determine whether $\sum_{k=1}^{\infty} \frac{\ln k}{k}$ converges or diverges. Justify your answer.

50. Which of the following statements about series is true?
    (a) If $\lim_{k \to \infty} u_k = 0$, then $\sum u_k$ converges.
    (b) If $\lim_{k \to \infty} u_k \neq 0$, then $\sum u_k$ diverges.
    (c) If $\sum u_k$ diverges, then $\lim_{k \to \infty} u_k \neq 0$.
    (d) $\sum u_k$ converges if and only if $\lim_{k \to \infty} u_k = 0$.
    (e) None of the preceding.

51. Determine whether $\sum_{k=1}^{\infty} \frac{\sqrt{k}}{k^2 + 1}$ converges or diverges. Justify your answer.

52. Determine whether $\sum_{k=1}^{\infty} \frac{k^2}{(2k^2 + 1)^2}$ converges or diverges. Justify your answer.

53. Determine whether $\sum_{k=1}^{\infty} \frac{1}{(k+3)(k+4)}$ converges or diverges. Justify your answer.

54. Determine whether $\sum_{k=1}^{\infty} \frac{1}{(k+3)(k-4)}$ converges or diverges. Justify your answer.

55. Determine whether $\sum_{k=1}^{\infty} \frac{1}{3k+2}$ converges or diverges. Justify your answer.

56. Determine whether $\sum_{k=1}^{\infty} \frac{1}{3^k + 2}$ converges or diverges. Justify your answer.

57. Determine whether $\sum_{k=1}^{\infty} \frac{\ln k}{k}$ converges or diverges. Justify your answer.

58. Determine whether $\sum_{k=1}^{\infty} \frac{k^2}{(k+2)(k+4)}$ converges or diverges. Justify your answer.

59. Determine whether $\sum_{k=1}^{\infty} \frac{k+1}{k^3+1}$ converges or diverges. Justify your answer.

60. Determine whether $\sum_{k=1}^{\infty} \frac{1}{1+\sqrt{k}}$ converges or diverges. Justify your answer.

61. Determine whether $\sum_{k=1}^{\infty} \frac{7k+2}{2k^5+7}$ converges or diverges. Justify your answer.

62. Determine whether $\sum_{k=1}^{\infty} \frac{1}{3^k - 2}$ converges or diverges. Justify your answer.

63. Determine whether $\sum_{k=1}^{\infty} \frac{4k-3}{k^3 - 5k - 7}$ converges or diverges. Justify your answer.

64. Determine whether $\sum_{k=1}^{\infty} \dfrac{1}{k^2 - 1}$ converges or diverges. Justify your answer.

65. Determine whether $\sum_{k=2}^{\infty} \dfrac{1}{k - \ln k}$ converges or diverges. Justify your answer.

66. Determine whether $\sum_{k=1}^{\infty} \dfrac{1}{3k^{3/2} + 1}$ converges or diverges. Justify your answer.

67. Determine whether $\sum_{k=1}^{\infty} \dfrac{k^2 + 3}{k(k+1)(k+2)}$ converges or diverges. Justify your answer.

68. Which of the following statements about $\sum_{k=2}^{\infty} \dfrac{1}{k \ln k}$ is true?
    (a) Converges because $\lim_{k \to \infty} \dfrac{1}{k \ln k} = 0$.
    (b) Converges because $\dfrac{1}{k \ln k} < \dfrac{1}{k}$.
    (c) Converges by ratio test.
    (d) Diverges by ratio test.
    (e) Diverges by integral test.

11.3 The Root Test; The Ratio Test

69. Determine whether $\sum_{k=1}^{\infty} \dfrac{k^2}{e^k}$ converges or diverges. Justify your answer by citing a relevant test.

70. Determine whether $\sum_{k=1}^{\infty} \dfrac{k}{2^k}$ converges or diverges. Justify your answer by citing a relevant test.

71. Determine whether $\sum_{k=1}^{\infty} \dfrac{k!}{10^{4k}}$ converges or diverges. Justify your answer by citing a relevant test.

72. Determine whether $\sum_{k=1}^{\infty} \dfrac{1}{2 + 3^{-k}}$ converges or diverges. Justify your answer by citing a relevant test.

73. Determine whether $\sum_{k=1}^{\infty} \left(\dfrac{k}{2k + 100} \right)^k$ converges or diverges. Justify your answer by citing a relevant test.

74. Determine whether $\sum_{k=1}^{\infty} \left(\dfrac{3k}{2k + 1} \right)^k$ converges or diverges. Justify your answer by citing a relevant test.

75. Determine whether $\sum_{k=5}^{\infty} \left(\dfrac{3k + 2}{2k - 9} \right)^k$ converges or diverges. Justify your answer by citing a relevant test.

76. Determine whether $\sum_{k=1}^{\infty} \dfrac{k!}{2^k}$ converges or diverges. Justify your answer by citing a relevant test.

77. Determine whether $\sum_{k=1}^{\infty} \dfrac{k^2}{2^k}$ converges or diverges. Justify your answer by citing a relevant test.

78. Determine whether $\sum_{k=1}^{\infty} \dfrac{1}{2k+9}$ converges or diverges. Justify your answer by citing a relevant test.

79. Determine whether $\sum_{k=1}^{\infty} \dfrac{e^k}{k!}$ converges or diverges. Justify your answer by citing a relevant test.

80. Determine whether $\sum_{k=1}^{\infty} \dfrac{10^k}{k!}$ converges or diverges. Justify your answer by citing a relevant test.

81. Determine whether $\sum_{k=1}^{\infty} \dfrac{k^3}{3^k}$ converges or diverges. Justify your answer by citing a relevant test.

82. Determine whether $\sum_{k=1}^{\infty} \dfrac{k!}{k^2}$ converges or diverges. Justify your answer by citing a relevant test.

83. Determine whether $\sum_{k=1}^{\infty} \left(\dfrac{\ln k}{k}\right)^k$ converges or diverges. Justify your answer by citing a relevant test.

84. Determine whether $\sum_{k=1}^{\infty} \dfrac{k^k}{k!}$ converges or diverges. Justify your answer by citing a relevant test.

85. Determine whether $\sum_{k=1}^{\infty} \dfrac{3^{2k}}{(2k)!}$ converges or diverges. Justify your answer by citing a relevant test.

11.4 Absolute and Conditional Convergence; Alternating Series

86. Determine whether $\sum_{k=1}^{\infty} \dfrac{(-1)^k}{k+\sqrt{k}}$ converges absolutely, converges conditionally, or diverges. Justify your answer by citing a relevant test.

87. Determine whether $\sum_{k=1}^{\infty} \dfrac{(-1)^{k+1}k+2}{k(k+1)}$ converges absolutely, converges conditionally, or diverges. Justify your answer by citing a relevant test.

88. Determine whether $\sum_{k=1}^{\infty} \dfrac{(-1)^{k+1}2^k}{3^k+1}$ converges absolutely, converges conditionally, or diverges. Justify your answer by citing a relevant test.

89. Determine whether $\sum_{k=2}^{\infty} \dfrac{(-1)^k \ln k}{k}$ converges absolutely, converges conditionally, or diverges. Justify your answer by citing a relevant test.

90. Determine whether $\sum_{k=1}^{\infty} \dfrac{(-1)^{k+1}2^k}{k^2}$ converges absolutely, converges conditionally, or diverges. Justify your answer by citing a relevant test.

91. Determine whether $\sum_{k=1}^{\infty} \dfrac{(-1)^k}{\sqrt{k}}$ converges absolutely, converges conditionally, or diverges. Justify your answer by citing a relevant test.

92. Determine whether $\sum_{k=1}^{\infty} \dfrac{(-1)^{k+1}}{3^k}$ converges absolutely, converges conditionally, or diverges. Justify your answer by citing a relevant test.

93. Determine whether $\sum_{k=1}^{\infty} \dfrac{(-1)^k k^2}{(2k^2+1)^2}$ converges absolutely, converges conditionally, or diverges. Justify your answer by citing a relevant test.

94. Determine whether $\sum_{k=1}^{\infty} \dfrac{(-1)^k k^3}{3^k}$ converges absolutely, converges conditionally, or diverges. Justify your answer by citing a relevant test.

95. Determine whether $\sum_{k=1}^{\infty} \dfrac{(-1)^k}{k\sqrt{k}}$ converges absolutely, converges conditionally, or diverges. Justify your answer by citing a relevant test.

96. Determine whether $\sum_{k=1}^{\infty} \dfrac{(-1)^k k}{k+2}$ converges absolutely, converges conditionally, or diverges. Justify your answer by citing a relevant test.

97. Determine whether $\sum_{k=1}^{\infty} \dfrac{(-1)^k 2}{e^k}$ converges absolutely, converges conditionally, or diverges. Justify your answer by citing a relevant test.

98. Determine whether $\sum_{k=1}^{\infty} \dfrac{(-1)^{k+1}(2k-1)}{k^3+1}$ converges absolutely, converges conditionally, or diverges. Justify your answer by citing a relevant test.

99. Determine whether $\sum_{k=1}^{\infty} \dfrac{(-1)^{k+1}}{3k+4}$ converges absolutely, converges conditionally, or diverges. Justify your answer by citing a relevant test.

100. Determine whether $\sum_{k=1}^{\infty} \dfrac{(-1)^{k+1} k^2}{2k+1}$ converges absolutely, converges conditionally, or diverges. Justify your answer by citing a relevant test.

101. Determine whether $\sum_{k=1}^{\infty} \dfrac{(-1)^k k!}{(2k+3)!}$ converges absolutely, converges conditionally, or diverges. Justify your answer by citing a relevant test.

102. Determine whether $\sum_{k=1}^{\infty} \dfrac{(-1)^k 2\sqrt{k}}{k^2+1}$ converges absolutely, converges conditionally, or diverges. Justify your answer by citing a relevant test.

103. Determine whether $\sum_{k=1}^{\infty} \dfrac{(-1)^k k^2}{e^k}$ converges absolutely, converges conditionally, or diverges. Justify your answer by citing a relevant test.

104. Estimate the error if the partial sum S_{20} is used to approximate the sum of the series $\sum_{k=0}^{\infty} \dfrac{(-1)^k}{\sqrt{3k+1}}$.

105. Find the smallest integer N such that S_N will approximate the sum of the series $\sum_{k=0}^{\infty} \dfrac{(-1)^k}{k^2+2}$ within 0.01.

11.5 Taylor Polynomials in x; Taylor Series in x

106. Find the Taylor polynomial $P_4(x)$ for $\ln(1+x)$.

107. Find the Taylor polynomial $P_4(x)$ for $\cos x \sin x$.

108. Find the Taylor polynomial $P_5(x)$ for $\ln(1+x)^{-2}$.

109. Find the Taylor polynomial $P_5(x)$ for $\sin^{-1} x$.

110. Find the Taylor polynomial $P_5(x)$ for $\cosh x$.

111. Determine $P_0(x), P_1(x), P_2(x), P_3(x)$ for $2x^3 + x^2 - 2x + 5$.

112. Determine $P_0(x), P_1(x), P_2(x), P_3(x)$ for $(x+2)^3$.

113. Use Taylor polynomials to estimate $\sin 1.3$ within 0.01.

114. Use Taylor polynomials to estimate $\ln 2.4$ within 0.01.

115. Use Taylor polynomials to estimate $e^{0.2}$ within 0.01.

116. Find the Lagrange form of the remainder R_3 for the function $f(x) = \sqrt{2-x}$.

117. Find the Lagrange form of the remainder R_{n+1} for the function $f(x) = e^{-3x}$.

11.6 Taylor Polynomials in $x - a$; Taylor Series in $x - a$

118. Expand $g(x) = 2x^4 - x^3 + 3x^2 - x + 1$ in powers of $x - 2$ and specify the values of x for which the expansion is valid.

119. Expand $g(x) = 1/x$ in powers of $x - 3$ and specify the values of x for which the expansion is valid.

120. Expand $g(x) = \ln(x - 1)$ in powers of $x - 2$ and specify the values of x for which the expansion is valid.

121. Expand $g(x) = e^{2x}$ in powers of $x - 3$ and specify the values of x for which the expansion is valid.

122. Expand $g(x) = \cos x$ in powers of $x - \pi/3$ and specify the values of x for which the expansion is valid.

123. Expand $g(x) = e^x \sin \pi x$ in powers of $x - 1$ and specify the values of x for which the expansion is valid.

124. Expand $g(x) = \tan x$ in powers of $x - \pi/3$.

125. Expand $g(x) = \ln x$ in powers of $x - 2$.

126. Expand $g(x) = e^{x^2}$ in powers of $x - 1$.

11.7 Power Series

127. Find the interval of convergence for $\sum_{k=1}^{\infty} \dfrac{k}{2^k}(x-1)^k$.

128. Find the interval of convergence for $\sum_{k=1}^{\infty} \dfrac{(-1)^k 2^k}{k^2}(x-2)^k$.

129. Find the interval of convergence for $\sum_{k=1}^{\infty} \frac{(-1)^k (x-2)^k}{k+1}$.

130. Find the interval of convergence for $\sum_{k=1}^{\infty} (-1) \frac{k x^k}{k}$.

131. Find the interval of convergence for $\sum_{k=1}^{\infty} \frac{2^k x^k}{\sqrt{k}}$.

132. Find the interval of convergence for $\sum_{k=1}^{\infty} \frac{2^k x^k}{3^k}$.

133. Find the interval of convergence for $\sum_{k=1}^{\infty} \frac{(x-1)^k}{2^k+1}$.

134. Find the interval of convergence for $\sum_{k=1}^{\infty} \frac{(x-3)^k}{k+1}$.

135. Find the interval of convergence for $\sum_{k=1}^{\infty} \frac{(x-3)^k}{(k+1)!}$.

136. Find the interval of convergence for $\sum_{k=1}^{\infty} \frac{k!}{(2k)!} x^k$.

137. Find the interval of convergence for $\sum_{k=1}^{\infty} k 3^k (x-2)^k$.

138. Find the interval of convergence for $\sum_{k=1}^{\infty} \frac{(-1)^k x^k}{e^k}$.

139. Find the interval of convergence for $\sum_{k=0}^{\infty} \left(\frac{x-1}{3}\right)^k$.

140. Find the interval of convergence for $\sum_{k=1}^{\infty} \frac{x^{k+1}}{2^k (k+1)}$.

141. Find the interval of convergence for $\sum_{k=1}^{\infty} \frac{2^k (x-3)^k}{k^2}$. For which values of x is the convergence absolute?

142. Find the interval of convergence for $\sum_{k=1}^{\infty} \frac{(-1)^k (x-1)^k}{3^k \sqrt[3]{k}}$.

143. Find the interval of convergence for $\sum_{k=1}^{\infty} \frac{k^2+k}{x^k}$.

144. Find the interval of convergence for $\sum_{k=1}^{\infty} \frac{(x+1)^k}{3^k k^2}$.

145. Find the interval of convergence for $\sum_{k=0}^{\infty} \frac{k! x^k}{2^k}$.

11.8 Differentiation and Integration of Power Series

146. Expand $\dfrac{1}{(1-x)^4}$ in powers of x, basing your calculation on the geometric series
$$\dfrac{1}{(1-x)} = 1 + x + x^2 + \cdots + x^n + \cdots$$

147. Expand $\ln(1 - 2x^2)$ in powers of x, basing your calculation on the geometric series
$$\dfrac{1}{(1-x)} = 1 + x + x^2 + \cdots + x^n + \cdots$$

148. Find $f^7(0)$ for $f(x) = x \sin x^2$.

149. Expand $e^x \sin x$ in powers of x.

150. Expand $e^{-x} \cos x$ in powers of x.

151. Expand $\dfrac{e^x}{1-x}$ in powers of x.

152. Expand $\dfrac{\cos x}{\sqrt{1+x}}$ in powers of x.

153. Expand $\coth x$ in powers of x.

154. Expand $\sec^2 x$ in powers of x.

155. Estimate $\displaystyle\int_0^1 \cos x^2 \, dx$ within 0.001.

156. Estimate $\displaystyle\int_0^1 \dfrac{\sin x}{x} \, dx$ within 0.001.

157. Estimate $\displaystyle\int_0^{1/2} \cos x^3 \, dx$ within 0.001.

158. Use a series to show $\displaystyle\lim_{x \to 0} \dfrac{\sin x}{x} = 1$.

159. Use a series to show $\displaystyle\lim_{x \to 0} \dfrac{\tan x}{x} = 1$ by first obtaining a series for $\tan x$.

160. Use a series to show $\displaystyle\lim_{x \to 0} \dfrac{e^x - 1}{\sin x} = 1$.

11.9 The Binomial Series

161. Expand $\sqrt{1 + x^3}$ in powers of x up to x^6.

162. Expand $\sqrt{1 - x^3}$ in powers of x up to x^6.

163. Expand $\dfrac{1}{\sqrt{1-x}}$ in powers of x up to x^4.

164. Expand $\dfrac{1}{\sqrt[3]{1-x}}$ in powers of x up to x^4.

165. Expand $\sqrt[5]{1-x}$ in powers of x up to x^4.

166. Expand $\dfrac{1}{\sqrt[5]{1+x}}$ in powers of x up to x^4.

167. Estimate $\sqrt{102}$ by using the first three terms of a binomial expansion, rounding off your answer to four decimal places.

168. Estimate $\sqrt[5]{28}$ by using the first three terms of a binomial expansion, rounding off your answer to four decimal places.

Answers to Chapter 11, Part B Questions

1. 32
2. 42
3. 44
4. −20
5. −180
6. $91/3^{10}$
7. $\dfrac{2}{e^4}(e^4 + e^3 + e^2 + e + 1) = 3.1426$
8. 65/16
9. $\sum_{k=0}^{10} 2^k$
10. $\sum_{k=1}^{6} (-1)^{k+1} 2^k x^{k+1}$
11. $\sum_{k=0}^{5} 3^{3-k} e^{-(2+k)}$
12. $\sum_{k=3}^{9} \dfrac{1}{k(k+1)}$
13. 1/2
14. 1/30
15. 9
16. 10/3
17. $\dfrac{3}{e-1}$
18. 1/4
19. diverges
20. 5/7
21. 1/5
22. 1/4
23. 1
24. 9/70
25. diverges
26. 4/15
27. 1/2
28. $\dfrac{21}{10^2} + \dfrac{21}{10^4} + \dfrac{21}{10^6} + \cdots = \dfrac{7}{33}$
29. $\dfrac{251}{10^3} + \dfrac{251}{10^6} + \dfrac{251}{10^9} + \cdots = \dfrac{251}{999}$
30. $\dfrac{315}{10^3} + \dfrac{315}{10^6} + \dfrac{315}{10^9} + \cdots = \dfrac{35}{111}$
31. $\sum_{k=0}^{\infty} x^{3k+1}$
32. $2\sum_{k=0}^{\infty} \dfrac{(-1)^k x^k}{3^{k+1}}$
33. diverges
34. converges
35. diverges
36. diverges
37. diverges
38. converges
39. diverges
40. converges
41. converges
42. diverges
43. $\dfrac{-50}{99}$
44. converges
45. converges
46. converges
47. converges
48. converges
49. diverges

Answers

50. b
51. converges
52. converges
53. converges
54. converges
55. diverges
56. converges
57. diverges
58. diverges
59. converges
60. diverges
61. converges
62. converges
63. converges
64. converges
65. diverges
66. converges
67. diverges
68. e
69. converges by ratio test
70. converges by ratio test
71. diverges by ratio test
72. diverges since $\lim_{k \to \infty} u_k \neq 0$
73. converges by root test
74. diverges by root test
75. diverges by root test
76. diverges by ratio test
77. diverges by ratio test
78. diverges by integral test
79. converges by ratio test
80. converges by ratio test
81. converges by ratio test
82. diverges by ratio test
83. converges by root test
84. diverges by ratio test
85. converges by ratio test
86. converges conditionally by limit comparison test with $\sum_{k=1}^{\infty} \frac{1}{k}$ and by alternating series test
87. converges conditionally by limit comparison test with $\sum_{k=1}^{\infty} \frac{1}{k}$ and by alternating series test
88. converges absolutely by comparison with geometric series $\sum_{k=1}^{\infty} \left(\frac{2}{3}\right)^k$
89. converges conditionally: converges by alternating series test but $\sum_{k=1}^{\infty} \frac{\ln k}{k}$ diverges by integral test
90. diverges by ratio test
91. converges conditionally: converges by alternating series test but $\sum_{k=1}^{\infty} \frac{1}{\sqrt{k}}$ divergent p series
92. converges absolutely by geometric series with $|r| = \frac{1}{3} < 1$
93. converges absolutely by comparison with p series $\frac{1}{4} \sum_{k=1}^{\infty} \frac{1}{k^2}$
94. converges absolutely by ratio test
95. converges absolutely: convergent p series
96. divergent since $\lim_{k \to \infty} \frac{k}{k+2} = 1 \neq 0$
97. converges absolutely by geometric series with $|r| = \frac{1}{e} < 1$

226   Calculus: One and Several Variables

98. converges absolutely by limit comparison test with convergent p series, $\sum_{k=1}^{\infty} \dfrac{1}{k^2}$

99. converges conditionally: converges by alternating series test but diverges by integral test

100. diverges by integral test

101. converges absolutely by ratio test

102. converges absolutely by limit comparison test with converging p series $2 \sum_{k=1}^{\infty} \dfrac{1}{k^{3/2}}$

103. converges absolutely by ratio test

104. $1/8$

105. $N = 9$

106. $x - \dfrac{x^2}{2} + \dfrac{x^3}{3} - \dfrac{x^4}{4}$

107. $x - \dfrac{2}{3}x^3$

108. $1 - 2x + 3x^2 - 4x^3 + 5x^4 - 6x^5$

109. $x - \dfrac{x^3}{6} + \dfrac{3}{40}x^5$

110. $1 + \dfrac{x^2}{2!} + \dfrac{x^4}{4!}$

111. $P_0(x) = 5$; $P_1(x) = 5 - 2x$; $P_2(x) = 5 - 2x + x^2$; $P_3(x) = 5 - 2x + x^2 + 2x^3$

112. $P_0(x) = 8$; $P_1(x) = 8 + 12x$; $P_2(x) = 8 + 12x + 6x^2$; $P_3(x) = 8 + 12x + 6x^2 + x^3$

113. 0.96

114. 0.88

115. 1.22

116. $R_3(x) = \dfrac{-x^3}{16(2-c)^{5/2}}$

117. $R_{n+1}(x) = \dfrac{(-3)^{n+1} e^{-3c}}{(n+1)!} x^{n+1}$

118. $35 + 63(x-2) + 45(x-2)^2 + 15(x-2)^3 + 2(x-2)^4$ valid for $-\infty < x < \infty$

119. $\sum_{k=0}^{\infty} (-1)^k \dfrac{1}{3^{k+1}} (x-3)^k$; valid for $0 < x < 6$

120. $\sum_{k=1}^{\infty} (-1)^{k+1} \dfrac{1}{k} (x-2)^k$; valid for $1 < x < 3$

121. $e^6 \sum_{k=0}^{\infty} \dfrac{2^k}{k!} (x-3)^k$; valid for $-\infty < x < \infty$

122. $\dfrac{1}{2} \sum_{k=0}^{\infty} (-1)^k \dfrac{\left(x - \dfrac{\pi}{3}\right)^{2k}}{2k!}$
$- \dfrac{\sqrt{3}}{2} \sum_{k=0}^{\infty} (-1)^k \dfrac{\left(x - \dfrac{\pi}{3}\right)^{2k+1}}{(2k+1)!}$
valid for $-\infty < x < \infty$

123. $-\pi e \left[(x-1) + \dfrac{2}{2!}(x-1)^2 \dfrac{(\pi^2 - 3)(x-1)^3}{3!} - 4 \dfrac{(\pi^2 - 1)(x-1)^4}{4!} \cdots \right]$
valid for $-\infty < x < \infty$

124. $\sqrt{3} + 4\left(x - \dfrac{\pi}{3}\right) + 4\sqrt{3}\left(x - \dfrac{\pi}{3}\right)^2$
$+ \dfrac{40}{3}\left(x - \dfrac{\pi}{3}\right)^3 + \dfrac{44}{3}\sqrt{3}\left(x - \dfrac{\pi}{3}\right) + \cdots$

125. $\ln 2 + \sum_{k=1}^{\infty} (-1)^{k+1} \dfrac{1}{2^k k} (x-2)^k$

126. $e \left[1 + 2(x-1) + 3(x-1)^2 + \dfrac{10}{3}(x-1)^3 + \dfrac{19}{6}(x-1)^4 + \cdots \right]$

127. $(-1, 3)$

128. $(3/2, 5/2)$

129. $(1, 3)$

130. $(-1, 1)$

131. $[-1/2, 1/2)$

132. $(-3/2, 3/2)$

133. $[0, 2)$

134. $[2, 4)$

135. $(-\infty, +\infty)$

Answers

136. $(-\infty, +\infty)$

137. $(5/3, 7/3)$

138. $(-e, e)$

139. $(-2, 4)$

140. $[-2, 2)$

141. $[5/2, 7/2]$

142. $(-2, 4]$

143. $(-\infty, -1) \cup (1, +\infty)$

144. $[-4, 2]$

145. $x = 0$

146. $\dfrac{1}{6} \displaystyle\sum_{k=1}^{\infty} k(k+1)(k+2)x^{k-1}$

147. $-\displaystyle\sum_{k=1}^{\infty} \dfrac{(2x^2)^k}{k}$

148. -840

149. $x + x^2 + \dfrac{x^3}{3} - \dfrac{x^4}{8} - \dfrac{x^5}{30} + \cdots$

150. $1 - x + \dfrac{x^3}{3} - \dfrac{x^4}{6} + \cdots$

151. $1 + 2x + \dfrac{5}{2}x^2 + \dfrac{8}{3}x^3 + \dfrac{65}{24}x^4 + \cdots$

152. $1 - \dfrac{1}{2}x - \dfrac{1}{8}x^2 - \dfrac{1}{16}x^3 + \dfrac{49}{384}x^4 + \cdots$

153. $\dfrac{1}{x} + \dfrac{1}{3}x - \dfrac{1}{45}x^3 + \dfrac{2}{945}x^5 + \cdots$

154. $1 + x^2 + \dfrac{2}{3}x^4 + \dfrac{17}{45}x^6 + \dfrac{62}{315}x^8 + \cdots$

155. 0.905

156. 0.946

157. 0.499

158. $\displaystyle\lim_{x \to 0} \dfrac{\sin x}{x} = \lim_{x \to 0}\left(1 - \dfrac{x^2}{3!} + \dfrac{x^4}{5!} - \dfrac{x^6}{7!} + \cdots\right) = 1$

159. $\displaystyle\lim_{x \to 0} \dfrac{\tan x}{x}$

    $= \displaystyle\lim_{x \to 0}\left(1 - \dfrac{x^2}{3} + \dfrac{2x^4}{15} - \dfrac{17x^6}{315} + \cdots\right) = 1$

160. $\displaystyle\lim_{x \to 0} \dfrac{e^x - 1}{\sin x} = \lim_{x \to 0} \dfrac{x + \dfrac{x^2}{2!} + \dfrac{x^3}{3!} + \dfrac{x^4}{4!} + \cdots}{x - \dfrac{x^3}{3!} + \dfrac{x^5}{5!} - \dfrac{x^7}{7!} + \cdots}$

    $= \displaystyle\lim_{x \to 0} \dfrac{1 + \dfrac{x}{2!} + \dfrac{x^2}{3!} + \dfrac{x^3}{4!} + \cdots}{1 - \dfrac{x^2}{3!} + \dfrac{x^4}{5!} - \dfrac{x^6}{7!} + \cdots} = 1$

161. $1 + \dfrac{x^3}{2} - \dfrac{x^6}{8}$

162. $1 - \dfrac{x^3}{2} - \dfrac{x^6}{8}$

163. $1 + \dfrac{x}{2} + \dfrac{3}{8}x^2 + \dfrac{5}{16}x^3 + \dfrac{35}{128}x^4$

164. $1 + \dfrac{1}{3}x + \dfrac{2}{9}x^2 + \dfrac{14}{81}x^3 + \dfrac{35}{243}x^4$

165. $1 + \dfrac{1}{5}x - \dfrac{2}{25}x^2 + \dfrac{6}{125}x^3 - \dfrac{21}{625}x^4$

166. $1 - \dfrac{1}{5}x + \dfrac{3}{25}x^2 - \dfrac{11}{125}x^3 + \dfrac{44}{625}x^4$

167. 10.0995

168. 1.9475

Part C

1. (a) Prove that if the series $\sum a_k$ converges and the series $\sum b_k$ diverges then $\sum (a_k + b_k)$ diverges. (11.1)

2. Show that $\sum_{k=1}^{\infty} kx^{k-1} = \frac{1}{(1-x)^2}$ for $|x| < 1$ (11.1)

    Hint: Verify S_n satisfies the identity
    $(1-x)^2 S_n = 1 - (n+1)x^n + nx^{n+1}$

3. Find the values of p for which the series $\sum_{k=2}^{\infty} \frac{\ln k}{k^p}$ converges. (11.2)

4. Let $\sum a_k$ be a series with nonnegative term. Prove that if $\sum a_k^2$ converges, then $\sum (a_k/k)$ converges. (11.2)

5. Consider the series $\frac{1}{2} + 1 + \frac{1}{8} + \frac{1}{4} + \frac{1}{32} + \cdots$ formed by rearranging a convergent geometric series: (a) use the root test to show the series converges; (b) show that the ratio test does not apply. (11.3)

6. Let r be a positive number. For what values of r (if any) does $\sum \frac{r^k}{k^r}$ converge? (11.3)

7. Verify that the series $1 - \frac{1}{2} + \frac{1}{2} - \frac{1}{3} + \frac{1}{3} - \frac{1}{2} + \frac{1}{3} - \frac{1}{4} + \frac{1}{3} - \frac{1}{4} + \frac{1}{3} - \frac{1}{4} + \cdots$ diverges and explain how this does not violate the theorem on alternating series. (11.4)

8. Let $\{a_k\}$ be a nonincreasing sequence of positive numbers that converge to 0. Does the alternating series $\sum (-1)^k a_k$ necessarily converge? (11.4)

9. Derive a series expansion in x for the given fraction and specify the numbers x for which the expansion is valid. Take $a > 0$ (11.5)
    $f(x) = \ln(1 - \alpha x)$

10. Find the Lagrange form of the remainder R_{n+1} for $f(x) = \sin 2x$. (11.5)

11. Show that e is irrational by following these steps:
    (1) Take the expansion $e = \sum_{k=0}^{\infty} \frac{1}{k!}$ (11.5)

    and show that the qth partial sum $S_g = \sum_{k=0}^{q} \frac{1}{k!}$

    satisfies the unequality $0 < q!(e - s_q) < \frac{1}{q}$ (11.5)
    (2) Show that $q!s_q$ is an integer and argue that, if e were of the form P/q then $q!(e - s_q)$ would be a positive integer less than 1

12. Expand $g(x)$
    $g(x) = (x-1)^n$ in powers of x (11.6)

13. Let $\sum a_k x^k$ be a power series and let r, $0 < r < \infty$, be its radius of convergence. Prove that if the series is absolutely convergent at one endpoint of its interval of convergence, then it is absolutely convergent at the other also. (11.7)

14. Let $\sum a_k x^k$ be a power series and let r, $0 < r < \infty$, be its radius of convergence. Prove that the power series $\sum a_k x^{2k}$ has radius of convergence $\sqrt{r}$. (11.7)

15. Reduce the differentiation formulas: $\dfrac{d}{dx}(\sinh x) = \cosh x$, $\dfrac{d}{dx}(\cosh x) = \sinh x$ from the expansions of $\sinh x$ and $\cosh x$ in powers of x. (11.7)

16. (a) Use the binomial series to find the Taylor series of $f(x) = \dfrac{1}{\sqrt{1+x^2}}$ in powers of x. (11.8)
    (b) Use the series for f in part (a) to find the Taylor series for $F(x) = \sinh^{-1} x$ and give the radius of convergence.

17. If $a > 1$, show that $\sum \dfrac{1}{a^{\ln n}}$ diverges if $a \leq e$ and converges for $a > e$. (11.2)

18. If k is any integer greater than 1, show that $\left(\dfrac{1}{n+1} + \dfrac{1}{n+2} + \cdots + \dfrac{1}{kn}\right) \to \ln k$. (11.2)

19. Use the ratio test to determine the behavior of $\sum \dfrac{1 \cdot 3 \cdot 5 \cdots (2n-1)}{2 \cdot 4 \cdot 6 \cdots (2n)}$. (11.3)

20. Show the series
$$\dfrac{1}{\sqrt{2}-1} - \dfrac{1}{\sqrt{2}+1} + \dfrac{1}{\sqrt{3}-1} - \dfrac{1}{\sqrt{3}+1} + \cdots$$ (11.4)
diverges

21. Find a power series representation for $\displaystyle\int \dfrac{\sin x}{x} dx$. (11.7)

22. Given $e^x = 1 + x + \dfrac{x^2}{2!} + \dfrac{x^3}{3!} + \cdots$

    $\sin x = x - \dfrac{x^3}{3!} + \dfrac{x^5}{5b} + \cdots$

    $\cos x = 1 - \dfrac{x^2}{2!} + \dfrac{x^4}{4!} - \cdots$

    assume $e^{ix} = 1 + (ix) + \dfrac{(ix)^2}{2!} + \dfrac{(ix)^3}{3!} + \cdots \quad i = r - 1$

    and $\sin(ix) = (ix) - \dfrac{(ix)^3}{3!} + \dfrac{(ix)^5}{5!} - \cdots$

    $\cos(ix) = 1 - \dfrac{(ix)^2}{2!} + \dfrac{(ix)^4}{4!} - \cdots$

    Show that (a) $e^{ix} = \cos x + i \sin x$ and
    (b) $e^{-ix} = \cos x - i \sin x$
    (c) let $x = \pi$ and derive $e^{\pi i} = \cos \pi + i \sin \pi \; e^{\pi i} = -1$ *Euler's formula

Answers to Chapter 11, Part C Questions

1. If $\sum(a_k + b_k)$ converges, then
   $\sum b_k = \sum(a_k + b_k) - \sum a_k$ would also converge.

2. Use induction to verify the hint, then
   $$S_n = \frac{1-(n+1)x^n + nx^{n+1}}{(1-x)^2} \to \frac{1}{(1-x)^2}$$
   Since $-(n+1)x^n + nx^{n+1} \to 0$ for $|x| < 1$.
   This last statement follows from observing that $nx^n \to 0$
   To see this, choose $\epsilon > 0$ so that $(1+\epsilon)|x| < 1$. Since $n^{1/n} \to 1$ there exists k so that $n^{1/n} < 1+\epsilon$ for $n \geq k$ then for $n \geq k$
   $|nx^n| = |n^{1/n}x|^n \leq ((1+\epsilon)|x|)^n \to 0$

3. If $p \leq 1$, $\sum \frac{\ln k}{k^p} > \sum \frac{1}{k^p}$ diverges
   if $p > 1$ then $\frac{p-1}{2} > 0$, so for large k,
   $\ln k < k^{\frac{p-1}{2}}$ then $\frac{\ln k}{k^p} < \frac{k^{\frac{p-1}{2}}}{k^p} = \frac{1}{k^{\frac{p+1}{2}}}$.
   Since $\frac{p+1}{2} > 1$, $\sum \frac{1}{k^{\frac{p+1}{2}}}$ converges.
   Hence so does $\sum \frac{\ln k}{k^p}$. So it converges iff $p > 1$.

4. Since $0 < \left(a_k - \frac{1}{k}\right)^2 < a_k^2 + \frac{1}{k^2}$, $\sum \left(a_k - \frac{1}{k}\right)^2$
   converges by comparison with $\sum a_k^2 + \sum \frac{1}{k^2}$
   But $\sum \left(a_k - \frac{1}{k}\right)^2 = \sum a_k^2 - 2\sum \frac{a_k}{k} + \sum \frac{1}{k^2}$
   so $\sum \frac{a_k}{k}$ must converge.

5. (a) $a_k = \begin{cases} \left(\frac{1}{2}\right)^k, & k \text{ is odd} \\ \left(\frac{1}{2}\right)^{k-2}, & k \text{ is even} \end{cases}$
   clearly $a_k^{1/k} \to \frac{1}{2} < 1$
   (b) $\lim_{k \to \infty} \frac{a_{k+1}}{a_k}$ doesn't exist
   since $\frac{a_{k+1}}{a_k} = \begin{cases} \frac{1}{8}k \text{ is even} \\ 2k \text{ is odd} \end{cases}$

6. by the root test $(a_k)^{1/k} = \frac{r}{k^{r/k}} = \frac{r}{(k^{1/k})^r} \Rightarrow r$
   Converges if $r < 1$ diverges if $r > 1$. If $r = 1$ we get $\sum \frac{1}{k}$ which diverges.

7. The series diverges because among the partial sums add all sums of the form $\frac{1}{2} + \frac{1}{3} + \frac{1}{4} + \cdots + \frac{1}{n}$
   Thus for instance
   $s_1 = \frac{1}{2}$, $s_5 = \frac{1}{2} + 3$, $s_{11} = \frac{1}{2} + \frac{1}{3} + \frac{1}{4}$ and so on.
   This doesn't violate the theorem on alternating series because, in the notation of the theorem it is not true that $\{a_k\}$ decreases.

8. yes

9. $\ln(1 - ax) = \sum_{k=1}^{\infty} \frac{(-1)^{k+1}}{k}(-ax)^k$
   $= -\sum_{k=1}^{\infty} \frac{a^k}{k} x^k, \left[-\frac{1}{a}, \frac{1}{a}\right]$

10. $R_{n+1}(x) = \frac{f^{n+1}(c)}{(n+1)!} x^{n+1}$
    $= \begin{cases} \frac{(-1)^{\frac{n-1}{2}} 2^{n+1} \cos 2c}{(n+1)!} x^{n+1} & n \text{ odd} \\ \frac{(-1)^{n/2} 2^{n+1} \sin 2c}{(n+1)!} x^{n+1} & \text{otherwise} \end{cases}$
    where c is between 0 and n.

11. (1) $0 < q!(e - s_q)$
    $= (q!) \sum_{k=q+1}^{\infty} \frac{1}{k!} = \frac{q!}{(q+1)!} + \frac{q!}{(q+2)!} + \cdots$
    $\leq \frac{1}{q+1} + \frac{1}{(q+1)^2} + \frac{1}{(q+b)^2} \cdots$
    $\leq \frac{1}{q+1}\left[1 + \frac{1}{q+1} + \frac{1}{(q+1)^2} + \cdots\right]$
    $= \frac{1}{q+1}\left[\frac{1}{1 - 1/(q+1)}\right] = \frac{1}{q}$
    (2) If e equaled p/q then $q!e$ would be an integer since $q!s_q$ is an integer, so would $q!(e - s_q)$
    But $0 < q!(e - s_q) < \frac{1}{q} < 1$ is impossible

12. $g(x) = (x - 1)^n = \sum_{k=0}^{n} \frac{n!}{(n-k)!k!} x^k (-1)^{n-k}$

13. $\sum |a_k r^k| = \sum |a_k(-r)^k|$

14. ratio test $\left|\frac{a_{k+1}}{a_k}\right| |x| \to \frac{|x|}{r}$
    so $\left|\frac{a_{k+1} x^{2(k+1)}}{a_k x^{2k}}\right| = \left|\frac{a_{k+1}}{a_k}\right| |x|^2 \to \frac{|x|^2}{r}$
    $\Rightarrow$ radius of convergence is $\sqrt{r}$

15. $\dfrac{d}{dx}(\sinh x) = \dfrac{d}{dx}\left[\displaystyle\sum_{k=0}^{\infty} \dfrac{x^{2k+1}}{(2k+1)!}\right]$

$= \displaystyle\sum_{k=0}^{\infty} \dfrac{(2k+1)x^{2k}}{(2k+1)!} = \displaystyle\sum_{k=0}^{\infty} \dfrac{x^{2k}}{(2k)!} = \cosh x$

$\dfrac{d}{dx}(\cosh x) = \dfrac{d}{dx}\left[\displaystyle\sum_{k=0}^{\infty} \dfrac{x^{2k}}{(2k)!}\right]$

$= \displaystyle\sum_{k=1}^{\infty} \dfrac{2kx^{2k-1}}{(2k)!} = \displaystyle\sum_{k=1}^{\infty} \dfrac{x^{2k-1}}{(2k-1)!} = \sinh x$

16. (a) $a = -\dfrac{1}{2}\dfrac{1}{\sqrt{1+x^2}}$

$f(x) = 1 - \dfrac{x^3}{2} + \dfrac{\left(-\frac{1}{2}\right)\left(-\frac{3}{2}\right)x^4}{2!} + \cdots$

$= 1 - \dfrac{1}{2}x^2 + \dfrac{3}{8}x^4 - \dfrac{5}{16}x^6 + \cdots$

(b) $\sinh^{-1} x = \displaystyle\int_0^x \dfrac{1}{\sqrt{1+t^2}}\, dt$

$= \displaystyle\int_0^x \left(1 - \dfrac{1}{2}t^2 + \dfrac{3}{8}x^4 - \dfrac{5}{16}x^6 \cdots\right) dt$

$= x - \dfrac{1}{6}x^3 + \dfrac{3}{40}x^4 - \dfrac{5}{112}x^7 + \cdots\quad r = 1$

17. $a^{\ln n} = (e^{\ln a})^{\ln n} = (e^{\ln n})^{\ln a} = n^{\ln a}$

$\displaystyle\sum \dfrac{1}{a^{\ln a}} = \displaystyle\sum \dfrac{1}{n^{\ln a}} \to$ p series

converges if $f(\ln a) > 1 \iff a > e$

18. Let $f(x) = \dfrac{1}{x}$; we have

$f(n+1) + f(n+2) + \cdots + f(kn)$

$\leq \displaystyle\int_n^{kn} f(x)\, dx$

$\leq f(n) + f(n+1) + \cdots + f(kn-1)$

$\displaystyle\int_n^{kn} f(x)\, dx = [\ln x]_n^{kn} = \ln(kn) - \ln(n)$

$= \ln\left(\dfrac{kn}{n}\right) = \ln k$

and when $n \to \infty$

$f(n+1) + f(n+2) + \cdots + f(kn)$

$= f(n) + f(n+1) + \cdots f(kn-1)$

$\to \dfrac{1}{n+1} + \dfrac{1}{n+2} + \cdots + \dfrac{1}{kn} \to \ln k$

by Squeeze theorem

19. $\displaystyle\sum \dfrac{1\cdot 3\cdot 5\cdots(2n-1)}{2\cdot 4\cdot 6\cdots(2n)}$

$= \bar{2}\dfrac{3}{2}\cdot\dfrac{5}{4}\cdot\dfrac{7}{6}\cdots\dfrac{2n-1}{2n-2}\cdot\dfrac{1}{2n}$

$> \displaystyle\sum(1\cdot 1\cdots)\dfrac{1}{2n} = \displaystyle\sum\dfrac{1}{2n}$ div.

so $\displaystyle\sum \dfrac{1\cdot 3\cdot 5\cdots(2n-1)}{2\cdot 4\cdot 6\cdots 2n}$ divergent

$\displaystyle\lim_{n\to\infty}\dfrac{a_{n+1}}{a_n} = \displaystyle\lim_{n\to\infty}\dfrac{1\cdot 3\cdot 5\cdots(2x+1)}{2\cdot 4\cdot 6\cdots(2x+2)}$

$\cdot\left(\dfrac{2\cdot 4\cdot 6\cdots 2x}{1\cdot 3\cdot 5\cdots(2x-1)}\right) = \displaystyle\lim_{n\to\infty}\dfrac{2n+1}{2n+2} = 1$

since $\lim = 1$ and $n+1 < a_n$
the ratio test does not determine the behavior

20. Since $\dfrac{1}{\sqrt{n-1}} - \dfrac{1}{\sqrt{n+1}} = \dfrac{2}{n-1}$

the $2n$(th) partial sum is

$\dfrac{1}{2-1} + \dfrac{1}{3-1} + \cdots + \dfrac{1}{n-1}$

$1 + \dfrac{1}{2} + \cdots + \dfrac{1}{n-1} + \dfrac{1}{n} - \dfrac{1}{n}$

which is unbounded.
Since the terms are not decreasing

$\left(\dfrac{1}{\sqrt{n+1}} < \dfrac{1}{\sqrt{n+1}-1}\text{ for } n \geq 2\right)$

it does not contradict the alternating series test.

21. $\displaystyle\int \dfrac{\sin x}{x}\, dx = \displaystyle\int \dfrac{1}{x}\displaystyle\sum_0^\infty (-1)^n \dfrac{x^{2n+1}}{(2n+1)!}\, dx$

$= \displaystyle\sum_0^\infty \dfrac{(-1)^n}{(2n+1)!}\displaystyle\int x^{2n}\, dx$

$= c + \displaystyle\sum \dfrac{(-1)^n}{(2n+1)!}\dfrac{x^{2n+1}}{2n+1}$

$= c + \displaystyle\sum(-1)^n \dfrac{x^{2n+1}}{(2n+1)(2n+1)!}$

$= c + x - \dfrac{1}{18}x^3 + \dfrac{1}{600}x^5 + \cdots$

22. Answers follow by substitution and
$i^2 = -1$
$i^3 = -1$
$i^4 = 1$

CHAPTER 12 Vectors
Part B

12.1 Cartesian Space Coordinates

1. Plot points $A(2, 7, 8)$ and $B(3, 9, 7)$ on a right-handed coordinate system. Then calculate the length of the line segment $\overline{AB}$ and find the midpoint.

2. Plot points $A(-3, -2, 4)$ and $B(9, 7, 2)$ on a right-handed coordinate system. Then calculate the length of the line segment $\overline{AB}$ and find the midpoint.

3. Plot points $A(-1, 1, 1)$ and $B(-1, 4, 4)$ on a right-handed coordinate system. Then calculate the length of the line segment $\overline{AB}$ and find the midpoint.

4. Find an equation for the plane through $(2, -1, -2)$ that is parallel to the xy-plane.

5. Find an equation for the plane through $(-3, 2, -1)$ that is perpendicular to the z-axis.

6. Find an equation for the plane through $(-2, -4, 3)$ that is parallel to the yz-plane.

7. Find an equation for the sphere centered at $(2, 1, 3)$ with radius 4.

8. Find an equation for the sphere that is centered at $(-4, 0, 6)$ and passes through $(2, 2, 3)$.

9. Find an equation for the sphere that is centered at $(5, 1, -4)$ and passes through $(3, -5, -1)$.

10. Find an equation for the sphere that has the line segment joining $(4, 3, 0)$ and $(2, 4, -4)$ as a diameter.

11. Find an equation for the sphere that is centered at $(-2, 1, 4)$ and is tangent to the plane $x = 2$.

12. The points $P(a, b, c)$ and $Q(3, 2, -1)$ are symmetric about the xy-plane. Find a, b, c.

13. The points $P(a, b, c)$ and $Q(-3, 2, -1)$ are symmetric about the yz-plane. Find a, b, c.

14. The points $P(a, b, c)$ and $Q(-3, -2, 1)$ are symmetric about the xz-plane. Find a, b, c.

15. The points $P(a, b, c)$ and $Q(1, 2, -4)$ are symmetric about the z-axis. Find a, b, c.

16. The points $P(a, b, c)$ and $Q(2, -1, 3)$ are symmetric about the plane $x = 2$. Find a, b, c.

17. The points $P(a, b, c)$ and $Q(-2, 1, -3)$ are symmetric about the plane $y = -3$. Find a, b, c.

18. The points $P(a, b, c)$ and $Q(4, 2, 2)$ are symmetric about the point $(0, 2, 1)$. Find a, b, c.

12.3 Vectors

19. Simplify $(3\,\mathbf{i} - \mathbf{j} + 2\,\mathbf{k}) - 2(\mathbf{i} - 2\,\mathbf{j} + \mathbf{k})$.

20. Simplify $2(\mathbf{i} - 3\,\mathbf{k}) - 3(2\,\mathbf{i} + \mathbf{j} - \mathbf{k})$.

234 Calculus: One and Several Variables

21. Calculate the norm of the vector $4\mathbf{i} - 3\mathbf{j}$.

22. Calculate the norm of the vector $3\mathbf{i} - \mathbf{j} + \mathbf{k}$.

23. Calculate the norm of $2(2\mathbf{i} - \mathbf{j} + \mathbf{k}) - (-2\mathbf{i} - \mathbf{j})$.

24. Let $\mathbf{a} = (-2, 3, 5), \mathbf{b} = (3, 5, -2), \mathbf{c} = (2, 1, 2), \mathbf{d} = (-3, 0, -1)$. Express $\mathbf{a} - 2\mathbf{b} + 2\mathbf{c} + 3\mathbf{d}$ as a linear combination of $\mathbf{i}, \mathbf{j}, \mathbf{k}$.

25. Given that $\mathbf{a} = (1, 2, 5)$ and $\mathbf{b} = (-1, 0, 3)$, calculate
    (a) $\|\mathbf{a}\|$
    (b) $\|\mathbf{b}\|$
    (c) $\|2\mathbf{a} - 3\mathbf{b}\|$
    (d) $\|3\mathbf{a} + \mathbf{b}\|$

26. Find α given that $3\mathbf{i} + 2\mathbf{j}$ and $-2\mathbf{i} + \alpha\mathbf{j}$ have the same length.

27. Find the unit vector in the direction of $2\mathbf{i} - \mathbf{j} + 2\mathbf{k}$.

28. Given that $\mathbf{a} = 3\mathbf{i} - 5\mathbf{k}$ and $\mathbf{b} = -\mathbf{i} + 2\mathbf{j} + 3\mathbf{k}$, find the unit vector in the direction of $\mathbf{a} - 2\mathbf{b}$.

29. Given that $\mathbf{a} = 2\mathbf{i} + 9\mathbf{j} + \mathbf{k}$ and $\mathbf{b} = \mathbf{i} + 7\mathbf{j} + 8\mathbf{k}$, find the unit vector in the direction of $2\mathbf{a} + \mathbf{b}$.

30. Find the vector of norm 2 in the direction of $3\mathbf{i} + 4\mathbf{j} + 2\mathbf{k}$.

31. Find the vector of norm 2 parallel to $5\mathbf{i} - 12\mathbf{j} + \mathbf{k}$.

12.4 The Dot Product

32. Simplify $(2\mathbf{a} \cdot 2\mathbf{b}) + \mathbf{a} \cdot (\mathbf{a} + 2\mathbf{b})$.

33. Simplify $(\mathbf{a} - 2\mathbf{b}) \cdot \mathbf{c} + \mathbf{b} \cdot (\mathbf{a} - \mathbf{c}) - 2\mathbf{a} \cdot (\mathbf{b} - 3\mathbf{c})$.

34. Taking $\mathbf{a} = \mathbf{i} + 2\mathbf{j}, \mathbf{b} = 2\mathbf{i} - \mathbf{j} + 3\mathbf{k}, \mathbf{c} = -2\mathbf{j} + \mathbf{k}$, calculate:
    (a) the three dot products $\mathbf{a} \cdot \mathbf{b}, \mathbf{a} \cdot \mathbf{c}, \mathbf{b} \cdot \mathbf{c}$
    (b) the cosines of the angles between these vectors
    (c) the component of $\mathbf{a}$ (i) in the $\mathbf{b}$ direction, (ii) in the $\mathbf{c}$ direction
    (d) the projection of $\mathbf{a}$ (i) in the $\mathbf{b}$ direction, (ii) in the $\mathbf{c}$ direction

35. Taking $\mathbf{a} = \mathbf{i} - 2\mathbf{j} + 4\mathbf{k}$ and $\mathbf{b} = 3\mathbf{i} - 2\mathbf{j} + \mathbf{k}, \mathbf{c} = -2\mathbf{j} - \mathbf{k}$, calculate:
    (e) the three dot products $\mathbf{a} \cdot \mathbf{b}, \mathbf{a} \cdot \mathbf{c}, \mathbf{b} \cdot \mathbf{c}$
    (f) the cosines of the angles between these vectors
    (g) the component of $\mathbf{a}$ (i) in the $\mathbf{b}$ direction, (ii) in the $\mathbf{c}$ direction
    (h) the projection of $\mathbf{a}$ (i) in the $\mathbf{b}$ direction, (ii) in the $\mathbf{c}$ direction

36. Find the angle between the vectors $2\mathbf{i} + 2\mathbf{j} - \mathbf{k}$ and $\mathbf{i} + 2\mathbf{j} + \mathbf{k}$.

37. Find the angle between the vectors $2\mathbf{i} + 3\mathbf{j} + \mathbf{k}$ and $3\mathbf{i} + \mathbf{j} + 8\mathbf{k}$.

38. Find the direction angles of the vector $\sqrt{2}\mathbf{i} - \mathbf{j} + \mathbf{k}$.

39. Find the direction angles of the vector $\sqrt{3}\mathbf{i} - 2\mathbf{k}$.

40. Find the unit vectors $\mathbf{u}$ that are perpendicular to both $\mathbf{i} + 2\mathbf{j} + \mathbf{k}$ and $\mathbf{i} - 2\mathbf{j} + 2\mathbf{k}$.

41. Find the cosine of the angle between $\mathbf{u} = 2\mathbf{i} - 2\mathbf{j} + \mathbf{k}$ and $\mathbf{v} = -\mathbf{i} + 4\mathbf{j} + 2\mathbf{k}$.

42. A 100 Newton force is applied along a rope making a 30° angle with the horizontal to pull a box a distance of 5 meters along the ground. What is the work done?

43. Find the work done by the force $\mathbf{F} = 3\mathbf{i} + 5\mathbf{j} + 2\mathbf{k}$ in moving an object from the point $P(2, 0, 2)$ to the point $Q(1, 4, 5)$.

12.5 The Cross Product

44. Calculate $(\mathbf{i} - \mathbf{k}) \times (\mathbf{j} + \mathbf{k})$.

45. Calculate $(\mathbf{j} \times \mathbf{k}) \cdot \mathbf{j}$.

46. Calculate $(\mathbf{k} \times \mathbf{j}) \times \mathbf{i}$.

47. Calculate $(\mathbf{i} - 4\mathbf{j} - 2\mathbf{k}) \times (2\mathbf{i} + \mathbf{j})$.

48. Calculate $[(\mathbf{i} + 2\mathbf{j} - \mathbf{k}) \times (\mathbf{i} + \mathbf{j} + \mathbf{k})] \times (\mathbf{i} + 2\mathbf{j} + 2\mathbf{k})$.

49. Calculate $(3\mathbf{i} - 4\mathbf{j} - 4\mathbf{k}) \times [(2\mathbf{i} - 6\mathbf{j}) \times (\mathbf{i} - 2\mathbf{j} + 2\mathbf{k})]$.

50. Calculate $(\mathbf{i} - \mathbf{j}) \cdot [(3\mathbf{i} - 4\mathbf{j}) \times (\mathbf{i} - 2\mathbf{j} + 2\mathbf{k})]$.

51. Calculate $(2\mathbf{i} + 3\mathbf{j} - 4\mathbf{k}) \cdot [(-\mathbf{i} + \mathbf{j} + 2\mathbf{k}) \times (\mathbf{i} - \mathbf{j} + \mathbf{k})]$.

52. Calculate $(3\mathbf{i} + 2\mathbf{k}) \times [(2\mathbf{i} + 2\mathbf{j} - \mathbf{k}) \times (\mathbf{i} - 2\mathbf{j} + 3\mathbf{k})]$.

53. Use a cross product to find the area of triangle PQR, $P(1, 2, 3)$, $Q(-1, 0, 1)$, $R(2, -2, -1)$.

54. Use a cross product to find the area of triangle PQR, $P(1, 1, 1)$, $Q(2, -1, 3)$, $R(2, 3, -4)$.

55. Find the volume of the parallelepiped with edges determined by $3\mathbf{i} - 4\mathbf{j} - \mathbf{k}, \mathbf{i} - 2\mathbf{j} + 2\mathbf{k}, \mathbf{i} + \mathbf{j}$.

56. Find the volume of the parallelepiped with vertices $A(0, 0, 0)$, $B(1, -1, 1)$, $C(2, 1, -2)$, and $D(-1, 2, -1)$.

57. Find the volume of the parallelepiped with edges determined by $\mathbf{i} + 2\mathbf{k}, 4\mathbf{i} + 6\mathbf{j} + 2\mathbf{k}, 3\mathbf{i} + 3\mathbf{j} - 6\mathbf{k}$.

58. Find the volume of the parallelepiped with edges determined by $2\mathbf{i} + \mathbf{k}, 3\mathbf{i} + 2\mathbf{j} + 5\mathbf{k}, -\mathbf{i} + 2\mathbf{k}$.

59. Find the area of the triangle with vertices $P(1, -2, 3)$, $Q(2, 4, 1)$, $R(2, 0, 1)$.

60. Find the area of the triangle with vertices $P(1, 2, 1)$, $Q(2, 4, 3)$, $R(5, -1, 4)$.

12.6 Lines

61. Which of the points $P(-1, 3, -1)$, $Q(3, 2, -1)$, $R(3, 0, -2)$ lie on the line l: $\mathbf{r}(t) = (2\mathbf{i} + \mathbf{j}) + t(3\mathbf{i} - 2\mathbf{j} + \mathbf{k})$?

62. Determine whether the lines are parallel.
    l_1: $\mathbf{r}_1(t) = (\mathbf{i} - 2\mathbf{k}) + t(\mathbf{i} - 2\mathbf{j} - 3\mathbf{k})$
    l_2: $\mathbf{r}_2(u) = (3\mathbf{i} + 2\mathbf{j} - 3\mathbf{k}) + u(\mathbf{i} + 2\mathbf{j} - \mathbf{k})$

63. Find a vector parametrization for the line that passes through $P(2, 3, 3)$ and is parallel to the line $\mathbf{r}(t) = (2\mathbf{i} - \mathbf{j}) + t\mathbf{k}$.

64. Find a vector parametrization for the line that passes through the origin and $P(3, 1, 8)$.

65. Find a vector parametrization for the line that passes through $P(4, 0, 5)$ and $Q(2, 3, 1)$.

66. Find a vector parametrization for the line that passes through $P(3, 3, 1)$ and $Q(4, 0, 2)$.

67. Find a set of scalar parametric equations for the line that passes through $P(1, 4, 6)$ and $Q(2, -1, 3)$.

68. Find a set of scalar parametric equations for the line that passes through $P(-3, -1, 0)$ and $Q(-1, 2, 1)$.

69. Find a set of scalar parametric equations for the line that passes through $P(4, -2, -1)$ and is perpendicular to the xy-plane.

70. Find a set of scalar parametric equations for the line that passes through $P(-1, 2, -3)$ and is perpendicular to the xz-plane.

71. Give a vector parametrization for the line that passes through $P(1, -2, 3)$ and is parallel to the line $3(x - 2) = 2(y + 2) = 5z$.

72. Find the point where l_1 and l_2 intersect and give the angle of intersection:
$l_1: x_1(t) = 3 - t, \ y_1(t) = 5 + 3t, \ z_1(t) = -1 - 4t$
$l_2: x_2(u) = 8 + 2u, \ y_2(u) = -6 - 4u, \ z_2(u) = 5 + u$

73. Where does the line that passes through $(1, 4, 2)$ and is parallel to $3\mathbf{i} + 2\mathbf{j} - 2\mathbf{k}$ intersect the xy-plane?

74. Where does the line that passes through $(3, 5, -1)$ and is parallel to $\mathbf{i} - \mathbf{j} + \mathbf{k}$ intersect the xz-plane?

75. Find scalar parametric equations for all lines that are perpendicular to the line $x(t) = 5 + 2t, \ y(t) = -5t$, $z(t) = -t$ and intersect the line at the point $P(-3, 2, 2)$.

76. Find the distance from $P(4, -3, 1)$ to the line through the origin parallel to $4\mathbf{i} - 3\mathbf{j} + \mathbf{k}$.

77. Find the distance from $P(3, -4, 1)$ to the line $\mathbf{r}(t) = 2\mathbf{i} - \mathbf{j} + t(\mathbf{i} - 2\mathbf{j} + 2\mathbf{k})$.

78. Find the standard vector parametrization for the line through $P(-1, 2, 4)$ parallel to $\mathbf{i} - 2\mathbf{j} + 3\mathbf{k}$.

79. Find the cosine of the angle between the line $x_1(t) = 2 + t, \ y_1(t) = 3 + t, \ z(t) = -1 + 2t$ and $x_2(u) = 2 + 2u, \ y_2(u) = 3 - u, \ z_2(u) = -1 + 3u$.

80. Find the cosine of the angle between the line $x(t) = 2t, \ y(t) = 3t, \ z(t) = t$ and the y-axis.

12.7 Planes

81. Which of the points $P(-2, 3, -1)$, $Q(2, 3, 4)$, $R(3, 4, 1)$ lie on the plane $2(x - 2) + 3(y - 2) - 2(z + 3) = 0$?

82. Which of the points $P(4, 1, 0)$, $Q(2, 1, -3)$, $R(4, 1, -2)$, $S(0, 2, -1)$ lie on the plane $\mathbf{N} \cdot (\mathbf{r} - \mathbf{r}_0) = 0$ if $\mathbf{N} = 2\mathbf{i} - 4\mathbf{j} + \mathbf{k}$ and $\mathbf{r}_0 = \mathbf{i} + 2\mathbf{j} - 3\mathbf{k}$?

83. Write an equation for the plane that passes through the point $P(2, 1, 3)$ and is perpendicular to $3\mathbf{i} + \mathbf{j} - 5\mathbf{k}$.

84. Write an equation for the plane that passes through the point $P(5, -2, -1)$ and is perpendicular to the plane $3x - y + 6z + 8 = 0$.

85. Find the unit normals for the plane $3x + 3y - 5z - 6 = 0$.

86. Write the equation of the plane $5x - 3y - 2z - 1 = 0$ in intercept form.

87. Where does the plane $4x + 3y - 2z + 4 = 0$ intersect the coordinate axes?

88. Find the angle between the planes $3(x-1) - 2(y-5) + 2(z+1) = 0$ and $2x + 5(y-1) + (z+4) = 0$.

89. Find the angle between the planes $x - 2y + 3z = 5$ and $2x + y - z = 7$.

90. Determine whether or not the vectors are coplanar: $\mathbf{i} + 2\mathbf{j} - 3\mathbf{k}, \mathbf{i} - 2\mathbf{j}, 4\mathbf{i} + \mathbf{j} - 2\mathbf{k}$.

91. Find an equation in x, y, z for the plane that passes through the points $P_1(1, 1, 1)$, $P_2(2, 4, 3)$, $P_3(-1, -2, -1)$.

92. Find a set of scalar parametric equations for the line formed by the two intersecting planes:
P_1: $3x - 2y + z = 0$, P_2: $8x + 2y + z - 11 = 0$.

93. Let l be the line determined by P_1, P_2, and let p be the plane determined by Q_1, Q_2, Q_3. Where, if anywhere, does l intersect p?
$P_1(2, 5, -2)$, $P_2(1, -2, 2)$; $Q_1(2, 1, -4)$, $Q_2(1, 2, 3)$, $Q_3(-1, 2, 1)$.

94. Find an equation in x, y, z for the plane that passes through $(1, 2, -3)$ and is perpendicular to the line
$x(t) = 1 + 2t, y(t) = 2 + t, z(t) = -3 - 5t$.

95. Find an equation in x, y, z for the plane that passes through $(2, 1, 5)$ and the line $x(t) = -1 + 3t, y(t) = -2, z(t) = 2 + 4t$.

96. Find a vector equation for the line through $(1, 1, 1)$ that is parallel to the line of intersection of the planes
$3x - 4y + 2z - 2 = 0$ and $4x - 3y - z - 5 = 0$.

97. Find parametric equations for the line through $(2, 0, -3)$ that is parallel to the line of intersection of the planes
$x + 2y + 3z + 4 = 0$ and $2x - y - z - 5 = 0$.

98. Find an equation for the plane that passes through $(3, 0, 1)$ and is perpendicular to the line $x(t) = 2t$, $y(t) = 1 - t, z(t) = 4 - 3t$.

99. Find an equation for the plane that contains the point $(-2, 1, 1)$ and the line
$\mathbf{r}(t) = 2\mathbf{i} + \mathbf{j} + \mathbf{k} + t(-\mathbf{i} + 4\mathbf{j} + 4\mathbf{k})$.

100. Find an equation for the plane that contains $P_1(1, 1, 1)$ and $P_2(-1, 2, 1)$ and is parallel to the line of intersection of the planes $2x + y - z - 4 = 0$ and $3x - y + z - 2 = 0$.

101. Find an equation for the plane that contains $P_1(3, 1, 2)$ and $P_2(-1, 2, -1)$ and is parallel to the line of intersection of the planes $2x - y - z - 2 = 0$ and $3x + 2y - 2z - 4 = 0$.

102. Sketch the graph of $20x + 12y + 15z - 60 = 0$.

103. Find the equation of the plane pictured below.

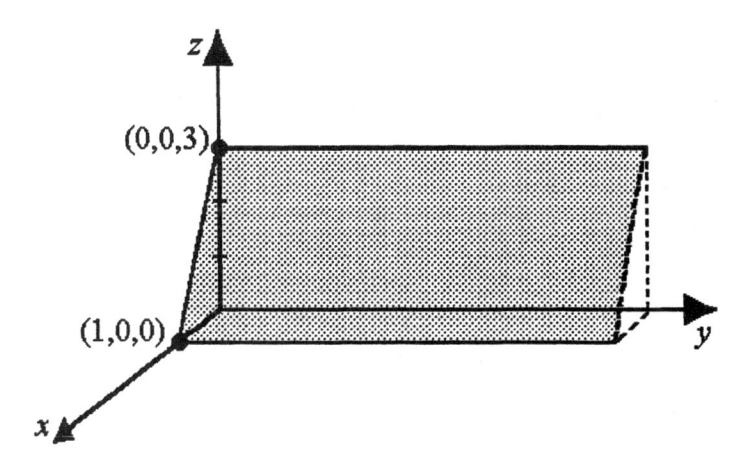

Answers to Chapter 12, Part B Questions

1. $\sqrt{6}; \left(\frac{5}{2}, 8, \frac{15}{2}\right)$

2. $\sqrt{229}; \left(3, \frac{5}{2}, 3\right)$

3. $3\sqrt{2}; \left(-1, \frac{5}{2}, \frac{5}{2}\right)$

4. $z = -2$

5. $z = -1$

6. $x = -2$

7. $(x-2)^2 + (y-1)^2 + (z-3)^2 = 16$

8. $(x+4)^2 + y^2 + (z-6)^2 = 49$

9. $(x-5)^2 + (y-1)^2 + (z+4)^2 = 49$

10. $(x-3)^2 + (y-7/2)^2 + (z+2)^2 = 21/4$

11. $(x+2)^2 + (y-1)^2 + (z-4)^2 = 16$

12. $(3, 2, 1)$

13. $(3, 2, -1)$

14. $(-3, 2, 1)$

15. $(-1, -2, -4)$

16. $(2, -1, 3)$

17. $(-2, -7, -3)$

18. $(-4, 2, 1)$

19. $\mathbf{i} + 3\mathbf{j}$

20. $-4\mathbf{i} - 3\mathbf{j} - 3\mathbf{k}$

21. 5

22. $\sqrt{11}$

23. $\sqrt{73}$

24. $-13\mathbf{i} - 5\mathbf{j} + 10\mathbf{k}$

25. (a) $\sqrt{30}$ (b) $\sqrt{10}$ (c) $\sqrt{42}$ (d) $2\sqrt{91}$

26. ± 3

27. $\frac{2}{3}\mathbf{i} + \frac{1}{3}\mathbf{j} + \frac{2}{3}\mathbf{k}$

28. $\frac{5}{\sqrt{162}}\mathbf{i} - \frac{4}{\sqrt{162}}\mathbf{j} - \frac{11}{\sqrt{162}}\mathbf{k}$

29. $\frac{1}{\sqrt{30}}\mathbf{i} + \frac{5}{\sqrt{30}}\mathbf{j} + \frac{2}{\sqrt{30}}\mathbf{k}$

30. $\frac{6}{\sqrt{29}}\mathbf{i} + \frac{8}{\sqrt{29}}\mathbf{j} + \frac{4}{\sqrt{29}}\mathbf{k}$

31. $\frac{10}{\sqrt{170}}\mathbf{i} - \frac{24}{\sqrt{170}}\mathbf{j} + \frac{2}{\sqrt{170}}\mathbf{k}$

32. $\mathbf{a} \cdot \mathbf{a} + 6\mathbf{a} \cdot \mathbf{b}$

33. $-\mathbf{a} \cdot \mathbf{b} + 7\mathbf{a} \cdot \mathbf{c} - 3\mathbf{b} \cdot \mathbf{c}$

34. (a) $0, -4, 5$
    (b) $\cos(\mathbf{a}, \mathbf{b}) = 0; \cos(\mathbf{a}, \mathbf{c}) = -4/5$
    $\cos(\mathbf{b}, \mathbf{c}) = \frac{\sqrt{70}}{14}$
    (c) (i) 0; (ii) $\frac{-4}{\sqrt{5}}$
    (d) (i) 0; (ii) $\frac{8}{5}\mathbf{j} + \frac{4}{5}\mathbf{k}$

35. (a) $11, -8, -5$
    (b) $\cos(\mathbf{a}, \mathbf{b}) = \frac{11}{\sqrt{21}\sqrt{14}}; \cos(\mathbf{a}, \mathbf{c}) = \frac{-8}{\sqrt{21}\sqrt{5}}$ $\cos(\mathbf{b}, \mathbf{c}) = \frac{-5}{\sqrt{14}\sqrt{5}}$
    (c) (i) $\frac{11}{\sqrt{14}}$; (ii) $\frac{-8}{\sqrt{5}}$
    (d) (i) $\left(\frac{33}{14}, \frac{-22}{14}, \frac{1}{14}\right)$; (ii) $\left(0, \frac{-16}{5}, \frac{8}{5}\right)$

36. $\approx 47.12°$ or 0.8225 radians

37. $\approx 58.12°$ or 1.014 radians

38. $\pi/4, 2\pi/3, \pi/3$

39. $\pi/6, 0, \pi/3$

40. $\frac{6}{\sqrt{53}}\mathbf{i} - \frac{1}{\sqrt{53}}\mathbf{j} - \frac{4}{\sqrt{53}}\mathbf{k}$ or

    $\frac{-6}{\sqrt{53}}\mathbf{i} + \frac{1}{\sqrt{53}}\mathbf{j} + \frac{4}{\sqrt{53}}\mathbf{k}$

Answers

41. $\dfrac{-8}{3\sqrt{21}}$

42. $250\sqrt{3}$ Joules

43. $W = 23$

44. $\mathbf{i} - \mathbf{j} + \mathbf{k}$

45. 0

46. 0

47. $2\mathbf{i} - 4\mathbf{j} + 9\mathbf{k}$

48. $-2\mathbf{i} - 7\mathbf{j} + 8\mathbf{k}$

49. $-12\mathbf{i} + 6\mathbf{j} - 54\mathbf{k}$

50. -2

51. 15

52. $14\mathbf{i} + 26\mathbf{j} - 21\mathbf{k}$

53. $5\sqrt{2}$

54. $\dfrac{1}{2}\sqrt{101}$

55. 17

56. 4

57. 54

58. 10

59. $2\sqrt{5}$

60. $\dfrac{\sqrt{290}}{2}$

61. $P(-1, 3, -1)$

62. No

63. $\mathbf{r}(t) = 2\mathbf{i} + 3\mathbf{j} + 3\mathbf{k} + t\mathbf{k}$

64. $\mathbf{r}(t) = t(3\mathbf{i} + \mathbf{j} + 8\mathbf{k})$

65. $\mathbf{r}(t) = 4\mathbf{i} + 5\mathbf{k} + t(-2\mathbf{i} + 3\mathbf{j} - 4\mathbf{k})$

66. $\mathbf{r}(t) = 3\mathbf{i} + 3\mathbf{j} + \mathbf{k} + t(\mathbf{i} - 3\mathbf{j} + \mathbf{k})$

67. $x(t) = 1 + t; y(t) = 4 - 5t; z(t) = 6 - 3t$

68. $x(t) = -3 + 2t; y(t) = -1 + 3t; z(t) = t$

69. $x(t) = 4; y(t) = -2; z(t) = -1 + t$

70. $x(t) = -1; y(t) = 2 + t; z(t) = -3$

71. $\mathbf{r}(t) = \mathbf{i} - 2\mathbf{j} + 3\mathbf{k} + t(10\mathbf{i} + 15\mathbf{j} + 6\mathbf{k})$

72. $(4, 2, 3); \theta = \cos^{-1}\left(\dfrac{-3}{\sqrt{273}}\right)$

73. $(4, 6, 0)$

74. $(8, 0, 4)$

75. $x(t) = -3 + t\,\mathbf{a}$
    $y(t) = 2 + t\,\mathbf{b}$
    $z(t) = 2 + t(2\,\mathbf{a} - 5\,\mathbf{b}); \mathbf{a}, \mathbf{b} \in \Re$

76. 0

77. $\sqrt{2}$

78. $\mathbf{r}(t) = -\mathbf{i} + 2\mathbf{j} + 4\mathbf{k} + t(\mathbf{i} - 2\mathbf{j} + 3\mathbf{k})$

79. $\dfrac{7}{2\sqrt{21}}$

80. $\dfrac{3}{\sqrt{14}}$

81. R

82. S

83. $3x + y - 5z + 8 = 0$

84. $3x + y + 6z - 7 = 0$

85. $\left(\dfrac{3}{\sqrt{43}}, \dfrac{3}{\sqrt{43}}, \dfrac{-5}{\sqrt{43}}\right), \left(\dfrac{-3}{\sqrt{43}}, \dfrac{-3}{\sqrt{43}}, \dfrac{5}{\sqrt{43}}\right)$

86. $\dfrac{x}{1/5} + \dfrac{y}{-1/3} + \dfrac{z}{-1/2} = 1$

87. $x = -1, y = -4/3, z = 2$

88. $\cos^{-1}\dfrac{2}{\sqrt{510}} \approx 84.92°$

89. $\cos^{-1}\dfrac{3}{2\sqrt{21}} \approx 70.89°$

90. No

91. $2y - 3z + 1 = 0$

92. $x = 1 - 4t, y = 3/2 + 5t, z = 22t$

93. $\left(\dfrac{92}{61}, \dfrac{95}{61}, \dfrac{-2}{61}\right)$

94. $2x + y - 5z - 19 = 0$

95. $10x - 3y - 9z + 28 = 0$

96. $\mathbf{i} + \mathbf{j} + \mathbf{k} + t(10\mathbf{i} + 11\mathbf{j} + 7\mathbf{k})$

97. $x = 2 + t, y = 7t, z = -3 - 5t$

98. $2x - y - 3z - 3 = 0$

99. $y - z = 0$

100. $x + 2y - 2z - 1 = 0$

101. $5x + 8y - 4z - 15 = 0$

102.

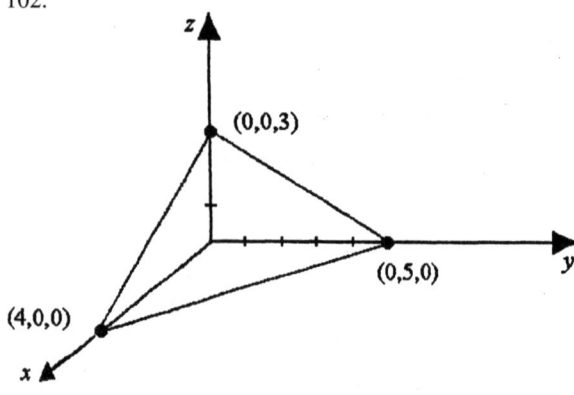

103. $3x + z = 3$

Part C

1. Find conditions on A, B, C, D such that the equation $x^2 + y^2 + z^2 + Ax + By + Cz + D = 0$ represents a sphere. (12.1)

2. (a) Show that, if $\mathbf{a}$ and $\mathbf{b}$ have the same direction, then $\|\mathbf{a} + \mathbf{b}\| = \|\mathbf{a}\| + \|\mathbf{b}\|$.
   (b) Does this equation necessarily hold if $\mathbf{a}$ and $\mathbf{b}$ are only parallel? (12.2) + (12.3)

3. Let P and Q be two points in space and let R be the point on $\overline{PQ}$ which is twice as far from P as it is from Q. Let $\mathbf{p} = \overrightarrow{OP}, \mathbf{q} = \overrightarrow{OQ}$, and $\mathbf{r} = \overrightarrow{OR}$. Prove that $\mathbf{r} = \frac{1}{3}\mathbf{p} + \frac{2}{3}\mathbf{pq}$. (12.2) + (12.3)

4. Under what conditions does $|\mathbf{a} \cdot \mathbf{b}| = \|\mathbf{a}\|\|\mathbf{b}\|$? (12.4)

5. Give an algebraic proof of Schwarz's inequality $|\mathbf{a} \cdot \mathbf{b}| \leq \|\mathbf{a}\|\|\mathbf{b}\|$. *Hint:* If $\mathbf{b} = \mathbf{0}$, the inequality is trivial, so assume $\mathbf{b} \neq \mathbf{0}$. Note that for any number λ we have $\|\mathbf{a} - \lambda\mathbf{b}\|^2 \geq 0$. First expand this inequality using the fact that $\|\mathbf{a} - \lambda\mathbf{b}\|^2 = (\mathbf{a} - \lambda\mathbf{b}) \cdot (\mathbf{a} - \lambda\mathbf{b})$. After collecting terms, make the special choice $\lambda = (\mathbf{a} \cdot \mathbf{b})/\|\mathbf{b}\|^2$ and see what happens. (12.4)

6. Let $\mathbf{a} = a_1 \mathbf{i} + a_2 \mathbf{j}$ and $\mathbf{b} = b_1 \mathbf{i} + b_2 \mathbf{j}$ be nonzero vectors in the xy-plane. Show that $\mathbf{a} \times \mathbf{b}$ is parallel to $\mathbf{k}$. (12.5)

7. If $\mathbf{a}, \mathbf{b}$, and $\mathbf{c}$ are mutually perpendicular, show that $\mathbf{a} \times (\mathbf{b} \times \mathbf{c}) = \mathbf{0}$. (12.5)

8. Suppose that the lines $l_1: \mathbf{r}(t) = \mathbf{r}_0 + t\,\mathbf{d}, l_2: \mathbf{R}(u) = \mathbf{R}_0 + u\,\mathbf{D}$ intersect at right angles. Show that the point of intersection is the origin iff $\mathbf{r}(t) \perp \mathbf{R}(u)$ for all real numbers t and u. (12.6)

9. Find the distance from the point $P(x_0, y_0, z_0)$ to the line $y = mx + b$ in the xy-plane. (12.6)

10. Let $\mathbf{a}, \mathbf{b}, \mathbf{c}$ be three nonzero vectors such that the angle between any pair of them is $\frac{1}{2}\pi$. Can these vectors be coplanar? (12.7)

11. $3x + 2y - 6 = 0$. (12.7)

Answers to Chapter 12, Part C Questions

1. $x^2 + y^2 + z^2 + Ax + By + Cz + D = 0$
   $$\Rightarrow \left(x + \frac{A}{2}\right)^2 + \left(y + \frac{B}{2}\right)^2 + \left(z + \frac{C}{2}\right)^2$$
   $$= \frac{A^2}{4} + \frac{B^2}{4} + \frac{C^2}{4} - D,$$
   so you get a sphere if $\frac{A^2}{4} + \frac{B^2}{4} + \frac{C^2}{4} - D > 0$.

2. (a) If $\mathbf{a}$ and $\mathbf{b}$ have the same direction $\mathbf{a} = \alpha \mathbf{b}$
   α constant >0. Thus
   $\|\mathbf{a} + \mathbf{b}\| = \|\alpha \mathbf{b} + \mathbf{b}\| = (\alpha + 1)\|\mathbf{b}\|$
   $= \alpha \|\mathbf{b}\| + \|\mathbf{b}\| = \|\alpha \mathbf{b}\| + \|\mathbf{b}\|$
   $= \|\mathbf{a}\| + \|\mathbf{b}\|$
   (b) No, for the choice $\mathbf{b} = -\mathbf{a} \neq 0$ $\|\mathbf{a} + \mathbf{b}\| = 0$
   but $\|\mathbf{a}\| + \|\mathbf{b}\| = 2\|\mathbf{a}\| > 0$

3. $\mathbf{r} - \mathbf{p} = 2(\mathbf{q} - \mathbf{r})$ $3\mathbf{r} = \mathbf{p} + 2\mathbf{q}$
   $$\mathbf{r} = \frac{1}{3}\mathbf{p} + \frac{2}{3}\mathbf{q}$$

4. $|\mathbf{a} \cdot \mathbf{b}| = \|\mathbf{a}\|\|\mathbf{b}\| |\cos\theta| = \|\mathbf{a}\|\|\mathbf{b}\|$
   iff $\theta = 0$, or $\theta = \pi$

5. Take λ arbitrary, $\mathbf{b} \neq 0$.
   $0 \leq \|\mathbf{a} - \lambda \mathbf{b}\|^2 = (\mathbf{a} - \lambda \mathbf{b}) \cdot (\mathbf{a} - \lambda \mathbf{b})$
   $= \mathbf{a} \cdot \mathbf{a} - \lambda(\mathbf{b} \cdot \mathbf{a})$
   $-\lambda(\mathbf{a} \cdot \mathbf{b}) + \lambda^2(\mathbf{b} \cdot \mathbf{b})$
   $= \|\mathbf{a}\|^2 - 2\lambda(\mathbf{a} \cdot \mathbf{b}) + \lambda^2\|\mathbf{b}\|^2$
   Setting $\lambda = (\mathbf{a} \cdot \mathbf{b})/\|\mathbf{b}\|^2$ we have
   $$0 \leq \|\mathbf{a}\|^2 - 2\frac{|(\mathbf{a} \cdot \mathbf{b})|^2}{\|\mathbf{b}\|^2} + \frac{|(\mathbf{a} \cdot \mathbf{b})|^2}{\|\mathbf{b}\|^2}$$
   $0 \leq \|\mathbf{a}\|^2\|\mathbf{b}\|^2 - |(\mathbf{a} \cdot \mathbf{b})|^2$.
   Thus $|(\mathbf{a} \cdot \mathbf{b})|^2 \leq \|\mathbf{a}\|^2\|\mathbf{b}\|^2$ and
   $|(\mathbf{a} \cdot \mathbf{b})| \leq \|\mathbf{a}\| \cdot \|\mathbf{b}\|$.

6. $\mathbf{a} \times \mathbf{b} = (a_1 b_2 - b_1 a_2)\mathbf{k}$

7. $\mathbf{b} \times \mathbf{c} \perp \mathbf{b}$ and $\mathbf{b} \times \mathbf{c} \perp \mathbf{c}$, so $\mathbf{b} \times \mathbf{c}$ must be parallel to $\mathbf{a}$. Hence $\mathbf{a} \times (\mathbf{b} \times \mathbf{c}) = 0$.

8. Since the two lines intersect, there exist numbers t_0 and u_0 such that
   $$\mathbf{r}(t_0) = \mathbf{R}(u_0)$$
   Suppose first that
   $$\mathbf{r}(t_0) = \mathbf{R}(u_0) = 0.$$
   Then
   $\mathbf{r}(t) = \mathbf{r}(t) - \mathbf{r}(t_0) = (\mathbf{r}_0 + t\mathbf{d}) - (\mathbf{r}_0 + t_0\mathbf{d})$
   $= (t - t_0)\mathbf{d}.$

   Similarly, $\mathbf{R}(u) = (u - u_0)\mathbf{D}$. Since $l_1 \perp l_2$, we have $\mathbf{d} \cdot \mathbf{D} = 0$ and thus
   $\mathbf{r}(t) \cdot \mathbf{R}(u) = (t - t_0)(u - u_0)(\mathbf{d} \cdot \mathbf{D}) = 0$
   for all t, u. Suppose now that
   $$\mathbf{r}(t_0) = \mathbf{R}(u_0) \neq 0$$
   Then
   $\mathbf{r}(t_0) \cdot \mathbf{R}(u_0) = \mathbf{r}(t_0) \cdot \mathbf{r}(t_0) = \|\mathbf{r}(t_0)\|^2 \neq 0$
   and it is therefore not true that
   $\mathbf{R}(t) \cdot \mathbf{R}(u) = 0$, for all t, u.

9. The line can be parameterized
   $\mathbf{r}(t) = b\mathbf{j} + t(\mathbf{i} + m\mathbf{j})$. Since the line contains the point $(0, b, 0)$
   $$d(P, l) = \frac{\|[x_0\mathbf{i} + (y_0 - b)\mathbf{j} + z_0\mathbf{k}] \times (\mathbf{i} + m\mathbf{j})\|}{\|\mathbf{i} + m\mathbf{j}\|}$$
   $$= \sqrt{\frac{(1 + m^2)z_0^2 + [y_0 - (mx_0 + b)]^2}{1 + m^2}}$$
   If $P(x_0, y_0, z_0)$ lies directly above or below the line, then $y_0 = mx_0 + b$ and $d(P, l)$ reduces to $\sqrt{z_0^2} = |z_0|$. This is evident geometrically.

10. No, each vector is normal to the plane of the other two. Another way to look at it: if they were coplanar, then together with a normal to that plane we would have four mutually perpendicular nonzero vectors. Three-space would then have four dimensions. Here is an algebraic argument:
    We are given that $\|\mathbf{a}\|, \|\mathbf{b}\|, \|\mathbf{c}\|$ are nonzero and $\mathbf{a} \cdot \mathbf{b} = \mathbf{a} \cdot \mathbf{c} = \mathbf{b} \cdot \mathbf{c} = 0$. The vectors $\mathbf{a}, \mathbf{b}, \mathbf{c}$ are coplanar iff there exist scalars s, t, u not all zero such that $s\mathbf{a} + t\mathbf{b} + u\mathbf{c} = 0$. Now assume that
    $$s\mathbf{a} + t\mathbf{b} + u\mathbf{c} = 0.$$
    The relation
    $0 = \mathbf{a} \cdot (s\mathbf{a} + t\mathbf{b} + u\mathbf{c}) = s\|\mathbf{a}\|^2 + t(\mathbf{a} \cdot \mathbf{b})$
    $+ u(\mathbf{a} \cdot \mathbf{c}) = s\|\mathbf{a}\|^2$
    gives $s = 0$. Similarly, we can show that $t = 0$ and $u = 0$. The vectors $\mathbf{a}, \mathbf{b}, \mathbf{c}$ cannot be coplanar.

11.

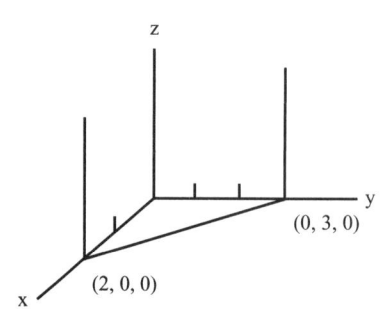

CHAPTER 13 Vector Calculus
Part B

13.1 Vector Functions

1. Differentiate $\mathbf{f}(t) = (3 - 4t)\mathbf{i} + 5t\mathbf{j} + (2 - 5t)\mathbf{k}$.

2. Differentiate $\mathbf{f}(t) = (1 - t)^{-1/2}\mathbf{i} + \sqrt{1 + 2t^2}\mathbf{j} - t^{3/2}\mathbf{k}$.

3. Differentiate $\mathbf{f}(t) = e^{-t}\mathbf{i} + \ln(2t^2 - t)\mathbf{j} + \sin^{-1} t\,\mathbf{k}$.

4. Differentiate $\mathbf{f}(t) = \sin^2 t^2\,\mathbf{i} + \cos 2t\,\mathbf{j} + t^2 e^{-2t}\,\mathbf{k}$.

5. Calculate $\int_1^3 \mathbf{f}(t)\,dt$ for $\mathbf{f}(t) = (1 + 2t^2)\mathbf{i} - 5t\,\mathbf{k}$.

6. Calculate $\int_{\pi/4}^{\pi/2} \mathbf{r}(t)\,dt$ for $\mathbf{r}(t) = t\sin 2t^2\,\mathbf{i} + \sin 3t\,\mathbf{j} + e^{2t}\,\mathbf{k}$.

7. Calculate $\int_0^1 \mathbf{g}(t)\,dt$ for $\mathbf{g}(t) = te^{-2t^2}\mathbf{i} + t^2 e^{4t^3}\mathbf{j} - \cos\dfrac{2t}{3}\,\mathbf{k}$.

8. Find $\lim_{t \to 1} \mathbf{f}(t)$ if it exists. $\mathbf{f}(t) = \ln t\,\mathbf{i} - \sqrt[3]{t}\,\mathbf{j} + e^{4t}\,\mathbf{k}$.

9. Find $\lim_{t \to 0} \mathbf{r}(t)$ if it exists. $\mathbf{r}(t) = t\cos t\,\mathbf{i} - e^{-t}\,\mathbf{j} + \dfrac{3t^2 - 2t + 1}{e^t}\,\mathbf{k}$.

10. Find a vector-valued function $\mathbf{f}$ that traces out the curve $16x^2 + 4y^2 = 64$ in (a) a counterclockwise direction and (b) a clockwise direction.

11. Find a vector-valued function $\mathbf{f}$ that traces out the curve $y = 2(x - 1)^2$ in a direction from (a) left to right and (b) right to left.

12. Find a vector-valued function $\mathbf{f}$ that traces out the directed line segment from $(2, -1, 3)$ to $(1, 4, -2)$.

13. Find $\mathbf{f}(t)$ given that $\mathbf{f}'(t) = (t^2 + 1)\mathbf{i} + 2t^2(1 + t^3)^{-1}\mathbf{j} + t^2 e^{t^3/2}\mathbf{k}$ and $\mathbf{f}(0) = 2\mathbf{i} - 3\mathbf{j} + \tfrac{1}{2}\mathbf{k}$.

14. Find $\mathbf{f}(t)$ given that $\mathbf{f}'(t) = 2t^2\mathbf{i} - 3(t + 1)\mathbf{j} + e^t(t + 1)\mathbf{k}$ and $\mathbf{f}(0) = \mathbf{i} - 2\mathbf{j} + 2\mathbf{k}$.

15. Sketch the curve represented by $\mathbf{r}(t) = 3t\,\mathbf{i} + t^3\,\mathbf{j}$ and indicate the orientation.

16. Sketch the curve represented by $\mathbf{r}(t) = 5\sin t\,\mathbf{i} + 3\cos t\,\mathbf{j}$ and indicate the orientation.

13.2 Differentiation Formulas

17. Find $\mathbf{f}'(t)$ and $\mathbf{f}''(t)$ for $\mathbf{f}(t) = e^t\mathbf{i} + e^{2t}\mathbf{j} + \mathbf{k}$.

18. Find $\mathbf{f}'(t)$ and $\mathbf{f}''(t)$ for $\mathbf{f}(t) = \sin t\,\mathbf{i} + \sinh 2t\,\mathbf{j} + \operatorname{sech} 2t\,\mathbf{k}$.

19. Find $\mathbf{f}'(t)$ and $\mathbf{f}''(t)$ for $\mathbf{f}(t) = \sqrt{t^2 + 2t}\,\mathbf{i} + \ln\sqrt{t^2 + 2t}\,\mathbf{j} + t\,\mathbf{k}$.

20. Find $\mathbf{f}'(t)$ and $\mathbf{f}''(t)$ for $\mathbf{f}(t) = t\,\mathbf{i} + \ln\cos 2t\,\mathbf{j} + \ln\sin 2t\,\mathbf{k}$.

21. Find $\mathbf{f}'(t)$ and $\mathbf{f}''(t)$ for $\mathbf{f}(t) = (\cos t + t\sin t)\,\mathbf{i} + (\sin t - t\cos t)\,\mathbf{j} + t^2\,\mathbf{k}$.

22. Find $\mathbf{f}'(t)$ and $\mathbf{f}''(t)$ for $\mathbf{f}(t) = (t^2\,\mathbf{i} + \cos t\,\mathbf{j}) \times (e^t\,\mathbf{j} + \sin t\,\mathbf{k})$.

23. Find $\mathbf{f}'(t)$ and $\mathbf{f}''(t)$ for $\mathbf{f}(t) = [(\sqrt{t}\,\mathbf{i} - t^{-3/2}\,\mathbf{j}) \cdot (\sin t\,\mathbf{j} - t\,\mathbf{k})]\,\mathbf{j}$.

24. Find the derivative $\dfrac{d^2}{dt^2}[e^t \sin^2 t\,\mathbf{i} + 2t^2\,\mathbf{j}]$.

25. Given $\mathbf{g}(t) = 2t\,\mathbf{i} + \dfrac{2}{3}t^3\,\mathbf{j} - (1+t^2)\,\mathbf{k}$ and $\mathbf{u}(t) = \dfrac{1}{4}t^2$, $\mathbf{f}(t) = t\,\mathbf{i} + \dfrac{1}{t}\,\mathbf{j} + e^{3t}\,\mathbf{k}$, find
    (a) $(\mathbf{f}+\mathbf{g})'(t)$
    (b) $(2\mathbf{f})'(t)$
    (c) $(u\mathbf{f})'(t)$
    (d) $(\mathbf{f} \cdot \mathbf{g})'(t)$
    (e) $(\mathbf{f} \times \mathbf{g})'(t)$
    (f) $(\mathbf{g} \times \mathbf{f})'(t)$
    (g) $(\mathbf{f}^{\circ} u)'(t)$

26. Find $\mathbf{r}(t)$ given that $\mathbf{r}'(t) = t\,\mathbf{i} - \mathbf{j} + e^{2t}\,\mathbf{k}$ and $\mathbf{r}(0) = \mathbf{i} - \dfrac{1}{4}\mathbf{j} + 2\,\mathbf{k}$.

27. Find $\mathbf{r}(t)$ given that $\mathbf{r}'(t) = \sin 2t\,\mathbf{i} + \cos 2t\,\mathbf{j} - t\,\mathbf{k}$ and $\mathbf{r}(0) = \mathbf{i} + \mathbf{j} + \mathbf{k}$.

28. Find $\mathbf{r}(t)$ given that $\mathbf{r}'(t) = \dfrac{1}{1+t^2}\,\mathbf{i} + 2\tan 2t\,\mathbf{j} + e^{-t}\,\mathbf{k}$ and $\mathbf{r}(0) = \mathbf{i} + \mathbf{j} + \mathbf{k}$.

29. Calculate $\mathbf{r}(t) \cdot \mathbf{r}'(t)$ and $\mathbf{r}(t) \times \mathbf{r}'(t)$ given that $\mathbf{r}(t) = (\sin t + t\cos t)\,\mathbf{i} + t\,\mathbf{j}$.

30. Find $\mathbf{f}'(\pi/3)$ for $\mathbf{f}(t) = t\,\mathbf{i} + \ln\sin 2t\,\mathbf{j} + \cos^2 2t\,\mathbf{k}$.

31. Find $\mathbf{f}'(t)$ for $\mathbf{f}(t) = \sin^{-1} 2t\,\mathbf{i} + \tan^{-1} 2t\,\mathbf{j}$.

13.3 Curves

32. Find the tangent to the vector $\mathbf{r}'(t)$ and the tangent line for $\mathbf{r}(t) = 3t\cos t\,\mathbf{i} + 3t\sin t\,\mathbf{j} + 4t\,\mathbf{k}$ at $t = \pi$.

33. Find the tangent to the vector $\mathbf{r}'(t)$ and the tangent line for $\mathbf{r}(t) = \sin^{-1}\dfrac{t}{2}\,\mathbf{i} + \tan^{-1} 3t\,\mathbf{j} - 3t\,\mathbf{k}$ at $t = 1$.

34. Find the tangent to the vector $\mathbf{r}'(t)$ and the tangent line for $\mathbf{r}(t) = 6\sin 2t\,\mathbf{i} + 6\cos 2t\,\mathbf{j} + \dfrac{2t^2}{\pi}\,\mathbf{k}$ at $t = \pi/4$.

35. Find the tangent to the vector $\mathbf{r}'(t)$ and the tangent line for $\mathbf{r}(t) = t^2\,\mathbf{i} + 4t^{-2}\,\mathbf{j} + (1-t^3)\,\mathbf{k}$ at $t = 1$.

36. Find the tangent to the vector $\mathbf{r}'(t)$ and the tangent line for $\mathbf{r}(t) = t^2\,\mathbf{i} + 2t\,\mathbf{j} + e^{2t}\,\mathbf{k}$ at $t = 1$.

37. Find the points on the curve $\mathbf{r}(t) = t\,\mathbf{j}$ at which $\mathbf{r}(t)$ and $\mathbf{r}'(t)$ are perpendicular.

38. Find the point at which the curves
    $\mathbf{r}_1(t) = (1-t)\,\mathbf{i} + (1+t)\,\mathbf{j} + (1-t)\,\mathbf{k}$ and
    $\mathbf{r}_2(u) = 2u^3\,\mathbf{i} + (1-u^2)\,\mathbf{j} + (1+u^2)\,\mathbf{k}$ intersect and find the angle of intersection.

39. Find a vector parametrization for the curve $x^2 = y - 1$, $x \geq 1$.

40. Find a vector parametrization for the curve $r = 2\cos 2\theta$, $\theta \in [0, \pi]$ (polar coordinates).

41. Find an equation in x and y for the curve $\mathbf{r}(t) = t^2\,\mathbf{i} + 2t\,\mathbf{j}$. Draw the curve. Does the curve have a tangent vector at the origin? If so, what is the unit tangent vector?

42. At $t = \pi/4$ find the unit tangent vector for the curve $\mathbf{r}(t) = \sin 2t\,\mathbf{i} + \cos 3t\,\mathbf{j} + \tan t\,\mathbf{k}$.

43. At $t = \pi/6$ find the unit tangent vector for the curve $\mathbf{r}(t) = \cos 2t\,\mathbf{i} + \sin 2t\,\mathbf{j} - 3t\,\mathbf{k}$.

44. Find the tangent vector $\mathbf{r}'(t)$ and the tangent line for $\mathbf{r}(t) = \sec 2t\,\mathbf{i} + \cos 2t\,\mathbf{j} + 2t\,\mathbf{k}$ at $t = 0$.

45. Find the unit tangent vector, the principal normal vector, and an equation in x, y, z for the osculating plane of the curve $\mathbf{r}(t) = t^2\,\mathbf{i} + 2t\,\mathbf{j} + t\,\mathbf{k}$ at $t = 1$.

46. Find the unit tangent to the curve $\mathbf{r}(t) = \ln t\,\mathbf{i} + t\,\mathbf{j} + \sin \pi t\,\mathbf{k}$ at $t = 2$.

47. Find the unit tangent vector, the principal normal vector, and an equation in x, y, z for the osculating plane of the curve $\mathbf{r}(t) = e^t \sin 2t\,\mathbf{i} + 2e^t \cos 2t\,\mathbf{j} + 2e^t\,\mathbf{k}$ at $t = 0$.

13.4 Arc Length

48. Find the length of the curve $\mathbf{r}(t) = 3\cos t\,\mathbf{i} + 3\sin t\,\mathbf{j} + t\,\mathbf{k}$ from $t = 0$ to $t = 2\pi$.

49. Find the length of the curve $\mathbf{r}(t) = 5t\,\mathbf{i} + 4\sin 3t\,\mathbf{j} + 4\cos 3t\,\mathbf{k}$ from $t = 0$ to $t = 2\pi$.

50. Find the length of the curve $\mathbf{r}(t) = \dfrac{t^3}{3}\,\mathbf{i} + \dfrac{t^2}{\sqrt{2}}\,\mathbf{j} + t\,\mathbf{k}$ from $t = 0$ to $t = 3$.

51. Find the length of the curve $\mathbf{r}(t) = 6\sin 2t\,\mathbf{i} + 6\cos 2t\,\mathbf{j} + 5t\,\mathbf{k}$ from $t = 0$ to $t = \pi$.

52. Find the length of the curve $\mathbf{r}(t) = \cos t\,\mathbf{i} + \sin t\,\mathbf{j} + t^{3/2}\,\mathbf{k}$ from $t = 0$ to $t = 20/3$.

53. Find the length of the curve $\mathbf{r}(t) = (2 + \cos 3t)\,\mathbf{i} + (3 - \sin t)\,\mathbf{j} + 4t\,\mathbf{k}$ from $t = 0$ to $t = 2\pi/3$.

54. Find the length of the curve $\mathbf{r}(t) = \sin^3 2t\,\mathbf{i} + \cos^3 2t\,\mathbf{j}$ from $t = 0$ to $t = \pi/4$.

55. Find the length of the curve $\mathbf{r}(t) = 2e^{-t}\,\mathbf{i} + (4 - 2t)\,\mathbf{j} + e^t\,\mathbf{k}$ from $t = 0$ to $t = 2$.

56. Find the curvature of $y = e^{2x}$.

57. Find the curvature of $y = \ln \sin 2x$.

58. Find the curvature of $y = 2x^2 - x + 1$.

59. Find the curvature of $y = 4\sin 3x$.

60. Find the radius of curvature of $y = x^2/4$ at $x = 1$.

61. Find the radius of curvature of $y^2 = 4x$ at $(1, 2)$.

62. Find the radius of curvature of $xy = 6$ at $x = 2$.

63. Find the radius of curvature of $y = e^x$ at $x = 0$.

64. Find the curvature of $\mathbf{r}(t) = 2\cos t\,\mathbf{i} + \cos 2t\,\mathbf{j}$ at $t = \pi/4$.

65. Find the curvature of $\mathbf{r}(t) = t^3 \mathbf{i} + 2t^2 \mathbf{j}$ at $t = 1$.

66. Find the curvature of $\mathbf{r}(t) = (t^2 + 1)\mathbf{i} + (t - 2)\mathbf{j}$ at $t = 2$.

67. Find the curvature of $\mathbf{r}(t) = 6\cos 2t\, \mathbf{i} + \sin 2t\, \mathbf{j} + 3t\, \mathbf{k}$ at $t = \pi$.

68. Find the curvature of $\mathbf{r}(t) = \sin t\, \mathbf{i} + \cos t\, \mathbf{j} + \ln\cos t\, \mathbf{k}$ at $t = 0$.

69. Find the curvature of $\mathbf{r}(t) = e^t \mathbf{i} + e^t \cos t\, \mathbf{j} + e^t \sin t\, \mathbf{k}$ at $t = 0$.

70. Interpret $\mathbf{r}(t)$ as the position of a moving object at time t. Find the curvature of $\mathbf{r}(t) = t^2 \mathbf{i} + \dfrac{1}{t}\mathbf{j}$ at $t = 1/2$ and determine the tangential and normal components of acceleration.

71. Interpret $\mathbf{r}(t)$ as the position of a moving object at time t. Find the curvature of $\mathbf{r}(t) = 2e^t \mathbf{i} + 2e^{-1}\mathbf{j}$ at $t = 0$ and determine the tangential and normal components of acceleration.

72. Find the curvature of $\mathbf{r}(t) = \ln t\, \mathbf{i} + t\, \mathbf{j}$ at $t = 2$.

73. Find the radius of curvature of $\mathbf{r}(t) = e^t \mathbf{i} + \sqrt{2}\, t\, \mathbf{j} + e^{-t}\mathbf{k}$ at $t = 0$.

74. Find the radius of curvature of $\mathbf{r}(t) = 4\sin t\, \mathbf{i} + (2t - \sin 2t)\mathbf{j} + \cos 2t\, \mathbf{k}$ at $t = \pi/2$.

75. Find the radius of curvature of $\mathbf{r}(t) = 2\cos t\, \mathbf{i} + 3\sin t\, \mathbf{j}$ at $t = \pi/2$.

76. Find the curvature of $x^2 + y^2 = 10x$ at $(2, -4)$.

77. Find the curvature of $y = 3\cosh x/3$ at $x = 0$.

13.5 Curvilinear Motion; Curvature

78. A particle moves so that $\mathbf{r}(t) = t^2 \mathbf{i} - 2t\, \mathbf{j}$. Find the velocity, speed, acceleration, and the magnitude of the acceleration at the time $t = 2$.

79. An object moves so that $\mathbf{r}(t) = 4\cos t\, \mathbf{i} + \sin t\, \mathbf{j}$. Sketch the curve and then compute and sketch the velocity and acceleration vectors at $t = \pi/2$.

80. An object moves so that $\mathbf{r}(t) = t^2 \mathbf{i} + t^3 \mathbf{j}$. Sketch the curve and then compute and sketch the velocity and acceleration vectors at $t = 1$.

81. An object moves so that $\mathbf{r}(t) = 6\cos 2t\, \mathbf{i} + 6\sin 2t\, \mathbf{j} + 5t\, \mathbf{k}$. Find the velocity, speed, acceleration, and the magnitude of the acceleration at the time $t = \pi$.

82. An object moves so that $\mathbf{r}(t) = 2t\, \mathbf{i} + 4\sin 3t\, \mathbf{j} + 4\cos 3t\, \mathbf{k}$. Find the velocity, speed, acceleration, and the magnitude of the acceleration at the time $t = \pi/2$.

83. An object moves so that $\mathbf{r}(t) = \cos t\, \mathbf{i} + \sin t\, \mathbf{j} + t^{3/2}\, \mathbf{k}$. Find the velocity, speed, acceleration, and the magnitude of the acceleration at the time $t = \pi/2$.

84. An object moves so that $\mathbf{r}(t) = e^t \mathbf{i} + e^t \cos t\, \mathbf{j} + e^t \sin t\, \mathbf{k}$, $t \geq 0$. Find
    (a) the initial velocity
    (b) the initial position
    (c) the initial speed
    (d) the acceleration throughout the motion
    (e) the acceleration at $t = 0$.

Vector Calculus   247

85. Find the force required to propel a particle of mass m so that $\mathbf{r}(t) = t^2\,\mathbf{i} + t^3\,\mathbf{j}$.

86. At each point $P(x(t), y(t), z(t))$ of its motion, an object of mass m is subject to a force $\mathbf{F}(t) = m\pi^2[2\cos\pi t\,\mathbf{i} + 3\sin\pi t\,\mathbf{j}]$. Given that $\mathbf{v}(0) = 2\,\mathbf{i} - 3\pi/2\,\mathbf{j} + \tfrac{1}{2}\,\mathbf{k}$ and $\mathbf{r}(0) = 3\,\mathbf{i} + 2\,\mathbf{j}$, find the following at time $t = 1$:
    (a) velocity
    (b) speed
    (c) acceleration
    (d) momentum
    (e) angular momentum
    (f) torque.

87. At each point $P(x(t), y(t), z(t))$ of its motion, an object of mass m is subject to a force $\mathbf{F}(t) = 2m\pi^2\left[\dfrac{3}{2}\cos\pi t\,\mathbf{i} - 2\sin\pi t\,\mathbf{j}\right]$. Given that $\mathbf{v}(0) = 3\,\mathbf{i} + 2\pi\,\mathbf{j} - \mathbf{k}$ and $\mathbf{r}(0) = 3\,\mathbf{j} - 2\,\mathbf{k}$, find the following at time $t = 1$:
    (a) velocity
    (b) speed
    (c) acceleration
    (d) momentum
    (e) angular momentum
    (f) torque.

88. Show that the position vector and the velocity vector of the particle whose position is given by $\mathbf{r}(t) = \sin t\cos t\,\mathbf{i} + \cos^2 t\,\mathbf{j} + \sin t\,\mathbf{k}$ are at right angles.

89. Solve the initial values problem: $\mathbf{F}(t) = m\,\mathbf{r}''(t) = m(e^t\,\mathbf{i} + e^{2t}\,\mathbf{j})$
    $\mathbf{r}_0 = \mathbf{r}(0) = 2\,\mathbf{i}$
    $\mathbf{v}_0 = \mathbf{v}(0) = 3\,\mathbf{j} + \mathbf{k}$

90. Solve the initial values problem: $\mathbf{F}(t) = m\,\mathbf{r}''(t) = m(3\cos 2t\,\mathbf{i} + 3\sin 2t\,\mathbf{j})$
    $\mathbf{r}_0 = \mathbf{r}(0) = \mathbf{i} - 2\,\mathbf{j}$
    $\mathbf{v}_0 = \mathbf{v}(0) = 2\,\mathbf{i} - \dfrac{2}{3}\,\mathbf{k}$

Answers to Chapter 13, Part B Questions

1. $-4\mathbf{i} + 5\mathbf{j} - 5\mathbf{k}$

2. $\frac{1}{2}(1-t)^{-3/2}\mathbf{i} + \frac{2t}{\sqrt{1+2t^2}}\mathbf{j} - \frac{3}{2}t^{1/2}\mathbf{k}$

3. $-e^{-t}\mathbf{i} + \frac{4t-1}{2t^2-t}\mathbf{j} - \frac{1}{\sqrt{1-t^2}}\mathbf{k}$

4. $4t\sin t^2 \cos t^2 \mathbf{i} - 2\sin 2t \mathbf{j} + 2te^{-2t}(1-t)\mathbf{k}$

5. $\frac{58}{3}\mathbf{i} - 20\mathbf{k}$

6. $-\frac{1}{4}\left(\cos\frac{\pi^2}{2} - \cos\frac{\pi^2}{8}\right)\mathbf{i} + \frac{\sqrt{2}}{6}\mathbf{j} + \frac{1}{2}\left(e^{\pi t} - e^{(\pi/2)t}\right)\mathbf{k}$

7. $\frac{1}{4}(1-e^{-2})\mathbf{i} + \frac{1}{12}(e^4-1)\mathbf{j} - \frac{3}{2}\left(\sin\frac{2}{3}\right)\mathbf{k}$

8. $-\mathbf{j} + e^4\mathbf{k}$

9. $-\mathbf{j} + \mathbf{k}$

10. (a) $\mathbf{f}(t) = 2\cos t\,\mathbf{i} + 4\sin t\,\mathbf{j}$
    (b) $\mathbf{g}(t) = 2\cos t\,\mathbf{i} - 4\sin t\,\mathbf{j}$

11. (a) $\mathbf{f}(t) = (t+1)\mathbf{i} + 2t^2\mathbf{j}$
    (b) $\mathbf{g}(t) = (1-t)\mathbf{i} + 2t^2\mathbf{j}$

12. $\mathbf{f}(t) = (2-t)\mathbf{i} + (-1+5t)\mathbf{j} + (3-5t)\mathbf{k}, t \in [0,1]$

13. $\mathbf{f}(t) = \left(\frac{t^3}{3} + t + 2\right)\mathbf{i} + \left(\frac{2}{3}\ln|1+t^3| - 3\right)\mathbf{j} + \left(\frac{2}{3}e^{t^3/2} - \frac{1}{6}\right)\mathbf{k}$

14. $\mathbf{f}(t) = \left(\frac{2}{3}t^2 + 1\right)\mathbf{i} - \left[3\left(\frac{t^2}{2} + t\right) + 2\right]\mathbf{j} + (te^t + 2)\mathbf{k}$

15.

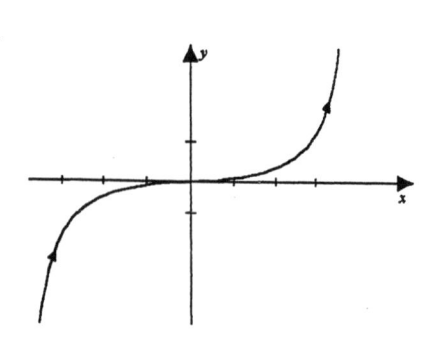

16.

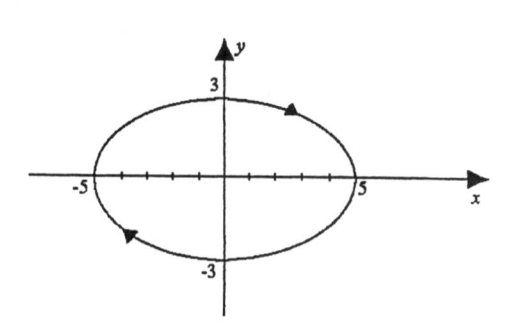

17. $\mathbf{f}'(t) = e^t\mathbf{i} + 2e^{2t}\mathbf{j}$
    $\mathbf{f}''(t) = e^t\mathbf{i} + 4e^{2t}\mathbf{j}$

18. $\mathbf{f}'(t) = \cos t\,\mathbf{i} + 2\cosh 2t\,\mathbf{j} - 2\,\text{sech}\,2t\tanh 2t\,\mathbf{k}$
    $\mathbf{f}''(t) = -\sin t\,\mathbf{i} + 4\sinh 2t\,\mathbf{j} + (4\,\text{sech}\,2t\tanh^2 2t - 4\,\text{sech}^3 2t)\mathbf{k}$

19. $\mathbf{f}'(t) = \frac{t+1}{\sqrt{t^2+2t}}\mathbf{i} - \frac{t+1}{\sqrt{t^2+2t}}\mathbf{j} + \mathbf{k}$

    $\mathbf{f}''(t) = \frac{-1}{(t^2+2t)^{3/2}}\mathbf{i} - \frac{t^2+t+1}{(t^2+2t)^2}\mathbf{j}$

20. $\mathbf{f}'(t) = \mathbf{i} - 2\tan 2t\,\mathbf{j} - 2\cot 2t\,\mathbf{k}$
    $\mathbf{f}''(t) = -4\sec^2 2t\,\mathbf{j} - 4\csc^2 2t\,\mathbf{k}$

21. $\mathbf{f}'(t) = t\cos t\,\mathbf{i} + t\sin t\,\mathbf{j} + 2t\,\mathbf{k}$
    $\mathbf{f}''(t) = (\cos t - t\sin t)\mathbf{i} + (\sin t + t\cos t)\mathbf{j} + 2\mathbf{k}$

22. $\mathbf{f}'(t) = \cos 2t\,\mathbf{i} - (2t\sin t + t^2\cos t)\mathbf{j} + e^t(t^2+2t)\mathbf{k}$
    $\mathbf{f}''(t) = -2\sin 2t\,\mathbf{i} - (2\sin t + 4t\cos t + t^2\sin t)\mathbf{j} + e^t(t^2+4t+2)\mathbf{k}$

23. $\mathbf{f}'(t) = t^{-3/2}\left(\frac{3\sin t}{2t} - \cos t\right)\mathbf{j}$

    $\mathbf{f}''(t) = t^{-3/2}\left(\frac{-15\sin t}{4t^2} + \frac{3\cos t}{t} + \sin t\right)\mathbf{j}$

24. $e^t(\sin^2 t + 2\sin 2t + 2\cos 2t)\mathbf{i} + 4\mathbf{j}$

25. (a) $3\mathbf{i} + \left(\frac{-1}{t^2} + 2t^2\right)\mathbf{j} + (3e^{3t} - 2t)\mathbf{k}$

    (b) $2\mathbf{i} - \frac{2}{t^2}\mathbf{j} + 6e^{3t}\mathbf{k}$

    (c) $\frac{3}{4}t^2\mathbf{i} + \frac{1}{4}\mathbf{j} + \left(\frac{3}{4}t^2 e^{3t} + \frac{t}{2}e^{3t}\right)\mathbf{k}$

    (d) $4t - \frac{4}{3}t - e^{3t}(3+2t+3t^2)$

(e) $\left(1 + \dfrac{1}{t^2} - 2t^2 e^{3t} - 2t^3 e^{3t}\right)\mathbf{i}$
$+ (2e^{3t} + 6te^{3t} + 1 + 3t^2)\mathbf{j} + \dfrac{8}{3}t^3 \mathbf{k}$

(f) $\left(-1 - \dfrac{1}{t^2} + 2t^2 e^{3t} + 2t^3 e^{3t}\right)\mathbf{i}$
$+ (-2e^{3t} - 6te^{3t} - 1 - 3t^2)\mathbf{j} - \dfrac{8}{3}t^3 \mathbf{k}$

(g) $\dfrac{t}{2}\mathbf{i} + \dfrac{8}{t^3}\mathbf{j} + \dfrac{3t}{2}e^{3t^2/4}\mathbf{k}$

26. $\mathbf{r}(t) = \left(\dfrac{t^2}{2} + 1\right)\mathbf{i} - \left(t + \dfrac{1}{4}\right)\mathbf{j} + \left(\dfrac{1}{2}e^{2t} + \dfrac{3}{2}\right)\mathbf{k}$

27. $\mathbf{r}(t) = \left(\dfrac{3}{2} - \dfrac{1}{2}\cos 2t\right)\mathbf{i} + \left(1 + \dfrac{1}{2}\sin 2t\right)\mathbf{j}$
$+ \left(1 - \dfrac{t^2}{2}\right)\mathbf{k}$

28. $\mathbf{r}(t) = (1 + \tan^{-1} t)\mathbf{i} + (1 - \ln\cos 2t)\mathbf{j}$
$+ (2 - e^{-t})\mathbf{k}$

29. $\mathbf{r}(t) \cdot \mathbf{r}'(t) = 3t\cos^2 t + \left(1 - \dfrac{t^2}{2}\right)\sin 2t$
$\mathbf{r}(t) \times \mathbf{r}'(t) = [(t^2 + 1)\sin t - t\cos t]\mathbf{k}$

30. $\mathbf{i} + \dfrac{2}{\sqrt{3}}\mathbf{j} + \sqrt{3}\mathbf{k}$

31. $\dfrac{2}{\sqrt{1-4t^2}}\mathbf{i} + \dfrac{2}{1+4t^2}\mathbf{j}$

32. $\mathbf{r}'(t) = 3(\cos^2 - t\sin t)\mathbf{i} + 3(\sin t + t\cos t)\mathbf{j} + 4\mathbf{k}$
tangent line: $-(3\pi + 3t)\mathbf{i} - 3\pi t\mathbf{j} + (4\pi + 4t)\mathbf{k}$

33. $\mathbf{r}'(t) = \dfrac{1}{\sqrt{4-t^2}}\mathbf{i} + \dfrac{3}{1+9t^2}\mathbf{j} - 3\mathbf{k}$
tangent line:
$\left(\dfrac{\pi}{6} - \dfrac{4}{\sqrt{15}}t\right) + \left(\tan^{-1} 3 + \dfrac{3}{10}t\right)\mathbf{j} - 3(1+t)\mathbf{k}$

34. $\mathbf{r}'(t) = 12\cos 2t\,\mathbf{i} - 12\sin 2t\,\mathbf{j} + \dfrac{4t}{\pi}\mathbf{k}$
tangent line: $6\mathbf{i} - 12t\mathbf{j} + \left(\dfrac{\pi}{8} + t\right)\mathbf{k}$

35. $\mathbf{r}'(t) = 2t\mathbf{i} - \dfrac{8}{t^3}\mathbf{j} - 3t^2\mathbf{k}$
tangent line: $(1+2t)\mathbf{i} + (4 - 8t)\mathbf{j} - 3t\mathbf{k}$

36. $\mathbf{r}(t) = 2t\mathbf{i} + 2\mathbf{j} + 2e^{2t}\mathbf{k}$
tangent line: $(1+t)\mathbf{i} + (2+t)\mathbf{j} + e^2(1+t)\mathbf{k}$

37. $(x, y) = (0, 0), (1/4, -1/2)$

38. $(2, 0, 2); \theta = \cos^{-1}\dfrac{-5}{\sqrt{33}} = 29.5°$

39. $\mathbf{r}(t) = t\mathbf{i} + (t^2 + 1)\mathbf{j}$

40. $\mathbf{r}(t) = t\,\mathbf{u}_\theta + (2 + \cos 2t)\,\mathbf{u}_r$

41. $x = y^2/4$, tangent vector at origin is $\mathbf{j}$

42. $\mathbf{T}(\pi/4) = \dfrac{-3}{\sqrt{17}}\mathbf{j} + \dfrac{2\sqrt{2}}{\sqrt{17}}\mathbf{k}$

43. $\mathbf{T} = \dfrac{-\sqrt{3}}{\sqrt{13}}\mathbf{i} + \dfrac{1}{\sqrt{13}}\mathbf{j} + \dfrac{3}{\sqrt{13}}\mathbf{k}$

44. $\mathbf{r}'(t) = (2\sec 2t\tan 2t)\mathbf{i} - 2\sin 2t\,\mathbf{j} + 2\mathbf{k}$
tangent line: $\mathbf{i} + \mathbf{j} + 2t\mathbf{k}$

45. $\mathbf{T}(1) = \dfrac{2}{3}\mathbf{i} + \dfrac{2}{3}\mathbf{j} + \dfrac{1}{3}\mathbf{k}$
$\mathbf{N}(1) = \dfrac{5}{3\sqrt{5}}\mathbf{i} + \dfrac{4}{3\sqrt{5}}\mathbf{j} + \dfrac{2}{3\sqrt{5}}\mathbf{k}$
osculating plane: $5x - 4y - 2z + 5 = 0$

46. $\mathbf{T}(2) = \dfrac{1}{\sqrt{5+4\pi^2}}\mathbf{i} + \dfrac{2}{\sqrt{5+4\pi^2}}\mathbf{j} + \dfrac{2\pi}{\sqrt{5+4\pi^2}}\mathbf{k}$

47. $\mathbf{T}(0) = \dfrac{1}{\sqrt{3}}\mathbf{i} + \dfrac{1}{\sqrt{3}}\mathbf{j} + \dfrac{1}{\sqrt{3}}\mathbf{k}$
$\mathbf{N}(0) = \dfrac{2}{\sqrt{14}}\mathbf{i} + \dfrac{3}{\sqrt{14}}\mathbf{j} + \dfrac{1}{\sqrt{14}}\mathbf{k}$
osculating plane: $2x - 3y + z + 4 = 0$

48. $2\pi\sqrt{10}$

49. 26π

50. 12

51. 13π

52. $56/3$

53. $10\pi/3$

54. $3/2$

55. $e^2 + 1 - 2e^{-2}$

56. $\dfrac{4e^{2x}}{(1+4e^{4x})^{3/2}}$

57. $\dfrac{4\csc^2 2x}{(1 + 4\cot^2 2x)^{3/2}}$

58. $\dfrac{4}{(2+8x+16x^2)^{3/2}}$

59. 0

60. $\dfrac{5\sqrt{5}}{4}$

61. $4\sqrt{2}$

62. $\dfrac{13\sqrt{13}}{12}$

63. $2\sqrt{2}$

64. $\dfrac{1}{3\sqrt{3}}$

65. $\dfrac{12}{125}$

66. $\dfrac{12}{17\sqrt{17}}$

67. $\dfrac{24}{169}$

68. $\sqrt{2}$

69. $\dfrac{\sqrt{2}}{3}$

70. $\dfrac{24}{17\sqrt{17}}, a_T = \dfrac{-62}{\sqrt{17}}, a_N = \dfrac{24}{\sqrt{17}}$

71. $\dfrac{1}{\sqrt{8}}, a_T = 0, a_N = \sqrt{8}$

72. $\dfrac{2}{5\sqrt{5}}$

73. $2\sqrt{2}$

74. $2\sqrt{2}$

75. 4/3

76. 1/5

77. 1/3

78. $\mathbf{v}(2) = 4\mathbf{i} - 2\mathbf{j}; \|\mathbf{v}(2)\| = 2\sqrt{5}$
    $\mathbf{a}(2) = 2\mathbf{i}; \|\mathbf{a}(2)\| = 2$

79. $\mathbf{v}(\pi/2) = -4\mathbf{i}; \mathbf{a}(\pi/2) = -\mathbf{j}$

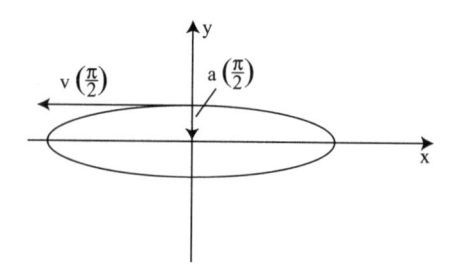

80. $\mathbf{v}(1) = 2\mathbf{i} + 3\mathbf{j}; \mathbf{a}(1) = 2\mathbf{i} + 6\mathbf{j}$

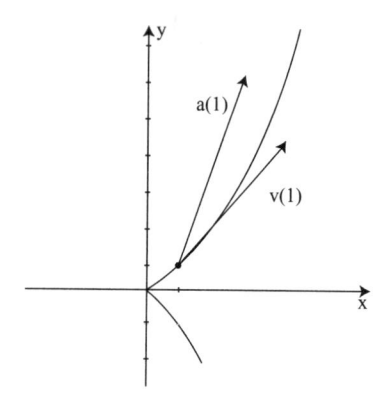

81. $\mathbf{v}(\pi) = 12\mathbf{j} + 5\mathbf{k}; \|\mathbf{v}(\pi)\| = 13$
    $\mathbf{a}(\pi) = -24\mathbf{i}; \|\mathbf{a}(\pi)\| = 24$

82. $\mathbf{v}(\pi/2) = 2\mathbf{i} + 12\mathbf{k}; \|\mathbf{v}(\pi/2)\| = 2\sqrt{37}$
    $\mathbf{a}(\pi/2) = 36\mathbf{j}; \|\mathbf{a}(\pi/2)\| = 36$

83. $\mathbf{v}(\pi/2) = -\mathbf{i} + \dfrac{3}{2}\sqrt{\dfrac{\pi}{2}}\mathbf{k}; \|\mathbf{v}(\pi/2)\| = \sqrt{1 + \dfrac{9}{8}\pi}$
    $\mathbf{a}(\pi/2) = -\mathbf{j} + \dfrac{3}{4}\sqrt{\dfrac{2}{\pi}}\mathbf{k}; \|\mathbf{a}(\pi/2)\| = \sqrt{1 + \dfrac{9}{8\pi}}$

84. (a) $\mathbf{i} + \mathbf{j} + \mathbf{k}$
    (b) (1, 1, 0)
    (c) $\sqrt{3}$
    (d) $\mathbf{a}(t) = e^t \mathbf{i} - 2e^t \sin t\, \mathbf{j} + 2e^t \cos t\, \mathbf{k}$
    (e) $\mathbf{a}(0) = \mathbf{i} + 2\mathbf{k}$

85. $\mathbf{F}(t) = 2m\mathbf{i} + 6mt\mathbf{j}$

86. (a) $2\mathbf{i} + 9\pi/2\,\mathbf{j} + {}^1/_2\,\mathbf{k}$
    (b) $\dfrac{1}{2}\sqrt{17 + 81\pi^2}$
    (c) $-2\pi^2\,\mathbf{i}$
    (d) $2m\mathbf{i} + \dfrac{9}{2}\pi m\,\mathbf{j} + \dfrac{m}{2}\mathbf{k}$

Answers

(e) $\left(1 - \dfrac{3\pi}{2}\right) m\,\mathbf{i} + \dfrac{1}{2} m\,\mathbf{j} + \left(\dfrac{75\pi}{2} - 4\right) m\,\mathbf{k}$

(f) $-m\pi^2\,\mathbf{j}$

87. (a) $3\,\mathbf{i} - 6\pi\,\mathbf{j} - \mathbf{k}$
    (b) $\sqrt{10 + 36\pi^2}$
    (c) $-3\pi^2\,\mathbf{i}$
    (d) $3m\,\mathbf{i} - 6\pi m\,\mathbf{j} + m\,\mathbf{k}$
    (e) $-(3 + 16\pi)m\,\mathbf{i} - (9 + 48\pi)m\,\mathbf{k}$
    (f) $9\pi^2 m\,\mathbf{j} + (9\pi^2 - 6\pi^3)m\,\mathbf{k}$

88. $\mathbf{r}(t) = \tfrac{1}{2}\sin 2t\,\mathbf{i} + \tfrac{1}{2}(1 + \cos 2t)\,\mathbf{j} + \cos t\,\mathbf{k}$
    $\mathbf{v}(t) = \cos 2t\,\mathbf{i} - \sin 2t\,\mathbf{j} + \sin t\,\mathbf{k}$
    $\mathbf{r}(t) \cdot \mathbf{v}(t) = 0$

89. $\mathbf{v}(t) = (e^t - 1)\,\mathbf{i} + \tfrac{1}{2}(e^{2t} + 5)\,\mathbf{j} + \mathbf{k}$
    $\mathbf{r}(t) = (e^t - t + 1)\,\mathbf{i} + \tfrac{1}{4}(e^{2t} + 10t - 1)\,\mathbf{j} + t\,\mathbf{k}$

90. $\mathbf{v}(t) = \left(\dfrac{3}{2}\sin 2t + 2\right)\mathbf{i} + \left(-\dfrac{3}{2}\cos 2t - \dfrac{3}{2}\right)\mathbf{j} - \dfrac{2}{3}\mathbf{k}$
    $\mathbf{r}(t) = \left(-\dfrac{3}{4}\cos 2t + 2t + \dfrac{7}{4}\right)\mathbf{i}$
    $\quad -\dfrac{3}{4}\left(\sin 2t - 2t + \dfrac{8}{3}\right)\mathbf{j} - \dfrac{2}{3}t\,\mathbf{k}$

Part C

1. No ϵ, δ's have surfaced so far, but they are still there at the heart of the limit process. Give an ϵ, δ characterization of $\lim_{t \to t_0} \mathbf{f}(t) = \mathbf{L}$. (13.1)

2. Assume that, as $t \to t_0$, $\mathbf{f}(t) \to \mathbf{L}$ and $\mathbf{g}(t) \to \mathbf{M}$. Show that $\mathbf{f}(t) \cdot \mathbf{g}(t) \to \mathbf{L} \cdot \mathbf{M}$. (13.1)

3. Derive the formula $(\mathbf{f} \cdot \mathbf{g})'(t) = [\mathbf{f}(t) \cdot \mathbf{g}'(t)] + [\mathbf{f}'(t) \cdot \mathbf{g}(t)]$ (13.2)
   (a) by appealing to components;
   (b) without appealing to components.

4. (a) Show that, if u is continuous at t_0 and $\mathbf{f}$ is continuous at $u(t_0)$, then the composition $\mathbf{f} \circ u$ is continuous at $u(t_0)$
   (b) Derive the chain rule for vector functions: $\dfrac{d\mathbf{f}}{dt} = \dfrac{d\mathbf{f}}{du}\dfrac{du}{dt}$. (13.2)

5. (a) Show that the curve $\mathbf{r}(t) = (t^2 - t + 1)\mathbf{i} + (t^3 - t + 2)\mathbf{j} + (\sin \pi t)\mathbf{k}$ intersects itself at $P(1, 2, 0)$ (13.3) by finding numbers $t_1 < t_2$ for which P is the tip of both $\mathbf{r}(t_1)$ and $\mathbf{r}(t_2)$.
   (b) Find the unit tangents at $P(1, 2, 0)$, first taking $t = t_1$, then taking $t = t_2$.

6. Let $\mathbf{r} = \mathbf{r}(t)$ be a differentiable vector-valued function. Prove that the tangent vector to the graph of $\mathbf{r}$ points in the direction of increasing t. (13.3)

7. Use vector methods to show that, if $y = f(x)$ has a continuous first derivative, then the length of the graph from $x = a$ to $x = b$ is given by the integral $\displaystyle\int_a^b \sqrt{1 + [f'(x)]^2}\, dx$. (13.4)

8. Let C_1 be the curve $\mathbf{r}(t) = (t - \ln t)\mathbf{i} + (t + \ln t)\mathbf{j}$, $1 \leq t \leq e$ and let C_2 be the graph of (13.4) $y = e^x$, $0 \leq x \leq 1$.
   Find a relation between the length of C_1 and the length of C_2.

9. Show that for an object of constant velocity the angular momentum is constant. (13.5)

10. (*Important*) A wheel is rotating about an axle with angular speed ω. Let $\boldsymbol{\omega}$ be the *angular velocity vector*, the vector of length ω that points along the axis of the wheel in such a direction that, observed from the tip of $\boldsymbol{\omega}$, the wheel rotates counterclockwise. Take the origin as the center of the wheel and let $\mathbf{r}$ be the vector from the origin to a point P on the rim of the wheel. Express the velocity $\mathbf{v}$ of P in terms of $\boldsymbol{\omega}$ and $\mathbf{r}$. (13.5)

11. Differentiating with respect to x, we have $\dfrac{d\phi}{dx} = \dfrac{1}{1 + (y')^2} \cdot \dfrac{d}{dx}(y') = \dfrac{y''}{1 + (y')^2}$. (13.7)
    We also know that the angular momentum $\mathbf{L}$ is constant and that $L = mr^2\dot\theta$. Use this fact to verify that
    $E = \dfrac{L^2}{2m}\left\{\dfrac{1}{r^2} + \dfrac{1}{r^4}\left(\dfrac{dr}{d\theta}\right)^2\right\} - \dfrac{m\rho}{r}$. Since E is a constant, this is a differential equation for r as a function of θ.

12. Show that the curvature of a polar curve $r = f(\theta)$ is given by $k = \dfrac{|[f(\theta)]^2 + 2[f'(\theta)]^2 - f(\theta)f''(\theta)|}{([f(\theta)]^2 + [f'(\theta)]^2)^{3/2}}$. (13.7)

13. Let $s(\theta)$ be the arc distance from the origin to the point $(x(\theta), y(\theta))$ along the exponential spiral $r = ae^{c\theta}$. (Take $a > 0$, $c > 0$.) Let $\rho(\theta)$ be the radius of curvature at that same point. Find an equation in s and ρ for that curve. (13.7)

Answers to Chapter 13, Part C Questions

1. For each $\epsilon > 0$ there exists $\delta > 0$ such that if $0 < |t - t_0| < \delta$, then $\|\mathbf{f}(t) - \mathbf{L}\| < \epsilon$.

2. $\|[\mathbf{f}(t) \cdot \mathbf{g}(t)] - [\mathbf{L} \cdot \mathbf{M}]\|$
   $= \|[\mathbf{f}(t) \cdot \mathbf{g}(t)] - [\mathbf{L} \cdot \mathbf{g}(t)] + [\mathbf{L} \cdot \mathbf{g}(t)] - [\mathbf{L} \cdot \mathbf{M}]\|$
   $= \|[(\mathbf{f}(t) - \mathbf{L}) \cdot \mathbf{g}(t)] + [\mathbf{L} \cdot (\mathbf{g}(t) - \mathbf{M})]\|$
   $\leq \|(\mathbf{f}(t) - \mathbf{L}) \cdot \mathbf{g}(t)\| + \|\mathbf{L} \cdot (\mathbf{g}(t) - \mathbf{M})\|$
   $\leq \|\mathbf{f}(t) - \mathbf{L}\|\|\mathbf{g}(t)\| + \|\mathbf{L}\|\|\mathbf{g}(t) - \mathbf{M}\|$

   by Schwarz's inequality
   As $t \to t_0$, the right side tends to
   $$(0)\|\mathbf{M}\| + \|\mathbf{L}\|(0) = 0.$$

3. (a) Routine
   (b) Write
   $$\frac{[\mathbf{f}(t+h) \cdot \mathbf{g}(t+h)] - [\mathbf{f}(t) \cdot \mathbf{g}(t)]}{h}$$
   so
   $$\left(\mathbf{f}(t+h) \cdot \left[\frac{\mathbf{g}(t+h) - \mathbf{g}(t)}{h}\right]\right)$$
   $$+ \left(\left[\frac{\mathbf{f}(t+h) - \mathbf{f}(t)}{h}\right] \cdot \mathbf{g}(t)\right)$$
   and take the limit as $h \to 0$. (Appeal to Theorem 13.1.3)

4. (a) and (b) can derived routinely by using components. An ϵ, δ derivation of (a) is also simple: Let $\epsilon > 0$. Since $\mathbf{f}$ is continuous at $u(t_0)$, there exists $\delta_1 > 0$ such that if
   $|x - u(t_0)| < \delta_1$, then $\|\mathbf{f}(x) - \mathbf{f}(u(t_0))\| < \epsilon$.
   Since u is continuous at t_0, there exists $\delta > 0$ such that
   if $|t - t_0| < \delta$, then $|u(t) - u(t_0)| < \epsilon$.
   Thus
   $|t - t_0| < \epsilon \Rightarrow |u(t) - u(t_0)| < \delta_1$
   $\Rightarrow \|\mathbf{f}(u(t)) - \mathbf{f}(u(t_0))\| < \epsilon$.

5. (a) $\mathbf{r}(0) = \mathbf{i} + 2\mathbf{j} = \mathbf{r}(1)$
   (b) $\mathbf{T}(0) = \dfrac{\mathbf{r}'(0)}{\|\mathbf{r}'(0)\|} = \dfrac{1}{\sqrt{\pi^2 + 2}}(-\mathbf{i} - \mathbf{j} + \pi\mathbf{k})$,
   $\mathbf{T}(1) = \dfrac{\mathbf{r}'(1)}{\|\mathbf{r}'(1)\|} = \dfrac{1}{\sqrt{\pi^2 + 5}}(\mathbf{i} + 2\mathbf{j} - \pi\mathbf{k})$

6. Since $\mathbf{r}'(t) = \lim\limits_{h \to 0} \dfrac{\mathbf{r}(t+h) - \mathbf{r}(t)}{h}$, for small positive h we have $\mathbf{r}'(t) \cong \dfrac{1}{h}[\mathbf{r}(t+h) - \mathbf{r}(t)]$, so $\mathbf{r}'(t)$ has the same direction as $\mathbf{r}(t+h) - \mathbf{r}(t)$, so $\mathbf{r}'(t)$ points in the direction of increasing t.

7. Parameterize the graph of f by setting
   $$\mathbf{r}(x) = x\hat{i} + f(x)\hat{j}.$$

8. $L_1 = \displaystyle\int_1^e \|\mathbf{r}'(t)\|\, dt = \int_1^e \sqrt{2 + \dfrac{2}{t^2}}\, dt;$
   $L_2 = \displaystyle\int_0^1 \sqrt{1 + e^{2x}}\, dx$
   Setting $t = e^x$, we get $L_1 = \sqrt{2} L_2$.

9. If $\mathbf{v}'(t) = 0$, then
   $\mathbf{L}'(t) = \mathbf{r}'(t) \times m\mathbf{v}(t) + \mathbf{r}(t) \times m\mathbf{v}'(t)$
   $= \mathbf{v}(t) \times m\mathbf{v}(t) + \mathbf{r}(t) \times 0 = 0$

10. Since $\mathbf{v}$ has magnitude $w\|\mathbf{r}\|$, lies in the plane of the wheel, and makes an angle of $90°$ counterclockwise with $\mathbf{r}$, we have $\mathbf{v} = \boldsymbol{\omega} \times \mathbf{r}$.

11. Using
    $$\dot{r} = \frac{dr}{d\theta}\dot{\theta} \quad \text{and} \quad \dot{\theta} = \frac{L}{mr^2}$$
    we have
    $$E + \frac{m\rho}{r} = \frac{1}{2}m(\dot{r}^2 + r^2\dot{\theta}^2)$$
    $$= \frac{1}{2}\left[\left(\frac{dr}{d\theta}\right)^2\dot{\theta}^2 + r^2\dot{\theta}^2\right]$$
    $$= \frac{1}{2}m\left[\left(\frac{dr}{d\theta}\right)^2\frac{L^2}{m^2r^4} + \frac{L^2}{m^2r^2}\right]$$
    $$= \frac{L^2}{2m}\left[\frac{1}{r^4}\left(\frac{dr}{d\theta}\right)^2 + \frac{1}{r^2}\right]$$
    and therefore
    $$E = \frac{L^2}{2m}\left[\frac{1}{r^2} + \frac{1}{r^4}\left(\frac{dr}{d\theta}\right)^2\right] - \frac{m\rho}{r}$$

12. Set $\mathbf{r}(\theta) = \cos\theta f(\theta)\mathbf{i} + \sin\theta f(\theta)\mathbf{j}$

13. $\mathbf{r}(\theta) = ae^{c\theta}\cos\theta\, \mathbf{i} + ae^{c\theta}\sin\theta\, \mathbf{j}$
    $s(\theta) = \displaystyle\int_0^\theta \sqrt{[x'(t)]^2 + [y'(t)]^2}\, dt$
    $= \displaystyle\int_0^\theta ae^{ct}\sqrt{1 + c^2}\, dt = \dfrac{a}{c}\sqrt{1 + c^2}\,e^{c\theta}$
    $f(\theta) = ae^{c\theta}, \quad f'(\theta) = ace^{c\theta}, \quad f''(\theta) = ac^2e^{c\theta}.$
    By Exercise 36,
    $\rho(\theta) = \dfrac{1}{k(\theta)}$
    $= \dfrac{([f(\theta)]^2 + [f'(\theta)]^2)^{3/2}}{|[f(\theta)]^2 + 2[f'(\theta)]^2 - f(\theta)f''(\theta)|}$
    $= a\sqrt{1 + c^2}\,e^{c\theta}$

CHAPTER 14 Functions of Several Variables
Part B

14.1 Elementary Examples

1. Find the domain and range of $f(x, y, z) = 2\tan^{-1}\dfrac{y}{x} + \ln(x^2 + z^2)$; find $f(1, 1, 1)$ and $f(1, -1, 1)$.

2. Find the domain and range of $f(x, y, z) = ze^x \cos y$; find $f(\ln 2, 0, -1)$.

3. Find the domain and range of $f(x, y, z) = yze^{\ln(x^2+y^2)}$; find $f(1, -1, 2)$ and $f(0, 1, 4)$.

4. Find the domain and range of $f(x, y) = \sin^{-1}+\left(\dfrac{x}{y}\right) + \tan^{-1}\left(\dfrac{x}{y}\right)$; find $f(1, 1)$.

5. Find the domain and range of $f(x, y) = 4 - \dfrac{2}{3}x - \dfrac{1}{2}y$; find $f(2, 3)$.

6. Find the domain and range of $f(x, y) = 1 - y^2$; find $f(2, 0)$.

7. Find the domain and range of $f(x, y, z) = \sqrt{4 - x^2 - y^2 - z^2}$; find $f(1, 1, 1/2)$.

8. Find the domain and range of $f(x, y) = \sqrt{\dfrac{x+y}{x-y}}$; find $f(3, 1)$.

9. Find the domain and range of $f(x, y) = 2xy - \dfrac{y}{x}$; if $x(t) = 2t$ and $y(t) = t^2$.

10. Determine a function f of x and y giving the volume of a cone of base diameter x and height y.

11. Determine a function f of x, y, and z giving the surface area of a box of length x, width y, and volume z.

14.2 A Brief Catalog of Quadratic Surfaces; Projections

12. Identify the surface $x^2 + 4y^2 + 9z^2 + 2x + 16y - 18z - 10 = 0$.

13. Identify the surface $5x^2 + 4y^2 + 20z^2 - 20x + 32y + 40z + 56 = 0$.

14. Identify the surface $9x^2 + 4y^2 - 54x - 16y - 36z + 277 = 0$.

15. Identify the surface $3x^2 - 2y^2 + 3z^2 + 30x - 8y - 24z + 131 = 0$.

16. Identify the surface $6x^2 + 4y^2 - 3z^2 + 36x - 16y - 6z + 55 = 0$.

17. Identify the surface $6x^2 + 4y^2 - 2z^2 - 6x - 4y + z = 0$.

18. Identify the surface $3x^2 - 2y^2 - z^2 - 6x + 8y - 2z + 6 = 0$.

19. Identify the surface $x^2 + y^2 - z^2 - 2x + 4y - 2z = 0$.

20. Sketch the cylinder $9x^2 + 4z^2 - 36 = 0$.

21. Identify and sketch the surface $z = 4x^2 + y^2$.

22. Identify and sketch the surface $z^2 = x^2 + y^2$.

23. Identify and sketch the surface $\dfrac{x^2}{4} + \dfrac{y^2}{9} + \dfrac{z^2}{16} = 1$.

24. Identify the surface and find the traces: $2x^2 + y^2 - 4z = 0$.

25. Identify the surface and find the traces: $x^2 - y^2 + z^2 + 2y = 1$.

26. Identify the surface and find the traces: $x^2 + y^2 + z - 5 = 0$.

27. Identify the surface and find the traces: $x^2 + 4y^2 + z^2 - 8y = 0$.

28. Write an equation for the surface obtained by revolving the parabola $y - 4z^2 = 0$ about the y-axis.

29. The planes $2x + y + z = 4$ and $x + y - z = 1$ intersect in a space curve C. Determine the projection of C onto the xy-plane.

30. The sphere $x^2 + (y-1)^2 + z^2 = 6$ and the hyperboloid $x^2 - y^2 + z^2 = 1$ intersect in a space curve C. Determine the projection of C onto the xy-plane.

31. The sphere $x^2 + (y-2)^2 + z^2 = 2$ and the cone $y^2 + z^2 = 5x^2$ intersect in a space curve C. Determine the projection of C onto the xy-plane.

32. The cone $x^2 + y^2 = 2z^2$ and the plane $y + 4z = 5$ intersect in a space curve C. Determine the projection of C onto the xy-plane.

14.3 Graphs; Level Curves and Level Surfaces

33. Sketch the graph of $f(x, y) = \sqrt{16 - x^2 - y^2}$.

34. Sketch the graph of $f(x, y) = \sqrt{16 - x^2 - 2y^2}$.

35. Identify the level curves and sketch some of them. $f(x, y) = x^2 + y^2$

36. Identify the level curves and sketch some of them. $f(x, y) = 4x^2 + y^2$

37. Identify the level curves and sketch some of them. $f(x, y) = \sqrt{\dfrac{x+y}{x-y}}$

38. Identify the level curves and sketch some of them. $f(x, y) = xy$

39. Identify the c-level surface for $f(x, y, z) = 2x - 3y + z$ and $c = 1$.

40. Identify the c-level surface for $f(x, y, z) = x^2 - y^2 + z^2$ and $c = 1$.

41. Identify the c-level surface for $f(x, y, z) = \dfrac{x^2}{9} + \dfrac{y^2}{4} + \dfrac{z^2}{2}$ and $c = 1$.

14.4 Partial Derivatives

42. Calculate the partial derivatives of $z = \dfrac{x}{y} \sin(xy^2)$.

43. Calculate the partial derivatives of $f(x, y) = x^y$.

44. Calculate the partial derivatives of $z = x^3 + xy - y\cos xy$.

45. Calculate the partial derivatives of $z = y^2 e^{-x} + y$.

46. Calculate the partial derivatives of $f(x, y) = \sqrt{16 - 4x^2 - 9y^2}$

47. Calculate the partial derivatives of $f(x, y) = \sin(x^2 y)$.

48. Calculate the partial derivatives of $f(x, y) = (1 + x^2 + y)^{5/3}$.

49. Find $f_x(1, 1)$ and $f_y(1, 1)$ given that $f(x, y) = y - e^{xy^2} + \sqrt{x^2 + 1}$.

50. Find $f_x(4, 2)$ and $f_y(4, 2)$ given that $f(x, y) = \ln(xy - 1) + e^y \sqrt{x}$.

51. Find $f_x(4, 2)$ and $f_y(4, 2)$ given that $f(x, y) = y\ln(x + y^2) + y^2 \sqrt{x}$.

52. Find $f_x(x, y)$ and $f_y(x, y)$ by forming the appropriate difference quotient and taking the limit as h tends to zero (Definition 14.4.1). $f(x, y) = xy^2$.

53. Find $f_x(x, y, z)$, $f_y(x, y, z)$, and $f_z(x, y, z)$ by forming the appropriate difference quotient and taking the limit as h tends to zero (Definition 14.4.2). $f(x, y, z) = x^2 yz$.

54. Let C be the curve of intersection of the surface $z = 3xy^2$ with the plane $y = 2$. Find an equation for the tangent line to C at the point $(1, 2, 12)$.

55. Let C be the curve of intersection of the surface $z = \sqrt{x^2 - y^2}$ with the plane $x = 3$. Find an equation for the tangent line to C at the point $(3, 1, 2\sqrt{2})$.

56. Use implicit differentiation to find $\dfrac{\partial z}{\partial x}$ and $\dfrac{\partial z}{\partial y}$ given that $x^2 z^2 - 2xyz + y^2 z^3 = 3$.

57. Use implicit differentiation to find $\dfrac{\partial z}{\partial x}$ at $(1, -2, 1)$ given that $x^3 z - 3xy^2 - (yz)^3 = -3$.

58. Use implicit differentiation to find $\dfrac{\partial z}{\partial x}$ and $\dfrac{\partial z}{\partial y}$ given that $x^2 + y^2 + z^2 - 2xyz = 5$.

59. Use implicit differentiation to find $\dfrac{\partial z}{\partial x}$ and $\dfrac{\partial z}{\partial y}$ given that $x^{1/3} - y^{1/3} + z^{1/3} = 16$.

60. Use implicit differentiation to find $\dfrac{\partial z}{\partial x}$ and $\dfrac{\partial z}{\partial y}$ given that $(x + y)^2 = (y - z)^3$.

61. Find $\dfrac{\partial z}{\partial x}$ and $\dfrac{\partial z}{\partial y}$ given that $x^3 z^2 - 2xyz^2 + z^3 y^2 = 2z$.

62. Evaluate $\dfrac{\partial z}{\partial x}$ and $\dfrac{\partial z}{\partial y}$ at $(3, 3, 2)$ given that $x^3 + y^3 + z^3 - 3xyz = 8$.

63. Evaluate $\dfrac{\partial z}{\partial x}$ and $\dfrac{\partial z}{\partial y}$ at $(1, 0, \pi/6)$ given that $x^2 \cos^2 z^3 - y^2 \sin z = \sin^2 2z$.

64. Verify that $x\dfrac{\partial z}{\partial x} + y\dfrac{\partial z}{\partial y} = z$ given that $z = x\sin\left(\dfrac{x}{y}\right) + ye^{y/x}$.

65. Verify that $x\dfrac{\partial z}{\partial x} + y\dfrac{\partial z}{\partial y} = 3z$ given that $z = x^3 + 2x^2 y - 3xy^2 + y^3$.

66. Verify that $4\dfrac{\partial z}{\partial x} - 3\dfrac{\partial z}{\partial y} = 0$ given that $z = (3x + 4y)^4$.

67. The area of a triangle is given by the formula $A = \frac{1}{2} bc \sin \theta$. At time t_0 we have $b_0 = 5$ cm, $c_0 = 10$ cm, and $\theta_0 = \pi/6$ radians.
   (a) Find the area of the triangle at time t_0.
   (b) Find the rate of change of the area with respect to b at time t_0 if c and θ remain constant.
   (c) Find the rate of change of the area with respect to θ at time t_0 if b and c remain constant.
   (d) Using the rate found in (c), calculate (by differentials) the approximate change in area if the angle is increased by one degree.
   (e) Find the rate of change of c with respect to b at time t_0 if the area and the angle are to remain constant.

14.5 Open Sets and Closed Sets

68. Specify the interior and the boundary of the set $\{(x, y) : 1 \le x \le 3, 2 \le y \le 4\}$. State whether the set is open, closed, or neither. Then sketch the set.

69. Specify the interior and the boundary of the set $\{(x, y) : 4 < x^2 + y^2 < 9\}$. State whether the set is open, closed, or neither. Then sketch the set.

70. Specify the interior and the boundary of the set $\{(x, y, z) : x^2 + y^2 \le 2, 0 \le z \le 2\}$. State whether the set is open, closed, or neither. Then sketch the set.

71. Specify the interior and the boundary of the set $\{(x, y, z) : x^2 + (y-1)^2 + z^2 < 1\}$. State whether the set is open, closed, or neither. Then sketch the set.

72. Specify the interior and the boundary of the set $\{y : 2 < y < 4\}$. State whether the set is open, closed, or neither.

73. Specify the interior and the boundary of the set $\{x : x \le 2\}$. State whether the set is open, closed, or neither.

74. Specify the interior and the boundary of the set $\{x : x > -1\}$. State whether the set is open, closed, or neither.

14.6 Limits and Continuity; Equality of Mixed Partials

75. Find the second partials of $f(x, y) = \sqrt{16 - 9x^2 - 4y^2}$.

76. Find the second partials of $f(x, y) = \cos(xy^2)$.

77. Find the second partials of $f(x, y) = (1 + x + y^2)^{4/3}$.

78. Find the second partials of $f(x, y, z) = 2 \tan^{-1} \dfrac{y}{x} + \ln(x^2 + z^2)$.

79. Find the second partials of $f(x, y, z) = xz^2 \cosh(\ln y)$.

80. Find the second partials of $f(x, y) = \dfrac{1}{x^2 + y^2 - 4}$.

81. Find the second partials of $f(x, y, z) = \ln(2x + 3y + 2z)$.

82. Find the second partials of $f(x, y) = (x^2 + xy)^{5/2}$.

83. Find the second partials of $f(x, y, z) = \sqrt{x^2 + y^2 + 2z}$.

84. Evaluate $\lim_{(x,y)\to(1,2)} (x^2 + 3y)$.

85. Evaluate $\lim_{(x,y)\to(1,1)} \dfrac{4+x-y}{3+x-3y}$.

86. Evaluate $\lim_{(x,y)\to(1,\pi/2)} x^3 \sin \dfrac{y}{x}$.

87. Evaluate $\lim_{(x,y)\to(0,0)} \dfrac{2x+y}{x^3+y^3}$.

88. Evaluate $\lim_{(x,y)\to(0,0)} \dfrac{\tan(x^2+y^2)}{x^2+y^2}$.

89. Evaluate $\lim_{(x,y)\to(1,1)} \dfrac{\sin^{-1}(xy-2)}{\tan^{-1}(3xy-4)}$.

90. Show that $\lim_{(x,y)\to(0,0)} \dfrac{x^2}{x^2+y^2}$ does not exist.

91. Show that the function $f(x,y) = e^{-2y} \cos 2x$ is harmonic.

92. For what value of $c > 0$ is $f(x,t) = \sin(2x+3t)$ a solution of the wave equation $\dfrac{\partial^2 f}{\partial t^2} - c^2 \dfrac{\partial^2 f}{\partial x^2} = 0$?

Answers to Chapter 14, Part B Questions

1. dom $(f) = \{(x, y, z) : x \neq 0\}$
    ran $(f) = (-\infty, \infty)$
    $f(1, 1, 1) = \pi/2 + \ln 2$
    $f(1, -1, 1) = -\pi/2 + \ln 2$

2. dom $(f) = \{(x, y, z) : x, y, z \in \Re\}$
    ran $(f) = (-\infty, \infty)$
    $f(\ln 2, 0, -1) = -2$

3. dom $(f) = \{(x, y, z) : x^2 + y^2 \neq 0\}$
    ran $(f) = (-\infty, \infty)$
    $f(1, -1, 2) = -4$
    $f(0, 1, 4) = 4$

4. dom $(f) = \{(x, y) : |x/y| \leq 1\}$
    ran $(f) = [-3\pi/4, 3\pi/4]$
    $f(1, 1) = 3\pi/4$

5. dom $(f) = \{(x, y) : x, y \in \Re\}$
    ran $(f) = (-\infty, \infty)$
    $f(2, 3) = 7/6$

6. dom $(f) = \{(x, y) : x, y \in \Re\}$
    ran $(f) = (-\infty, 1]$
    $f(2, 0) = 1$

7. dom $(f) = \{(x, y, z) : x^2 + y^2 + z^2 \leq 4\}$
    ran $(f) = [0, 2]$
    $f(1, 1, 1/2) = \dfrac{\sqrt{7}}{2}$

8. dom $(f) = \{(x, y) : -x \leq y \leq x, x > 0;$
    or $x \leq y \leq -x, x < 0\}$
    ran $(f) = [0, \infty)$
    $f(3, 1) = \sqrt{2}$

9. dom $(f) = \{(x, y) : x \neq 0\} = \{t : t \neq 0\}$
    ran $(f) = (-\infty, \infty)$

10. $f(x, y) = \dfrac{1}{12}\pi x^2 y$

11. $f(x, y, z) = 2xy + 2\dfrac{z}{y} + 2\dfrac{z}{x}$

12. ellipsoid

13. ellipsoid

14. elliptic paraboloid

15. hyperboloid of two sheets

16. hyperboloid of one sheet

17. hyperboloid of one sheet

18. hyperboloid of one sheet

19. hyperboloid of one sheet

20. $\dfrac{x^2}{4} + \dfrac{z^2}{9} = 1$

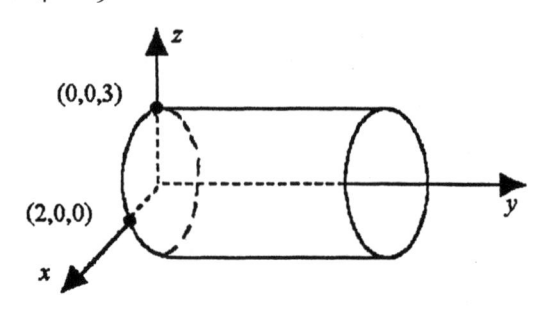

21. $z = \dfrac{x^2}{1/4} + y^2$; elliptic paraboloid

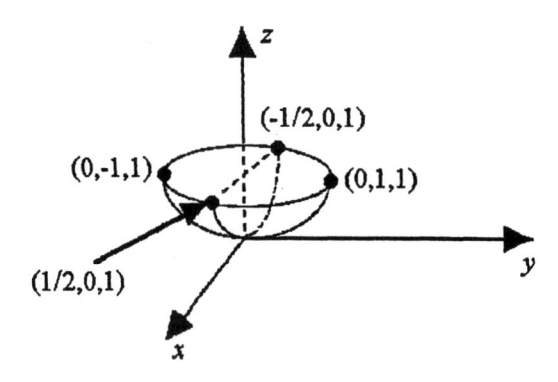

22. circular cone

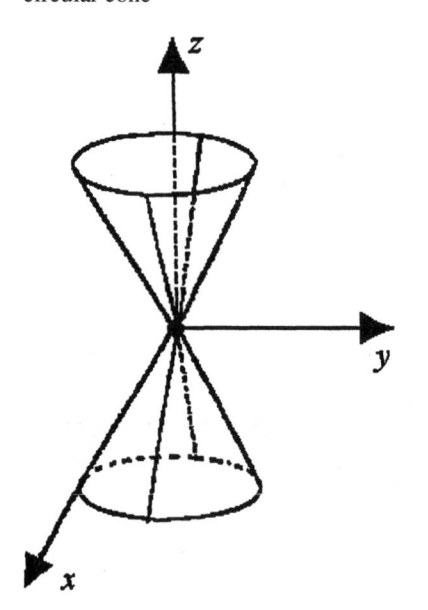

23. hyperboloid of one sheet

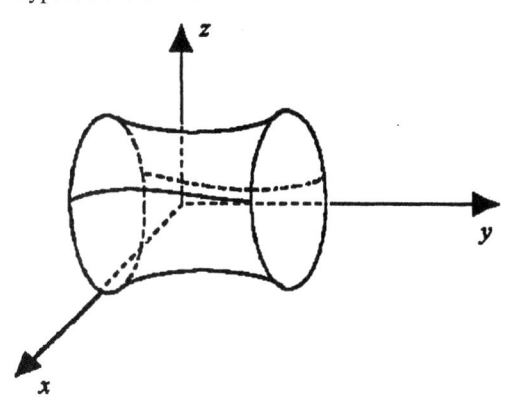

24. elliptic paraboloid
    traces: origin, $z = \dfrac{y^2}{4}$, $z = \dfrac{x^2}{2}$

25. circular cone
    traces: $z = \pm(y - 1)$, $x^2 + z^2 = 1$, $x = \pm(y - 1)$

26. circular paraboloid
    traces: $z = 5 - y^2$, $z = 5 - x^2$, $x^2 + y^2 = 5$

27. ellipsoid
    traces: $(y - 1)^2 + z^2/4 = 1$, origin,
    $x^2/4 + (y - 1)^2 = 1$

28. $4x^2 + 4z^2 = y$

29. $3x + 2y = 5$

30. $x = 0$

31. $x^2 = \dfrac{2}{3}\left(y - \dfrac{1}{2}\right)$

32. $\dfrac{x^2}{25/7} = \dfrac{(y + 5/7)^2}{200/49} = 1$; an ellipse

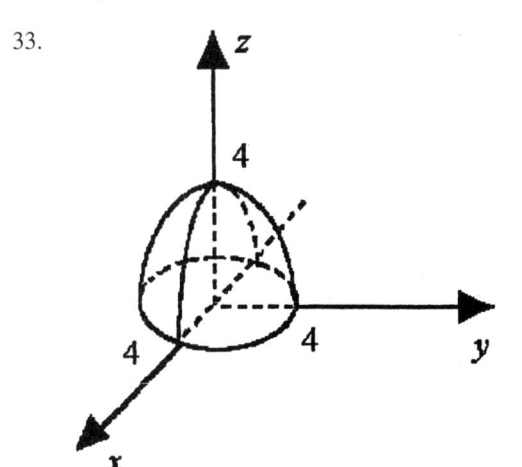

33.

34.

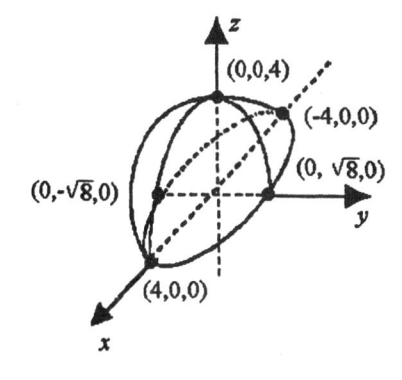

35. The level curves for $z \geq 0$ are concentric circles in the xy plane.

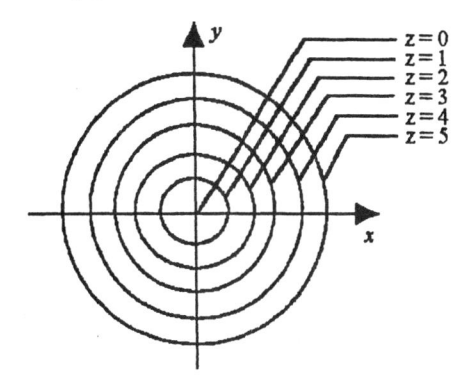

36. The level curves for $z \geq 0$ are concentric ellipses in the xy plane.

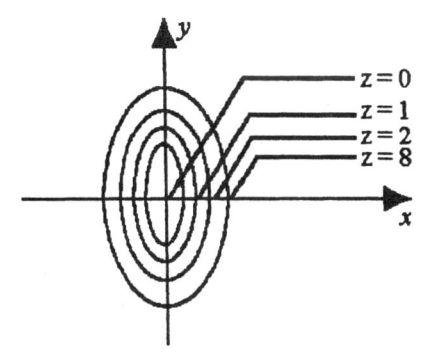

37. The level curves are straight lines through the origin.

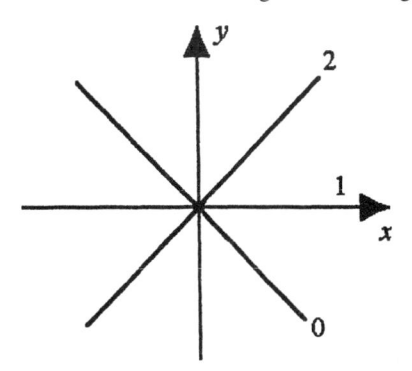

38. hyperbolas

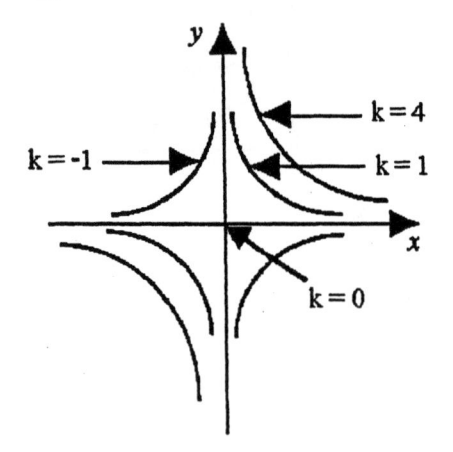

39. plane

40. hyperboloid of one sheet

41. ellipsoid

42. $\dfrac{\partial z}{\partial x} = z_x = xy\cos(xy^2) + \dfrac{1}{y}\sin(xy^2)$

$\dfrac{\partial z}{\partial y} = z_y = 2x^2\cos(xy^2) - \dfrac{x}{y^2}\sin(xy^2)$

43. $f_x = yx^{y-1}; f_y = x^y \ln x$

44. $z_x = 3x^2 + y + y^2 \sin xy$
$z_y = x + xy \sin xy - \cos xy$

45. $z_x = -y^2 e^{-x}$
$z_y = 2ye^{-x} + 1$

46. $f_x = \dfrac{-4x}{\sqrt{16 - 4x^2 - 9y^2}}$

$f_y = \dfrac{-9x}{\sqrt{16 - 4x^2 - 9y^2}}$

47. $f_x = 2xy \cos(x^2 y)$
$f_y = x^2 \cos(x^2 y)$

48. $f_x = \dfrac{10x}{3}(1 + x^2 + y^2)^{2/3}$

$f_x = \dfrac{5}{3}(1 + x^2 + y^2)^{2/3}$

49. $f_x(1, 1) = -e + \dfrac{\sqrt{2}}{2}$
$f_y(1, 1) = 1 - 2e$

50. $f_x(4, 2) = \dfrac{2}{7} + \dfrac{e^2}{4}; f_y(4, 2) = \dfrac{4}{7} + 2e^2$

51. $f_x(4, 2) = 5/4; f_y(4, 2) = \ln 8 + 9$

52. $f_x(x, y) = y^2; f_y(x, y) = 2xy$

53. $f_x(x, y, z) = 2xyz; f_y(x, y, z) = x^2 z$
$f_z(x, y, z) = x^2 y$

54. $\mathbf{r}(t) = t\,\mathbf{i} + 2\,\mathbf{j} + 12t\,\mathbf{k}$

55. $\mathbf{r}(t) = 3\,\mathbf{i} + t\,\mathbf{j} + \dfrac{\sqrt{2}}{4}(9 - t)\,\mathbf{k}$

56. $\dfrac{\partial z}{\partial x} = \dfrac{2yz - 2xz^2}{2x^2 z - 2xy + 3y^2 z^2}$

$\dfrac{\partial z}{\partial y} = \dfrac{2xz - 2yz^3}{2x^2 z - 2xy + 3y^2 z^2}$

57. $\dfrac{\partial z}{\partial x} = \dfrac{9}{25}$

58. $\dfrac{\partial z}{\partial x} = \dfrac{x - yz}{xy - z}, \dfrac{\partial z}{\partial y} = \dfrac{xz - y}{z - xy}$

59. $\dfrac{\partial z}{\partial x} = -\left(\dfrac{z}{x}\right)^{2/3}, \dfrac{\partial z}{\partial y} = -\left(\dfrac{z}{y}\right)^{2/3}$

60. $\dfrac{\partial z}{\partial x} = \dfrac{-2(x + y)}{3(y - z)^2}, \dfrac{\partial z}{\partial y} = \dfrac{1 - 2(x + y)}{3(y - z)^2}$

61. $\dfrac{\partial z}{\partial x} = \dfrac{3x^2 z^2 - 2yz^2}{2 - 2x^3 z + 4xyz - 3z^2 y^2}$

$\dfrac{\partial z}{\partial y} = \dfrac{3z^3 y - 2yz^2}{2 - 2x^3 z + 4xyz - 3z^2 y^2}$

62. $\dfrac{\partial z}{\partial x} = \dfrac{3}{5}, \dfrac{\partial z}{\partial y} = \dfrac{3}{5}$

63. $\dfrac{\partial z}{\partial x} = \dfrac{1}{\sqrt{3}}, \dfrac{\partial z}{\partial y} = 0$

64. $x\left(\sin\dfrac{x}{y} + \dfrac{x}{y}\cos\dfrac{x}{y} - \dfrac{y^2}{x^2}e^{y/x}\right)$

$+ y\left(\dfrac{-x^2}{y^2}\cos\dfrac{x}{y} + e^{y/x} + \dfrac{y}{x}e^{y/x}\right)$

$= x\sin\dfrac{x}{y} + ye^{y/x} = z$

65. $x(3x^2 + 4xy + 3y^2) + y(2x^2 + 6xy + 3y^2) = z$

66. $4[12(3x + 4y)^3] - 3[16(3x + 4y)^3] = 0$

67. (a) $25/2$ cm^2
    (b) $5/2$ cm^2/cm
    (c) $\dfrac{25\sqrt{3}}{2}$ cm^2/rad
    (d) $\dfrac{5\pi\sqrt{3}}{72}$ cm^2
    (e) $-c/b$

68. Interior: $\{(x, y) : 1 < x < 3, 2 < y < 4\}$
    Boundary: $\{(x, y) : 1 \le x \le 3, y = 2 \text{ or } y = 4\}$
    $\cup \{(x, y) : 2 < y < 4, x = 1 \text{ or } x = 3\}$
    closed set

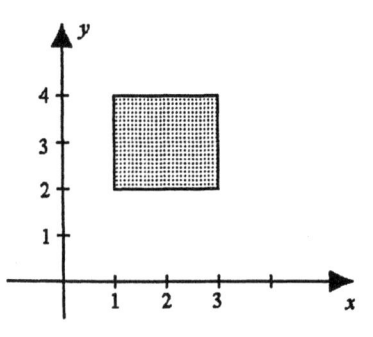

69. Interior: $\{(x, y) : 4 < x^2 + y^2 < 9\}$
    Boundary: $\{(x, y) : x^2 + y^2 = 4\}$
    $\cup \{(x, y) : x^2 + y^2 = 9\}$
    open set

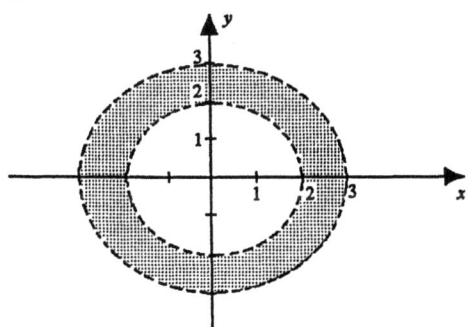

70. Interior: $\{(x, y, z) : x^2 + y^2 < 2, 0 < z < 2\}$
    Boundary: $\{(x, y, z) : x^2 + y^2 = 2, 0 < z < 2\}$
    $\cup \{(x, y, z) : x^2 + y^2 \le 2, z = 0 \text{ or } 2\}$
    closed set

71. Interior: $\{(x, y) : x^2 + (y-1)^2 + z^2 < 1\}$
    Boundary: $\{(x, y) : x^2 + (y-1)^2 + z^2 = 1\}$
    open set

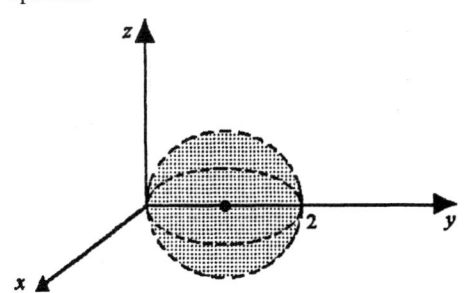

72. Interior: $\{y : 2 < y < 4\}$
    Boundary: $y = 2, y = 4$
    open set

73. Interior: $\{x : x < 2\}$
    Boundary: $x = 2$
    closed set

74. Interior: $\{x : x > -1\}$
    Boundary: $x = -1$
    open set

75. $f_{xx} = \dfrac{36y^2 - 144}{(16 - 9x^2 - 4y^2)^{3/2}}$
    $f_{yy} = \dfrac{36x^2 - 64}{(16 - 9x^2 - 4y^2)^{3/2}}$
    $f_{xy} = \dfrac{-36xy}{(16 - 9x^2 - 4y^2)^{3/2}}$

76. $f_{xx} = -y^4 \cos(xy^2)$
    $f_{yy} = -4x^2 y \cos(xy^2) - 2x \sin(xy^2)$
    $f_{xy} = f_{yx} = -2xy^3 \cos(xy^2) - 2y \sin(xy^2)$

77. $f_{xx} = \dfrac{4}{9}(1 + x + y^2)^{-2/3}$
    $f_{yy} = \dfrac{24 + 24x + 40y^2}{9(1 + x + y^2)^{2/3}}$
    $f_{xy} = f_{yx} = \dfrac{8y}{9(1 + x + y^2)^{2/3}}$

78. $f_{xx} = \dfrac{4xy}{(x^2 + y^2)^2} + \dfrac{2z^2 - 2x^2}{(x^2 + y^2)^2}$
    $f_{yy} = \dfrac{-4xy}{(x^2 + y^2)^2}, \; f_{zz} = \dfrac{2x^2 - 2z^2}{(x^2 + z^2)^2}$
    $f_{xy} = \dfrac{-2x^2 + 2y^2}{(x^2 + y^2)^2}, \; f_{xz} = \dfrac{-4xz}{(x^2 + z^2)^2}$
    $f_{yz} = 0$

79. $f_{xx} = 0, \; f_{xy} = \dfrac{z^2}{y} \sinh(\ln y)$
    $f_{yy} = \dfrac{xz^2}{y^2}[\cosh(\ln y) - \sinh(\ln y)]$
    $f_{xz} = 2z \cosh(\ln y), \; f_{zz} = 2x \cosh(\ln y)$
    $f_{yz} = \dfrac{2xz}{y}[\sinh(\ln y)]$

80. $f_{xx} = \dfrac{6x^2 - 2y^2 + 8}{(x^2 + y^2 - 4)^3}, \; f_{yy} = \dfrac{6y^2 - 2x^2 + 8}{(x^2 + y^2 - 4)^3}$
    $f_{xy} = \dfrac{8xy}{(x^2 + y^2 - 4)^3}$

81. $f_{xx} = \dfrac{-4}{(2x+3y+2z)^2}$, $f_{yy} = \dfrac{-9}{(2x+3y+2z)^2}$
$f_{zz} = \dfrac{-4}{(2x+3y+2z)^2}$, $f_{xy} = \dfrac{-6}{(2x+3y+2z)^2}$
$f_{xz} = \dfrac{-4}{(2x+3y+2z)^2}$, $f_{yz} = \dfrac{-6}{(2x+3y+2z)^2}$

82. $f_{xx} = \dfrac{15}{4}(x^2+xy)^{1/2}(2x+y)^2 + 5(x^2+xy)^{3/2}$
$f_{xy} = \dfrac{15x}{4}(x^2+xy)^{1/2}(2x+y) + \dfrac{5}{2}(x^2+xy)^{3/2}$
$f_{yy} = \dfrac{15x^2}{4}(x^2+xy)^{1/2}$

83. $f_{xx} = \dfrac{y^2+2z}{(x^2+y^2+2z)^{3/2}}$,
$f_{yy} = \dfrac{x^2+2z}{(x^2+y^2+2z)^{3/2}}$
$f_{xy} = \dfrac{-xy}{(x^2+y^2+2z)^{3/2}}$,
$f_{xz} = \dfrac{-x}{(x^2+y^2+2z)^{3/2}}$
$f_{yz} = \dfrac{-y}{(x^2+y^2+2z)^{3/2}}$

84. 7

85. 4

86. 1

87. does not exist

88. 1

89. 2

90. along $y=0$ $\lim\limits_{(x,y)\to(0,0)} \dfrac{x^2}{x^2} = 1$
along $y=x$ $\lim\limits_{(x,y)\to(0,0)} \dfrac{x^2}{2x^2} = \dfrac{1}{2}$
limit does not exist in either case

91. $\dfrac{\partial^2 f}{\partial x^2} = -4e^{-2y}\cos 2x = -\dfrac{\partial^2 f}{\partial y^2}$
so $\dfrac{\partial^2 f}{\partial y^2} + \dfrac{\partial^2 f}{\partial y^2} = 0$

92. $c = 3/2$

Part C

1. Let $f(x, y) = \cos \pi x \sin \pi y$ and $g(x, y, z) = \cos \pi x \sin \pi y$. Determine the domain and range of each of these functions and compare your results. (14.1)

2. Determine a function f of three variables whose value at (x, y, z) is (14.1)
   (a) The surface area of a box with no top whose sides have lengths x, y, and z.
   (b) The angle between the vectors $\mathbf{i} + \mathbf{j}$ and $x\mathbf{i} + y\mathbf{j} + z\mathbf{k}$.
   (c) The volume of the parallelepiped whose sides are the vectors $\mathbf{i}$, $\mathbf{i} + \mathbf{j}$, and $x\mathbf{i} + y\mathbf{j} + z\mathbf{k}$.

3. The hyperbola $c^2 y^2 - b^2 z^2 - b^2 c^2 = 0$ is revolved about the z-axis. Find an equation for the resulting surface. (14.2)

4. The given surfaces intersect in a space curve C. Determine the projection of C onto the xy-plane. The cylinder $y^2 + z - 4 = 0$ and the paraboloid $x^2 + 3y^2 = z$. (14.2)

5. Identify the c-level surfaces of $f(x, y, z) = 9x^2 - 4y^2 + 36z^2$ taking (i) $c < 0$, (ii) $c = 0$, (iii) $c > 0$. (14.3)

6. $f(x, y) = (x^2 + y) \ln [2 - x + e^y]$; $P(2, 1)$. (14.3)

7. The surface $z = \sqrt{4 - x^2 - y^2}$ is a hemisphere of radius 2 centered at the origin. (14.3)
   (a) Find equations for the tangent line l_1 to the curve of intersection of the hemisphere with the plane $x = 1$ at the point $(1, 1, \sqrt{2})$.
   (b) Find equations for the tangent line l_2 to the curve of intersection of the hemisphere with the plane $y = 1$ at the point $(1, 1, \sqrt{2})$.
   (c) The tangent lines l_1 and l_2 determine a plane. Find an equation for this plane. As you might expect, this plane is tangent to the surface at the point $(1, 1, \sqrt{2})$.

8. The law of cosines for a triangle can be written $a^2 = b^2 + c^2 - 2bc \cos \theta$.
   At time t_0 we have $b_0 = 10$ inches, $c_0 = 15$ inches, $\theta_0 = \frac{1}{3}\pi$ radians. (14.3)
   (a) Find a_0.
   (b) Find the rate of change of a with respect to b at time t_0 if c and θ remain constant.
   (c) Using the rate found in (b), calculate (by differentials) the approximate change in a if b is decreased by 1 inch.
   (d) Find the rate of change of a with respect to θ at time t_0 if b and c remain constant.
   (e) Find the rate of change of c with respect to θ at time t_0 if a and b remain constant.

9. Given that $x = r \cos \theta$ and $y = r \sin \theta$, find
$$\frac{\partial x}{\partial r}\frac{\partial y}{\partial \theta} - \frac{\partial x}{\partial \theta}\frac{\partial y}{\partial r}.$$ (14.4)

10. Let $\emptyset$ be the empty set. Let X be the real line, the entire plane, or, in the three-dimensional case, all of three-space. For each subset A of X, let $X - A$ be the set of all points $\mathbf{x} \in X$ such that $\mathbf{x} \notin A$. (14.5)
    (a) Show that $\emptyset$ is both open and closed.
    (b) Show that X is both open and closed. (It can be shown that $\emptyset$ and X are the only subsets of X that are both open and closed.)
    (c) Let U be a subset of X. Show that U is open iff $X - U$ is closed.
    (d) Let F be a subset of X. Show that F is closed iff $X - F$ is open.

11. Verify that
$$\frac{\partial^2 f}{\partial y \partial x} = \frac{\partial^2 f}{\partial x \partial y}$$ (14.6)

given that
(a) $f(x, y) = g(x) + h(y)$ with g and h differentiable.
(b) $f(x, y) = g(x)h(y)$ with g and h differentiable.
(c) $f(x, y)$ is a polynomial in x and y.
Hint: Check each term $x^m y^n$ separately.

12. Show that the following functions do not have a limit at (0, 0): (14.6)

(a) $f(x, y) = \dfrac{x^2 - y^2}{x^2 + y^2}$.

(b) $f(x, y) = \dfrac{y^2}{x^2 + y^2}$.

Answers to Chapter 14, Part C Questions

1. dom (f) = the entire plane, ran $(f) = [-1, 1]$
   dom (g) = all of space, ran $(g) = [-1, 1]$

2. (a) $f(x, y, z) = xy + 2xz + 2yz$
   (b) $f(x, y, z) = \cos^{-1} \dfrac{(\mathbf{i} + \mathbf{j}) \cdot (x\mathbf{i} + y\mathbf{j} + z\mathbf{k})}{||\mathbf{i} + \mathbf{j}||\,||x\mathbf{i} + y\mathbf{j} + z\mathbf{k}||}$
   $= \cos^{-1} \dfrac{x + y}{\sqrt{2}\sqrt{x^2 + y^2 + z^2}}$
   (c) $f(x, y, z) = [\mathbf{i} \times (\mathbf{i} + \mathbf{j})] \cdot (x\mathbf{i} + y\mathbf{j} + z\mathbf{k}) = z$

3. $c^2 x^2 + c^2 y^2 - b^2 z^2 = b^2 c^2$ (hyperboloid of revolution, one sheet)

4. $y^2 + (x^2 + 3y^2) = 4 \Longrightarrow x^2 + 4y^2 = 4$, an ellipse.

5. (i) hyperboloid of two sheets
   (ii) quadric cone
   (iii) hyperboloid of one sheet

6. $(x^2 + y)\ln(2 - x + e^y) = 5$

7. $f(x, y) = (4 - x^2 - y^2)^{1/2}$,
   $f_x(x, y) = -x(4 - x^2 - y^2)^{-1/2}$,
   $f_y(x, y) = -y(4 - x^2 - y^2)^{-1/2}$
   (a) $f_y(1, 1) = -\dfrac{\sqrt{2}}{2} \Longrightarrow x = 1$,
   $z - \sqrt{2} = -\dfrac{\sqrt{2}}{2}(y - 1)$
   (b) $f_x(1, 1) = -\dfrac{\sqrt{2}}{2} \Longrightarrow y = 1$,
   $z - \sqrt{2} = -\dfrac{\sqrt{2}}{2}(x - 1)$
   (c) l_1 and l_2 have direction vectors
   $\mathbf{j} - \dfrac{\sqrt{2}}{2}\mathbf{k}, \mathbf{i} - \dfrac{\sqrt{2}}{2}\mathbf{k}$, respectively.
   The normal to the plane is
   $\left(\mathbf{j} - \dfrac{\sqrt{2}}{2}\mathbf{k}\right) \times \left(\mathbf{i} - \dfrac{\sqrt{2}}{2}\mathbf{k}\right)$
   $= -\dfrac{\sqrt{2}}{2}\mathbf{i} - \dfrac{\sqrt{2}}{2}\mathbf{j} - \mathbf{k}$,
   so the tangent plane is
   $-\dfrac{\sqrt{2}}{2}(x - 1) - \dfrac{\sqrt{2}}{2}(y - 1) - (z - \sqrt{2}) = 0$,
   or $(x - 1) + (y - 1) + \sqrt{2}(z - \sqrt{2}) = 0$

8. (a) $a_0 = \left(b_0^2 + c_0^2 - 2 b_0 c_0 \cos \theta_0\right)^{1/2} = 5\sqrt{7}$
   (b) $\dfrac{\partial a}{\partial b} = (2b - 2c \cos \theta)\left(\dfrac{1}{2}\right)$
   $\times (b^2 + c^2 - 2bc \cos \theta)^{-1/2} = \dfrac{\sqrt{7}}{14}$
   (c) $a \cong a_0 + \dfrac{\sqrt{7}}{14}(b - b_0) = 5\sqrt{7} + \dfrac{\sqrt{7}}{14} \cdot (-1)$
   decreases by about $\dfrac{\sqrt{7}}{14}$ inches.
   (d) $\dfrac{\partial a}{\partial \theta} = 2bc \sin \theta \left(\dfrac{1}{2}\right)$
   $\times (b^2 + c^2 - 2bc \cos \theta)^{-1/2} = \dfrac{15}{7}\sqrt{21}$

9. $\dfrac{\partial x}{\partial r}\dfrac{\partial y}{\partial \theta} - \dfrac{\partial x}{\partial \theta}\dfrac{\partial y}{\partial r} = \cos\theta(r\cos\theta) - (-r\sin\theta)(\sin\theta)$
   $= r(\cos^2\theta + \sin^2\theta)$
   $= r$

10. (a) ϕ is open because it contains no boundary points,
    ϕ is closed because it contains its boundary (the boundary is empty).
    (b) X is open because it contains a neighborhood of each of its points,
    X is closed because it contains its boundary (the boundary is empty).
    (c) Suppose that U is open. Let x be a boundary point of $X - U$. Then every neighborhood of x contains points from $X - U$. The point x cannot be in U because U contains a neighborhood of each of its points. Thus $x \in X - U$. This shows that $X - U$ contains its boundary and is therefore closed.
    Suppose now that $X - U$ is closed. Let x be a point of U. If no neighborhood of x lies entirely in U, then every neighborhood of x contains points from $X - U$. This makes x a boundary point of $X - U$ and, since $X - U$ is closed, places x in $X - U$. This contradiction shows that some neighborhood of x lies entirely in U. Thus U contains a neighborhood of each of its points and is therefore open.
    (d) Set $U = X - F$ and note that $F = X - U$. By (c) $F = X - U$ is closed iff $X - F = U$ is open.

11. (a) mixed partials are 0
    (b) mixed partials are $g'(x)h'(y)$
    (c) by the Hint, mixed partials for each term $x^m y^n$ are $mn x^{m-1} y^{n-1}$

12. (a) as (x, y) tends to $(0, 0)$ along the x-axis, $f(x, y) = f(x, 0) = 1$ tends to 1;
    as (x, y) tends to $(0, 0)$ along the line $y = x$, $f(x, y) = f(x, x) = 0$ tends to 0;
    (b) as (x, y) tends to $(0, 0)$ along the x-axis, $f(x, y) = f(x, 0) = 0$ tends to 0;
    as (x, y) tends to $(0, 0)$ along the line $y = x$, $f(x, y) = f(x, x) = \dfrac{1}{2}$ tends to $\dfrac{1}{2}$.

CHAPTER 15 Gradients; Extreme Values; Differentials
Part B

15.1 Differentiability and Gradient

1. Find the gradient of $f(x, y) = x^2 e^{xy}$.

2. Find the gradient of $f(x, y) = 2x^3 + 5xy - 2y^2$.

3. Find the gradient of $f(x, y) = xy \sin xy^2$.

4. Find the gradient of $f(x, y, z) = 2x^2 yz$.

5. Find the gradient of $f(x, y, z) = \dfrac{1}{\sqrt{x+y+z}}$.

6. Find the gradient of $f(x, y, z) = x^3 y + yz^2 - xy^2 z$.

7. Find the gradient of $f(x, y, z) = e^y \ln xz$.

8. Find the gradient of $f(x, y) = (x^2 - y) \cos(x + y)$.

9. Find the gradient of $f(x, y, z) = x e^{x+y+z}$.

10. Find the gradient of $f(x, y) = \sqrt{x^2 + y^2}$.

11. Find the gradient vector at $(1, 2)$ of $f(x, y) = 3x^2 - 2xy + 4y^2$.

12. Find the gradient vector at $(1, 2)$ of $f(x, y) = \dfrac{2x + y}{x^2 + y^2}$.

13. Find the gradient vector at $(\tfrac{1}{2}, 1)$ of $f(x, y) = (x + y) \sin \pi x$.

14. Find the gradient vector at $(1, -1, 2)$ of $f(x, y, z) = e^x \sin(y + z^2)$.

15. Find the gradient vector at $(1, 1, 2)$ of $f(x, y, z) = \ln(x + y + z^2)$.

16. Calculate (a) $\nabla(\ln r^3)$; (b) $\nabla(\sin 2r)$; (c) $\nabla(e^{2r^2 + r})$ where $r = \sqrt{x^2 + y^2 + z^2}$

15.2 Gradients and Directional Derivatives

17. Find the directional derivative of $f(x, y) = e^x \sin y$ at $(0, \pi/3)$ in the direction of $5\mathbf{i} - 2\mathbf{j}$.

18. Find the directional derivative of $f(x, y) = \ln \sqrt[3]{x^2 + y^2}$ at $(3, 4)$ in the direction of $4\mathbf{i} + 3\mathbf{j}$.

19. Find the directional derivative of $f(x, y) = \dfrac{x^2}{16} + \dfrac{y^2}{9}$ at $(4, 3)$ in the direction of $\mathbf{i} + \mathbf{j}$.

20. Find the directional derivative of $f(x, y) = e^x \cos y$ at $(2, \pi)$ in the direction of $2\mathbf{i} + 3\mathbf{j}$.

21. Find the directional derivative of $f(x, y) = 3xy^2 - 4x^3 y$ at $(1, 2)$ in the direction of $3\mathbf{i} + 4\mathbf{j}$.

270   Calculus: One and Several Variables

22. Find the directional derivative of $f(x, y) = e^x \sin \pi y$ at $(0, 1/3)$ toward the point $(3, 7/3)$.

23. Find the directional derivative of $f(x, y) = x \tan^{-1} y/x$ at $(1, 1)$ in the direction of $\mathbf{a} = 2\mathbf{i} - \mathbf{j}$.

24. Find the rate of change of $f(x, y) = \dfrac{2x}{x - y}$ at $(1, 0)$ in the direction of a vector at an angle of $60°$ from the positive x-axis.

25. Find the rate of change of $f(x, y) = \dfrac{x + y}{2x - y}$ at $(1, 1)$ in the direction of a vector at an angle of $150°$ from the positive x-axis.

26. Find the rate of change of $f(x, y) = 2xy - \dfrac{y}{x}$ at $(1, 2)$ in the direction of a vector at an angle of $120°$ from the positive x-axis.

27. Find the directional derivative of $f(x, y, z) = x \sin(\pi yz) + yz \tan(\pi x)$ at $(1, 2, 3)$ in the direction of $2\mathbf{i} + 6\mathbf{j} - 9\mathbf{k}$.

28. Find the directional derivative of $f(x, y, z) = x^2 - 2y^2 + z^2$ at $(3, 3, 1)$ in the direction of $2\mathbf{i} + \mathbf{j} - \mathbf{k}$.

29. Find the directional derivative of $f(x, y, z) = x^2 y^3 + \sqrt{xz}$ at $(1, -2, 3)$ in the direction of $5\mathbf{j} + \mathbf{k}$.

30. Find the directional derivative of $f(x, y, z) = x^2 y + xy^2 + z^2$ at $(1, 1, 1)$ toward the point $(3, 1, 2)$.

31. The temperature, T, at a point (x, y) on a semicircular plate is given by $T(x, y) = 3x^2 y - y^3 + 273°$ Celsius.
    (a) Find the temperature at $(1, 2)$.
    (b) Find the rate of change of temperature at $(1, 2)$ in the direction of $\mathbf{i} - 2\mathbf{j}$.
    (c) Find a unit vector in the direction in which the temperature increases most rapidly at $(1, 2)$ and find this maximum rate of increase in temperature at $(1, 2)$.

32. The temperature, T, at a point (x, y) in the xy-plane is given by $T(x, y) = xy - x$. Find a unit vector in the direction in which the temperature increases most rapidly at $(1, 1)$ and find this maximum rate of increase in temperature at $(1, 1)$.

33. Find the unit vector in the direction in which $f(x, y, z) = 4e^{xy} \cos z$ decreases most rapidly at $(0, 1, \pi/4)$ and find the rate of decrease of f in that direction.

34. Find the unit vector in the direction in which $f(x, y, z) = \ln(1 + x^2 + y^2 - z^2)$ increases most rapidly at $(1, -1, 1)$ and find the rate of increase of f in that direction.

15.3   The Mean-Value Theorem; Chain Rules

35. Find the rate of change of $f(x, y) = xy^2$ with respect to t along the curve $\mathbf{r}(t) = e^t \mathbf{i} + t^2 \mathbf{j}$.

36. Find the rate of change of $f(x, y, z) = x^2 + y^2 + z^2$ with respect to t along the curve $\mathbf{r}(t) = t\mathbf{i} + t^2 \mathbf{j} - t^3 \mathbf{k}$.

37. Find the rate of change of $f(x, y, z) = xy + yz - xz$ with respect to t along the curve $\mathbf{r}(t) = 2 \cos \omega t \, \mathbf{i} + 3 \sin \omega t \, \mathbf{j} - 2\omega t \, \mathbf{k}$.

38. Find the rate of change of $f(x, y, z) = x^2 \cos(y + z)$ with respect to t along the curve $\mathbf{r}(t) = t\mathbf{i} - \sin t \, \mathbf{j} - t^2 \mathbf{k}$.

39. Use the chain rule to find $\dfrac{\partial u}{\partial t}$ if $u = \sqrt{x^2 + y^2}$; $x = e^t$, $y = \sin t$.

Gradients; Extreme Values; Differentials 271

40. Use the chain rule to find $\dfrac{\partial u}{\partial t}$ if $u = y^2 e^x$; $x = \cos t$, $y = t^3$.

41. Use the chain rule to find $\dfrac{\partial u}{\partial t}$ if $u = xy$ and $x = e^x \cos x$.

42. Use the chain rule to find $\dfrac{\partial u}{\partial t}$ if $u = xy$ and $x = y \sin y$.

43. Use the chain rule to find $\dfrac{\partial u}{\partial t}$ at $t = 1$ if $u = x^3 y^2$; $x = t^2 + 1$, $y = t^3 + 2$.

44. Use the chain rule to find $\dfrac{\partial \omega}{\partial t}$ at $\omega = \tan^{-1}(xyz)$ and $x = t^2$, $y = t^3$, $z = t^{-4}$.

45. Use the chain rule to find $\dfrac{\partial \omega}{\partial t}$ at $\omega = \sin xy + \ln xz + z$ and $x = e^t$, $y = t^2$, $z = 1$.

46. At $t = 0$, the position of a particle on a rectangular membrane is given by $P(x, y) = \sin \pi x/3 \sin \pi y/5$. Find the rate at which P changes if the particle moves from $(3/4, 15/4)$ in a direction of a vector at an angle of $30°$ from the positive x-axis.

47. Use the chain rule to find $\dfrac{\partial z}{\partial s}$ and $\dfrac{\partial z}{\partial t}$ if $z = x \sin y$, $x = se^t$, $y = se^{-t}$.

48. Use the chain rule to find $\dfrac{\partial z}{\partial u}$ and $\dfrac{\partial z}{\partial v}$ if $z = x^2 \tan y$, $x = u^2 + v^3$, $y = \ln(u^2 + v^2)$.

49. Use the chain rule to find $\dfrac{\partial z}{\partial r}$ and $\dfrac{\partial z}{\partial \theta}$ if $z = \dfrac{xy}{x^2 + y^2}$, $x = r \cos \theta$.

50. Use the chain rule to find $\dfrac{\partial z}{\partial u}$ and $\dfrac{\partial z}{\partial v}$ if $z = x \cos y + y \sin x$, $x = uv^2$, $y = u + v$.

51. Use the chain rule to find $\dfrac{\partial z}{\partial s}$ and $\dfrac{\partial z}{\partial t}$ if $z = x^2 + y^3$, $x = s + t$, $y = s - t$.

52. Use the chain rule to find $\dfrac{\partial z}{\partial u}$ and $\dfrac{\partial z}{\partial v}$ if $z = x^3 + xy + y^2$, $x = 2u + v$, $y = u - 2v$.

53. Verify that $z = f(x^3 - y^2)$ satisfies the equation $2y \dfrac{\partial z}{\partial x} + 3x^2 \dfrac{\partial z}{\partial y} = 0$.

54. Verify that $z = f(y/x)$ satisfies the equation $x \dfrac{\partial z}{\partial x} + y \dfrac{\partial z}{\partial y} = 0$.

55. Use the chain rule to find $\dfrac{\partial \omega}{\partial r}$ and $\dfrac{\partial \omega}{\partial s}$ if $\omega = \ln(x^2 + y^2 + 2z)$, $x = r + s$, $y = r - s$, $z = 2rs$.

56. Use the chain rule to find $\dfrac{\partial \omega}{\partial r}, \dfrac{\partial \omega}{\partial \theta}$, and $\dfrac{\partial \omega}{\partial z}$ if $\omega = xy + yz$, $x = r \cos \theta$, $y = r \sin \theta$, $z = z$.

57. Use the chain rule to find $\dfrac{\partial \omega}{\partial r}$ and $\dfrac{\partial \omega}{\partial s}$ if $\omega = \ln(x^2 + y^2 + z^2)$, $x = e^r \cos s$, $y = e^r \sin s$, $z = e^s$.

58. Use the chain rule to find $\dfrac{\partial \omega}{\partial t}$ if $\omega = x^2 + y^2 + z^2$, $x = e^t \cos t$, $y = e^t \sin t$, $z = e^t$.

59. Use the chain rule to find $\dfrac{\partial \omega}{\partial t}, \dfrac{\partial \omega}{\partial u}$, and $\dfrac{\partial \omega}{\partial v}$ if $\omega = 2x + y - z$, $x = t^2 + u^2$, $y = u^2 + v^2$, $z = v^2 + t^2$.

272 Calculus: One and Several Variables

60. Use the chain rule to find $\dfrac{\partial \omega}{\partial u}$ and $\dfrac{\partial \omega}{\partial v}$ if $\omega = 2x - 3y + z$, $x = u \sin v$, $y = v \sin u$, $z = \sin u \sin v$.

61. Use the chain rule to find $\dfrac{\partial \omega}{\partial r}$ and $\dfrac{\partial \omega}{\partial s}$ if $\omega = \sqrt{x^2 + y^2 + z^2}$, $x = r \cos s$, $y = r \sin s$, $z = r \tan s$.

62. Use the chain rule to find $\dfrac{\partial \omega}{\partial r}$ and $\dfrac{\partial \omega}{\partial s}$ if $\omega = x^2 + y^2 + z^2$, $x = r \cos s$, $y = r \sin s$, $z = rs$.

63. Evaluate $\dfrac{\partial z}{\partial x}$ and $\dfrac{\partial z}{\partial y}$ at $(½, -1, 2)$ if $e^{xz} + \ln(yz + 3) = y + 1 + e$ and z is a differentiable function of x and y.

64. Use the chain rule to find $\dfrac{\partial z}{\partial u}$ and $\dfrac{\partial z}{\partial v}$ if $z = x^2 y^3 + x \sin y$, $x = u^2$, $y = uv$.

65. Show that $u = \dfrac{1}{\sqrt{x^2 + y^2 + z^2}}$ satisfies $\dfrac{\partial^2 z}{\partial x^2} + \dfrac{\partial^2 u}{\partial y^2} + \dfrac{\partial^2 u}{\partial z^2} = 0$.

66. Let $z = f(x, y)$ with $x = r \cos \theta$ and $y = r \sin \theta$. Show that $\left(\dfrac{\partial z}{\partial r}\right)^2 + \dfrac{1}{r^2}\left(\dfrac{\partial z}{\partial \theta}\right)^2 = \left(\dfrac{\partial z}{\partial x}\right)^2 + \left(\dfrac{\partial z}{\partial y}\right)^2$.

15.4 The Gradient as a Normal; Tangent Lines and Tangent Planes

67. Write an equation for the plane tangent to the surface $4x^2 + 9y^2 + z = 17$ at the point $(-1, 1, 4)$.

68. Write an equation for the plane tangent to the surface $z = e^x \sin \pi y$ at the point $(2, 1, 0)$.

69. Write an equation for the plane tangent to the surface $z = e^{-x} y^2 + y$ at the point $(0, 2, 6)$.

70. Write an equation for the plane tangent to the surface $z = x^2 + y^2$ at the point $(2, -1, 5)$.

71. Write an equation for the plane tangent to the surface $z = xe^{\sin y}$ at the point $(2, \pi, 2)$.

72. Write an equation for the plane tangent to the surface $z = 3x^2 + 2y^2$ at the point $(2, -1, 4)$.

73. Find a normal vector at the point indicated. Write an equation for the normal line and an equation for the plane tangent to the surface $z = \dfrac{y^2}{3} - x$ at the point $(0, 0, 0)$.

74. Find a normal vector at the point indicated. Write an equation for the normal line and an equation for the plane tangent to the surface $zx^2 - xy^2 - yz - 18 = 0$ at the point $(0, -2, 3)$.

75. Find a normal vector at the point indicated. Write an equation for the normal line and an equation for the plane tangent to the surface $xyz + x + y + z - 3 = 0$ at the point $(1, -2, 2)$.

76. Find a normal vector at the point indicated. Write an equation for the normal line and an equation for the plane tangent to the surface $\dfrac{x^2}{4} + \dfrac{y^2}{9} + \dfrac{z^2}{36} = 1$ at the point $(2, 3, 6)$.

77. The surfaces $2x^2 - 2y^2 + z^3 = 4$ and $3x^2 - 2y^2 + 3z = 1$ intersect in a curve that passes through the point $(1, 2, 2)$. What are the equations of the respective tangent planes to the two surfaces at this point?

78. Find a point on the surface $z = 16 - 4x^2 - y^2$ at which the tangent plane is perpendicular to the line $x = 3 + 4t$, $y = 2t$, $z = 2 - t$.

Gradients; Extreme Values; Differentials   273

79. Find a point on the surface $z = 9 - x^2 - y^2$ at which the tangent plane is parallel to the plane $2x + 3y + 2z = 6$.

15.5 Local Extreme Values and 15.6 Absolute Extreme Values

80. Find the stationary points and determine the local extreme values for the function $f(x, y) = 2x^2 - 3y + y^2$.

81. Find the stationary points and determine the local extreme values for the function $f(x, y) = x^2 - 2x - y^3$.

82. Find the stationary points and determine the local extreme values for the function $f(x, y) = x^2 + xy + y^2 - 2x - 2y + 6$.

83. Find the stationary points and determine the local extreme values for the function $f(x, y) = x^2 - xy + y^2 - 2x - 2$.

84. Find the stationary points and determine the local extreme values for the function $f(x, y) = x^3 + 3x - 2y$.

85. Find the stationary points and determine the local extreme values for the function $f(x, y) = 3xy - 5x^2 - y^2 + 5x - 2y$.

86. Find the stationary points and determine the local extreme values for the function $f(x, y) = x^2 - xy + y^2 - 5x + 2$.

87. Find the stationary points and determine the local extreme values for the function $f(x, y) = 2x^2 + y^2 + 4y$.

88. Find the stationary points and determine the local extreme values for the function $f(x, y) = x^2 - xy + y^2 + 5x + 1$.

89. Find the stationary points and determine the local extreme values for the function $f(x, y) = y \sin 2x$, $-\pi/2 < x < \pi/2$.

90. Find the absolute extreme values taken on by $f(x, y) = \sqrt{x^2 + y^2}$ on the set $D = \{(x, y): 2 \le x \le 4, 1 \le y \le 4\}$.

91. Find the absolute extreme values taken on by $f(x, y) = 2x^2 - 3y^2$ on the set $D = \{(x, y): -2 \le x \le 2\}$.

92. Find the absolute extreme values taken on by $f(x, y) = (x - 2)^2 + (y - 1)^2$ on the set $D = \{(x, y): x^2 + y^2 \le 1\}$.

93. Find the absolute extreme values taken on by $f(x, y) = (x - 2)^2 + y^2$ on the set $D = \{(x, y): 0 \le x \le 1, x^2 \le y \le 4x\}$.

94. Find the absolute extreme values taken on by $f(x, y) = (x - y)^2$ on the set $D = \{(x, y): 0 \le x \le 2, 0 \le y \le 4 - x\}$.

95. Find the distance from the point $(1, 1, 1)$ to the sphere $x^2 + y^2 + z^2 = 4$.

96. Find the distance from the point $(\tfrac{1}{2}, 1, \tfrac{1}{2})$ to the sphere $(x - 1)^2 + y^2 + z^2 = 2$.

97. Find the stationary points and the local extreme values for $f(x, y) = 5xy - 7x^2 - y^2 + 3x - 6y + 2$.

98. Find the stationary points and the local extreme values for $f(x, y) = x^3 + y^2 - 12x - 6y + 7$.

99. Find the stationary points and the local extreme values for $f(x, y) = x^2 + 3xy + 3y^2 - 6x + 3y - 6$.

100. Find the stationary points and the local extreme values for $f(x, y) = x^2 - xy + y^2 + 2x + 2y - 4$.

101. Find the stationary points and the local extreme values for $f(x, y) = x^2 - 2y^2 - 6x + 8y + 3$.

102. Find the stationary points and the local extreme values for $f(x, y) = x^2 + 3xy + y^2 - 10x - 10y$.

103. Find the stationary points and the local extreme values for $f(x, y) = 2x^2 + y^2 - 4x - 6y$.

104. Find the stationary points and the local extreme values for $f(x, y) = x^3 - 9xy + y^3$.

105. Find the stationary points and the local extreme values for $f(x, y) = x^2 + \frac{1}{3}y^3 - 2xy - 3y$.

106. A rectangular box, open at the top, is to contain 256 cubic inches. Find the dimensions of the box for which the surface area is a minimum.

107. Find the point on the plane $2x - 3y + z = 19$ that is closest to $(1, 1, 0)$.

108. Find the shortest distance to $2x + y - z = 5$ from $(1, 1, 1)$.

109. An open rectangular box containing 18 cubic inches is to be constructed so that the base material costs 3 cents per square inch, the front face costs 2 cents per square inch, and the sides and back each cost 1 cent per square inch. Find the dimensions of the box for which the cost of construction will be a minimum.

110. Find the points on $z^2 = x^2 + y^2$ that are closest to $(2, 2, 0)$.

111. Find the point on $x + 2y + z = 1$ that is closest to the origin.

112. Find the maximum product of x, y, and z where x, y, and z are positive numbers such that $4x + 3y + z = 108$.

113. Find the maximum sum of $9x + 5y + 3z$ if x, y, and z are positive numbers such that $xyz = 25$.

114. Find the maximum product of x^2yz if x, y, and z are positive numbers such that $3x + 2y + z = 24$.

15.7 Maxima and Minima with Side Conditions

115. Maximize xy on the ellipse $\frac{x^2}{4} + \frac{y^2}{9} = 1$.

116. Minimize xy on the ellipse $\frac{x^2}{4} + \frac{y^2}{9} = 1$.

117. Maximize x^2y on the circle $x^2 + y^2 = 4$.

118. Minimize x^2y on the circle $x^2 + y^2 = 4$.

119. Maximize xyz^2 on the sphere $x^2 + y^2 + z^2 = 4$.

120. Use the Lagrange multiplier method to find the point on the surface $z = xy + 1$ that is closest to the origin.

121. Use the Lagrange multiplier method to find the point on the plane $x + 2y + z = 1$ that is closest to the point $(1, 1, 0)$.

122. Use the Lagrange multiplier method to find three positive numbers whose sum is 12 for which x^2yz is a maximum.

123. Use the Lagrange multiplier method to find the maximum for $x^2 + y^2 + z^2$ if $x + 2y + 2z = 12$.

124. An open rectangular box is to contain 256 cubic inches. Use the Lagrange multiplier method to find the dimensions of the box that uses the least amount of material.

125. An open rectangular box containing 18 cubic inches is constructed of material costing 3 cents per square inch for the base, 2 cents per square inch for the front face, and 1 cent per square inch for the sides and back. Use the Lagrange multiplier method to find the dimensions of the box for which the cost of construction is a minimum.

126. Use the Lagrange multiplier method to find the maximum possible volume for a rectangular box inscribed in the ellipsoid $2x^2 + 3y^2 + 4z^2 = 12$.

127. Use the Lagrange multiplier method to find the maximum possible volume for a rectangular box inscribed in the ellipsoid $2x^2 + 3y^2 + 6z^2 = 18$.

128. Use the Lagrange multiplier method to find the point on the plane $2x - 3y + z = 19$ that is closest to $(1, 1, 0)$.

129. Use the Lagrange multiplier method to find the shortest distance from $(1, 1, 1)$ to $2x + y - z = 5$.

130. Use the Lagrange multiplier method to find the points on $z^2 = x^2 + y^2$ that are closest to $(2, 2, 0)$.

131. Use the Lagrange multiplier method to find three positive numbers whose sum is 36 and whose product is as large as possible.

132. Use the Lagrange multiplier method to find three positive numbers whose product is 64 and whose sum is as small as possible.

133. Use the Lagrange multiplier method to find three positive numbers x, y, and z whose product is as large as possible given that $2x + 2y + z = 84$.

134. The base of a rectangular box costs three times as much per square foot as do the sides and top. Use the Lagrange multiplier method to find the dimensions of the box with least cost if the box is to contain 54 cubic feet.

15.8 Differentials

135. Find the differential df given that $f(x, y, z) = e^{2z} \sqrt[3]{x^2 + y^2}$.

136. Find the differential df given that $f(x, y, z) = x^2 + 3xy - 2y^2 + 3xz + z^2$.

137. Find the differential df given that $f(x, y, z) = z^4 - 3yz^2 + y \sin z$.

138. Find the differential df given that $f(x, y, z) = 3x^2 + y^2 + z^2 - 3xy + 4xz - 15$.

139. Find the differential df given that $f(x, y, z) = x \sin^{-1} y + x^2 y$.

140. Compute Δu and du for $u = 2x^2 + 3xy - y^2$ at $x = 2$, $y = -2$, $\Delta x = -0.2$, $\Delta y = 0.1$.

141. The radius and height of a right-circular cylinder are measured with errors of at most 0.1 inches. If the height and radius are measured to be 10 inches and 2 inches, respectively, use differentials to approximate the maximum possible error in the calculated value of the volume.

142. The power consumed in an electrical resistor is given by $P = E^2/R$ watts. Suppose $E = 200$ volts and $R = 8$ ohms, approximate the change in power if E is decreased by 5 volts and R is decreased by 0.20 ohm.

143. Let $f(x, y) = \sqrt{x + 2y}$. Use a total differential to approximate the change in $f(x, y)$ as (x, y) varies from $(3, 5)$ to $(2.98, 5.1)$.

144. The legs of a right triangle are measured to be 6 and 8 inches with a maximum error of 0.10 inches in each measurement. Use differentials to estimate the maximum possible error in the calculated value of the hypotenuse and the area of the triangle.

145. Find the differential df given that $f(x, y) = \ln \sqrt[3]{1+xy}$.

15.9 Reconstructing a Function from Its Gradient

146. Determine whether the vector function $x^2 y^3 \mathbf{i} + x^3 y^2 \mathbf{j}$ is the gradient $\nabla f(x, y)$ of a function everywhere defined. If so, find such a function.

147. Determine whether the vector function $e^{2x} \mathbf{i} - e^{-2y} \mathbf{j}$ is the gradient $\nabla f(x, y)$ of a function everywhere defined. If so, find such a function.

148. Determine whether the vector function $\dfrac{x}{y^2} \mathbf{i} - \left(\dfrac{x^2}{y^3} + y^2\right) \mathbf{j}$ is the gradient $\nabla f(x, y)$ of a function everywhere defined. If so, find such a function.

149. Determine whether the vector function $\sin(x+y-2z)\mathbf{i} - \sin(x+y-2z)\mathbf{j} + 2\cos(x+y-2z)\mathbf{k}$ is the gradient $\nabla f(x, y, z)$ of a function everywhere defined. If so, find such a function.

150. Determine whether the vector function $(3x^2 y e^{z^2} - z)\mathbf{i} - (x^3 e^{z^2} + 1)\mathbf{j} + (2x^3 y z e^{z^2} - x)\mathbf{k}$ is the gradient $\nabla f(x, y, z)$ of a function everywhere defined. If so, find such a function.

151. Determine whether the vector function $\left(\dfrac{2x}{x^2+y^2+1} + \dfrac{yz}{1+x^2 y^2 z^2}\right)\mathbf{i} + \left(\dfrac{2y}{x^2+y^2+1} + \dfrac{xz}{1+x^2 y^2 z^2}\right)\mathbf{j} + \dfrac{xy}{1+x^2 y^2 z^2}\mathbf{k}$ is the gradient $\nabla f(x, y, z)$ of a function everywhere defined. If so, find such a function.

Answers to Chapter 15, Part B Questions

1. $(2x + x^2y)e^{xy}\mathbf{i} + x^3e^{xy}\mathbf{j}$

2. $(6x^2 + 5y)\mathbf{i} + (5x - 4y)\mathbf{j}$

3. $[y\sin(xy^2) + xy^3\cos(xy^2)]\mathbf{i} + [x\sin(xy^2) + 2x^2y^2\cos(xy^2)]\mathbf{j}$

4. $4xyz\mathbf{i} + 2x^2z\mathbf{j} + 2x^2y\mathbf{k}$

5. $\dfrac{-1}{2(x+y+z)^{3/2}}(\mathbf{i}+\mathbf{j}+\mathbf{k})$

6. $(3x^2y - y^2z)\mathbf{i} + (x^3 + z^2 - 2xyz)\mathbf{j} + (2yz - xy^2)\mathbf{k}$

7. $\dfrac{e^y}{x}\mathbf{i} + e^y \ln xz\, \mathbf{j} + \dfrac{e^y}{z}\mathbf{k}$

8. $[2x\cos(x+y) - (x^2-y)\sin(x+y)]\mathbf{i} - [\cos(x+y) - (x^2-y)\sin(x+y)]\mathbf{j}$

9. $xe^{x+y+z}(\mathbf{i}+\mathbf{j}+\mathbf{k}) + e^{x+y+z}\mathbf{i}$

10. $\dfrac{x}{\sqrt{x^2+y^2}}\mathbf{i} + \dfrac{y}{\sqrt{x^2+y^2}}\mathbf{j}$

11. $2\mathbf{i} + 14\mathbf{j}$

12. $\dfrac{2}{25}\mathbf{i} - \dfrac{11}{25}\mathbf{j}$

13. $\mathbf{i}+\mathbf{j}$

14. $e\sin 3\,\mathbf{i} + e\cos 3\,\mathbf{j} + 4e\cos 3\,\mathbf{k}$

15. $\dfrac{1}{6}\mathbf{i} + \dfrac{1}{6}\mathbf{j} + \dfrac{2}{3}\mathbf{k}$

16. (a) $\dfrac{3}{r^2}(x\mathbf{i} + y\mathbf{j} + z\mathbf{k})$
    (b) $\dfrac{2}{r}\cos 2r\,(x\mathbf{i} + y\mathbf{j} + z\mathbf{k})$
    (c) $e^{2r^2+r}\left(4 + \dfrac{1}{r}\right)(x\mathbf{i} + y\mathbf{j} + z\mathbf{k})$

17. $\dfrac{5\sqrt{3}-2}{2\sqrt{29}}$

18. $\dfrac{16}{125}$

19. $\dfrac{7}{6\sqrt{2}}$

20. $\dfrac{-2e^2}{\sqrt{13}}$

21. $-4/5$

22. $\dfrac{3\sqrt{3}+2\pi}{2\sqrt{13}}$

23. $\dfrac{\pi}{2\sqrt{5}} - \dfrac{3}{2\sqrt{5}}$

24. $\sqrt{3}$

25. $\dfrac{3 + 3\sqrt{3}}{2}$

26. $\dfrac{\sqrt{3}-6}{2}$

27. $\dfrac{12\pi}{11}$

28. $\dfrac{-\sqrt{6}}{3}$

29. $\dfrac{60}{\sqrt{26}} + \dfrac{1}{2\sqrt{78}}$

30. $\dfrac{8\sqrt{5}}{5}$

31. (a) $271\,°C$
    (b) $6\sqrt{5}$
    (c) $\mathbf{u} = \dfrac{4}{5}\mathbf{i} - \dfrac{3}{5}\mathbf{j};\ 5$

32. $\mathbf{u} = \mathbf{j};\ 1$

33. $\mathbf{u} = \dfrac{-\sqrt{2}}{2}\mathbf{i} - \dfrac{\sqrt{2}}{2}\mathbf{k};\ 4$

34. $\mathbf{u} = \dfrac{1}{\sqrt{3}}(\mathbf{i} - \mathbf{j} - \mathbf{k});\ \sqrt{3}$

35. $e^t(t^4 + 4t^3)$

36. $2t + 4t^3 + 6t^5$

37. $6\omega\cos 2\omega t - 4\omega^2 t - 6\omega^2 t\cos\omega t - 6\omega\sin\omega t - 4\omega\cos\omega t$

38. $2t\cos(t^2 - \sin t) + (t^2\cos t - 2t^3)\sin(t^2 - \sin t)$

278 Calculus: One and Several Variables

39. $\dfrac{e^{2t} + \cos t \sin t}{\sqrt{e^{2t} + \sin^2 t}}$

40. $6t^5 e^{\cos t} - t^6 e^{\cos t} \sin t$

41. $e^x \cos x + xe^x \cos x - xe^x \sin x$

42. $y^2 \cos y + 2y \sin y$

43. 360

44. $\dfrac{1}{1 + t^2}$

45. $t^2 e^t \cos t^2 e t + 2te^t \cos t^2 e^t + 1$

46. $\dfrac{(5\sqrt{3} - 3)\pi}{60}$

47. $\dfrac{\partial z}{\partial s} = e^t \sin(se^{-t}) + s \cos(se^{-t})$

$\dfrac{\partial z}{\partial t} = se^t \sin(se^{-t}) + s^2 \cos(se^t)$

48. $\dfrac{\partial z}{\partial u} = \dfrac{4u(u^2 + v^3)\tan \ln(u^2 + v^2)}{u^2 + v^2}$
$+ \dfrac{2u(u^2 + v^3)\sec^2 \ln(u^2 + v^2)}{u^2 + v^2}$

$\dfrac{\partial z}{\partial v} = \dfrac{6v^2(u^2 + v^3)\tan \ln(u^2 + v^2)}{u^2 + v^2}$
$+ \dfrac{2v(u^2 + v^3)\sec^2 \ln(u^2 + v^2)}{u^2 + v^2}$

49. $\dfrac{\partial z}{\partial r} = 0; \ \dfrac{\partial z}{\partial \theta} = \cos 2\theta$

50. $\dfrac{\partial z}{\partial v} = 2uv \cos(u + v) + 2uv(u + v)\cos uv^2$
$- uv^2 \sin(u + v) + \sin uv^2$

$\dfrac{\partial z}{\partial u} = v^2 \cos(u + v) + v^2(u + v)\cos uv^2$
$- uv^2 \sin(u + v) + \sin uv^2$

51. $\dfrac{\partial z}{\partial s} = 2(s + t) + 3(s - t)^2;$

$\dfrac{\partial z}{\partial t} = 2(s + t) - 3(s - t)^2$

52. $\dfrac{\partial z}{\partial u} = 6(2u + v)^2 + 6u - 7v$

$\dfrac{\partial z}{\partial v} = 3(2u + v)^2 - 7u + 4v$

53. $2y \cdot 3x^2 f'(x^3 - y^2) + 3x^2(-2y) f'(x^3 - y^2) = 0$

54. $x \dfrac{\partial z}{\partial x} + y \dfrac{\partial z}{\partial y} = \dfrac{-y}{x} f'\left(\dfrac{y}{x}\right) + \dfrac{y}{x} f'\left(\dfrac{y}{x}\right) = 0$

55. $\dfrac{\partial \omega}{\partial r} = \dfrac{2}{r + s}; \ \dfrac{\partial \omega}{\partial s} = \dfrac{2}{r + s}$

56. $\dfrac{\partial \omega}{\partial r} = r \sin 2\theta + z \sin \theta$

$\dfrac{\partial \omega}{\partial \theta} = r^2 \cos 2\theta + rz \cos \theta$

$\dfrac{\partial \omega}{\partial z} = r \sin \theta$

57. $\dfrac{\partial \omega}{\partial r} = \dfrac{2e^{2r}}{e^{2r} + e^{2r}}; \ \dfrac{\partial \omega}{\partial s} = \dfrac{2e^{2s}}{e^{2r} + e^{2s}}$

58. $4e^{2t}$

59. $\dfrac{\partial \omega}{\partial t} = 2t, \ \dfrac{\partial \omega}{\partial u} = 6u, \ \dfrac{\partial \omega}{\partial v} = 0$

60. $\dfrac{\partial \omega}{\partial r} = 2u \cos v - 3 \sin u \sin u \cos v$

$\dfrac{\partial \omega}{\partial u} = 2 \sin v - 3v \cos u + \cos u \sin v$

61. $\dfrac{\partial \omega}{\partial r} = \sec s; \ \dfrac{\partial \omega}{\partial s} = \tan s \sec s$

62. $\dfrac{\partial \omega}{\partial r} = 2r + 2rs^2; \ \dfrac{\partial \omega}{\partial s} = 2r^2 s$

63. $\dfrac{\partial z}{\partial x} = \dfrac{4e}{2 - e}; \ \dfrac{\partial z}{\partial y} = \dfrac{2e}{2 - e}$

64. $\dfrac{\partial z}{\partial u} = 7u^6 v^3 + 2u \sin uv + u^2 v \cos uv$

$\dfrac{\partial z}{\partial v} = 3u^7 v^2 + u^3 \cos uv$

65. $\dfrac{\partial^2 u}{\partial x^2} + \dfrac{\partial^2 u}{\partial y^2} + \dfrac{\partial^2 u}{\partial z^2}$
$= \dfrac{2x^2 - 2x^2 + 2y^2 - 2y^2 + 2z^2 - 2z^2}{(x^2 + y^2 + z^2)^{3/2}} = 0$

66. $f_x^2(\cos^2 \theta + \sin^2 \theta) + f_y^2(\cos^2 \theta + \sin^2 \theta)$
$= \left(\dfrac{\partial z}{\partial x}\right)^2 + \left(\dfrac{\partial z}{\partial y}\right)^2$

67. $8x - 18y - z + 30 = 0$

68. $\pi e^2 y + z - \pi e^2 = 0$

69. $4x - 5y + z + 4 = 0$

70. $4x - 2y - z - 5 = 0$

71. $x - 2y - z + 2\pi = 0$

Answers

72. $12x - 4y - z - 14 = 0$

73. normal vector: $\mathbf{i} + \mathbf{k}$
    normal line: $x = t, z = t$
    tangent plane: $x + z = 0$

74. normal vector: $4\mathbf{i} + 3\mathbf{j} - 2\mathbf{k}$
    normal line: $x = 4t, y = -2 + 3t, z = 3 - 2t$
    tangent plane: $4x + 3y - 2z + 12 = 0$

75. normal vector: $3\mathbf{i} - 3\mathbf{j} + \mathbf{k}$
    normal line: $x = 1 + 3t, y = -2 - 3t, z = 2 + t$
    tangent plane: $3x - 3y + z - 11 = 0$

76. normal vector: $3\mathbf{i} + 2\mathbf{j} - \mathbf{k}$
    normal line: $x = 2 + 3t, y = 3 + 2t, z = 6 - t$
    tangent plane: $3x - 2y + z - 6 = 0$

77. $4x - 3y + 6z - 10 = 0$
    $6x - 8y + 3z + 4 = 0$

78. $(-1/2, -1, 14)$

79. $(1/2, 3/4, 131/16)$

80. stationary point: $(0, 3/2, -9/4)$ local minimum

81. stationary point: $(1, 0, -1)$ local minimum

82. stationary point: $(2/3, 2/3, 14/3)$ local minimum

83. stationary point: $(4/3, 2/3, -10/3)$ local minimum

84. No stationary points

85. stationary point: $(4/11, -5/11, 15/11)$ local maximum

86. stationary point: $(10/3, -5/3, -44/3)$ local minimum

87. stationary point: $(1, -2, -4)$ local minimum

88. stationary point: $(-10/3, -5/3, -22/3)$ local minimum

89. stationary point: $(-\pi/2, 0), (-\pi/4, 0), (0, 0), (\pi/4, 0), (\pi/2, 0)$ no local extrema

90. absolute minimum at $(0, 0, 0)$
    absolute maximum at $(4, 4, 4\sqrt{2})$

91. absolute maximum at $(-2, 0, 8)$ and $(2, 0, 8)$ no absolute minimum

92. absolute maximum at $\left(\dfrac{-2}{\sqrt{5}}, \dfrac{-1}{\sqrt{5}}, 6 + 2\sqrt{5}\right)$
    absolute minimum at $\left(\dfrac{2}{\sqrt{5}}, \dfrac{1}{\sqrt{5}}, 6 - 2\sqrt{5}\right)$

93. absolute maximum at $(1, 4, 17)$
    absolute minimum at $(0.83512, 0.6974, 1.8433)$

94. absolute maximum at $(0, 4, 16)$
    absolute minimum at $(x, x, 0), 0 \le x \le 2$

95. $x = y = z = \dfrac{2\sqrt{3}}{3}$, distance: $2 - \sqrt{3}$

96. $\dfrac{2\sqrt{2} - \sqrt{6}}{2}$

97. $f(-8, -23) = 59$, relative maximum

98. $f(2, -3) = -32$, relative minimum;
    $(-2, -3)$ is a saddle point

99. $f(15, -8) = -63$, relative minimum

100. $f(-2, -2) = -8$, relative minimum

101. $(3, 2)$ is a saddle point

102. $(2, 2)$ is a saddle point

103. $f(1, 3) = -11$, relative minimum

104. $f(3, 3) = -27$, relative minimum;
     $(0, 0)$ is a saddle point

105. $f(3, 3) = -9$, relative minimum;
     $(-1, -1)$ is a saddle point

106. 8 in. by 8 in. by 4 in.

107. $[27/7, -23/7, 10/7]$

108. $\dfrac{\sqrt{6}}{2}$

109. 2 in. by 3 in. by 3 in.

110. $(1, 1, \sqrt{2})$ and $(1, 1, -\sqrt{2})$

111. $(1/6, 1/3, 1/6)$

112. $x = 9, y = 12, z = 36$; product $= 3888$

113. $x = 5/3, y = 3, z = 5$; sum $= 45$

114. $x = 4, y = 3, z = 6$; product $= 288$

115. $(x, y) = \left(-\sqrt{2}, \dfrac{-3}{\sqrt{2}}\right)$ or $\left(-\sqrt{2}, \dfrac{-3}{\sqrt{2}}\right)$; $xy = 3$

116. $(x, y) = \left(\sqrt{2}, \dfrac{-3}{\sqrt{2}}\right)$ or $\left(-\sqrt{2}, \dfrac{-3}{\sqrt{2}}\right)$; $xy = -3$

117. $(x, y) = \left(\dfrac{-2\sqrt{2}}{\sqrt{3}}, \dfrac{-2}{\sqrt{3}}\right)$ or $\left(\dfrac{2\sqrt{2}}{\sqrt{3}}, \dfrac{-2}{\sqrt{3}}\right)$;

$x^2 y = \dfrac{-16\sqrt{3}}{9}$

118. $(x, y) = \left(\dfrac{-2\sqrt{2}}{\sqrt{3}}, \dfrac{-2}{\sqrt{3}}\right)$ or $\left(\dfrac{2\sqrt{2}}{3}, \dfrac{-2}{\sqrt{3}}\right)$;

$x^2 y = \dfrac{-16\sqrt{3}}{9}$

119. $(x, y, z) = (-1, -1, \pm\sqrt{2})$, $(1, 1, \pm\sqrt{2}); xyz^2 = 2$

120. $(0, 0, 1)$

121. $(2/3, 1/3, -1/3)$

122. $x = 6, y = 3, z = 3$; product$=324$

123. $x = 4/3, y = 8/3, z = 8/3$; sum $= 16$

124. 8 in. by 8 in. by 4 in.

125. 2 in. by 3 in. by 3 in.

126. $x = \sqrt{2}, y = \dfrac{2}{\sqrt{3}}, z = 1$; volume $= \dfrac{16\sqrt{6}}{3}$

127. $x = \sqrt{2}, y = \sqrt{3}, z = 1$; volume $= 8\sqrt{6}$

128. $[27/7, -23/7, 10/7]$

129. $\dfrac{\sqrt{6}}{2}$

130. $(1, 1, \sqrt{2})$ and $(1, 1, -\sqrt{2})$

131. $x = y = z = 12$; product $= 1728$

132. $x = y = z = 4$; sum $= 12$

133. $x = 14, y = 14, z = 28$; product $= 5488$

134. 3 ft. by 3 ft. by 6 ft.

135. $\dfrac{2xe^{2z}}{3(x^2 + y^2)^{2/3}} dx + \dfrac{2ye^{2z}}{3(x^2 + y^2)^{2/3}} dy$
$+ 2e^{2z}(x^2 + y^2)^{1/3} dz$

136. $(2x + 3y + 3z) dx + (3x - 4y) dy + (3x + 2z) dz$

137. $0\, dx + (\sin z - 3z^2) dy + (4z^3 - 6yz$
$+ y \cos z) dz$

138. $(6x - 3y + 4z) dx + (2y - 3x) dy + (2z + 4x) dz$

139. $(\sin^{-1} y + 2xy) dx + \left(\dfrac{x}{\sqrt{1 - y^2}} + x^2\right) dy$

140. $\Delta u = 0.61; du = 0.60$

141. 4.4π

142. decreased by 125 watts

143. ≈ 0.025

144. error in hypotenuse $= 0.14$; error in area $= 0.7$

145. $\dfrac{y}{3(1 + xy)} dy + \dfrac{y}{3(1 + xy)} dy$

146. $f(x, y) = \dfrac{x^3}{3} y^3 + C$

147. $f(x, y) = \dfrac{1}{2} e^{2x} + \dfrac{1}{2} e^{2y} + C$

148. It is the gradient of $f(x, y) = \dfrac{x^2}{2y^2} - \dfrac{y^3}{3} + C$, but f is only defined for $y \neq 0$.

149. not a gradient

150. $f(x, y, z) = x^3 y e^{z^2} - xz - y + C$

151. $f(x, y, z) = \ln(x^2 + y^2 + 1) + \tan^{-1}(xyz) + C$

Part C

1. Find the gradient of the function $f(x, y) = \frac{1}{2}x^2 + 2xy + y^2$. (15.1)

2. Show that, if g is $o(\mathbf{h})$, then $\lim_{\mathbf{h} \to 0} g(\mathbf{h}) = 0$. (15.1)

3. Verify that, if g is continuous at $\mathbf{x}$, then (15.2)
   (a) $g(\mathbf{x} + \mathbf{h}) o(\mathbf{h}) = o(\mathbf{h})$ and
   (b) $[g(\mathbf{x} + \mathbf{h}) - g(\mathbf{x})] \nabla f(\mathbf{x}) \cdot \mathbf{h} = o(\mathbf{h})$.

4. Assume that $\nabla f(\mathbf{x})$ and $\nabla g(\mathbf{x})$ exist, and that $g(\mathbf{x}) \neq 0$. Derive the quotient rule (15.2)
$$\nabla \left[\frac{f(\mathbf{x})}{g(\mathbf{x})} \right] = \frac{g(\mathbf{x}) \nabla f(\mathbf{x}) - f(\mathbf{x}) \nabla g(\mathbf{x})}{g^2(\mathbf{x})}.$$

5. (*Important*) Set $r = \|\mathbf{r}\|$ where $\mathbf{r} = x\mathbf{i} + y\mathbf{j} + z\mathbf{k}$. If f is a continuously differentiable function of r, then (15.3)

   (15.3.11) $\quad \boxed{\nabla[f(r)] = f'(r)\dfrac{\mathbf{r}}{r}.}$

   Derive this formula.

6. (*A chain rule for vector-valued functions*) Suppose that $\mathbf{u}(x, y) = u_1(x, y)\mathbf{i} + u_2(x, y)\mathbf{j}$ where $x = x(t), y = y(t)$. (15.3)
   (a) Show that

   (15.3.12) $\quad \boxed{\dfrac{d\mathbf{u}}{dt} = \dfrac{\partial \mathbf{u}}{\partial x}\dfrac{dx}{dt} + \dfrac{\partial \mathbf{u}}{\partial y}\dfrac{dy}{dt},}$

   where $\dfrac{\partial \mathbf{u}}{\partial x} = \dfrac{\partial u_1}{\partial x}\mathbf{i} + \dfrac{\partial u_2}{\partial x}\mathbf{j}$ and $\dfrac{\partial \mathbf{u}}{\partial y} = \dfrac{\partial u_1}{\partial y}\mathbf{i} + \dfrac{\partial u_2}{\partial y}\mathbf{j}$.
   (b) Let $\mathbf{u} = e^x \cos y\, \mathbf{i} + e^x \sin y\, \mathbf{j}$ where $x = \frac{1}{2}t^2, y = \pi t$.

7. Show that in the case of a surface of the form $z = xf(x/y)$ with f continuously differentiable, all the tangent planes share a point in common. (15.4)

8. Show that, for all planes tangent to the surface $x^{2/3} + y^{2/3} + z^{2/3} = a^{2/3}$, the sum of the squares of the intercepts is the same. (15.4)

9. Given that $0 < a < b$, find the absolute maximum value taken on by the function (15.5)
$$f(x, y) = \frac{xy}{(a+x)(x+y)(b+y)} \text{ on the open square } \{(x, y) : a < x < b, a < y < b\}.$$

10. Find the absolute maximum value of $f(x, y) = \dfrac{(ax + by + c)^2}{x^2 + y^2 + 1}$. (15.5)

11. Let $f(x, y) = ax^2 + bxy + cy^2$, taking $abc \neq 0$. (15.5)
    (a) Find the discriminant D.
    (b) Find the stationary points and local extreme values if $D \neq 0$.
    (c) Suppose that $D = 0$. Find the stationary points and the local and absolute extreme values given that
        (i) $a > 0, c > 0$.   (ii) $a < 0, c < 0$.

12. (a) Determine the maximum value of $f(x, y, z) = (xyz)^{1/3}$ given that x, y, and z are nonnegative numbers and $x + y + z = k$, k a constant. (15.6)

    (b) Use the result in (a) to show that if x, y, and z are nonnegative numbers, then $(xyz)^{1/3} \leq \dfrac{x + y + z}{3}$.

    Note: $(xyz)^{1/3}$ is the *geometric mean* of x, y, z.

13. Find the volume of the largest rectangular box that can be inscribed in the ellipsoid $4x^2 + 9y^2 + 36z^2 = 36$ if the edges of the box are parallel to the coordinate axes. (15.6)

14. Estimate by a differential the change in the volume of a right circular cylinder if the height is increased from 12 to 12.2 inches and the radius is decreased from 8 to 7.7 inches. (15.7)

15. The dimensions of a rectangular box with a top are length $= 4$ feet, width $= 2$ feet, height $= 3$ feet. It has a coat of paint $\dfrac{1}{16}$ inch thick. Estimate the amount of paint (cubic inches) on the box. (15.7)

16. Given that g and its first and second partials are everywhere continuous, find the general solution of the differential equation $\nabla f(x, y) = e^{g(x,y)}[g_x(x, y)\mathbf{i} + g_y(x, y)\mathbf{j}]$. (15.8)

    Theorem 15.8.2 has a three-dimensional analog. In particular we can show that, if P, Q, R are continuously differentiable on an open rectangular box S, then the vector function $P(x, y, z)\mathbf{i} + Q(x, y, z)\mathbf{j} + R(x, y, z)\mathbf{k}$ is a gradient on S iff
    $$\frac{\partial P}{\partial y} = \frac{\partial Q}{\partial x}, \quad \frac{\partial P}{\partial z} = \frac{\partial R}{\partial x}, \quad \frac{\partial Q}{\partial z} = \frac{\partial R}{\partial y} \quad \text{throughout } S.$$

17. Verify that every vector function of the form $\mathbf{h}(\mathbf{r}) = kr^n \mathbf{r}$ (k constant, n an integer) is a gradient. (15.8)

Answers to Chapter 15, Part C Questions

1. $\nabla f = (x+2y)\mathbf{i} + (2x+2y)\mathbf{j}$

2. $\lim_{h \to 0} g(h) = \lim_{h \to 0} \left(\|\mathbf{h}\| \frac{g(\mathbf{h})}{\|\mathbf{h}\|} \right)$
   $= \left(\lim_{h \to 0} \|\mathbf{h}\| \right) \left(\lim_{h \to 0} \frac{g(\mathbf{h})}{\|\mathbf{h}\|} \right)$
   $= (0)(0) = (0).$

3. (a) $\frac{g(\mathbf{x}+\mathbf{h})\,o(\mathbf{h})}{\|\mathbf{h}\|} = g(\mathbf{x}+\mathbf{h})\frac{o(\mathbf{h})}{\|\mathbf{h}\|} \to g(\mathbf{x}) \cdot 0 = 0$

   (b) $\frac{|[g(\mathbf{x}+\mathbf{h}) - g(\mathbf{x})]\nabla f(\mathbf{x})] \cdot \mathbf{h}|}{\|\mathbf{h}\|}$
   $\leq \frac{\|[g(\mathbf{x}+\mathbf{h}) - g(\mathbf{x})]\nabla f(\mathbf{x})\|\|\mathbf{h}\|}{\|\mathbf{h}\|}$
   by Schwarz's inequality
   $= |g(\mathbf{x}+\mathbf{h}) - g(\mathbf{x})| \cdot \|\nabla f(\mathbf{x})\| \to 0.$

4. $\nabla\left(\frac{f}{g}\right) = \frac{\partial}{\partial x}\left(\frac{f}{g}\right)\mathbf{i} + \frac{\partial}{\partial y}\left(\frac{f}{g}\right)\mathbf{j} + \frac{\partial}{\partial z}\left(\frac{f}{g}\right)\mathbf{k}$
   $= \frac{\frac{\partial f}{\partial x}g - f\frac{\partial g}{\partial x}}{g^2}\mathbf{i} + \frac{\frac{\partial f}{\partial y}g - f\frac{\partial g}{\partial y}}{g^2}\mathbf{j}$
   $+ \frac{\frac{\partial f}{\partial z}g - f\frac{\partial g}{\partial z}}{g^2}\mathbf{k}$
   $= \frac{g(\mathbf{x})\nabla f(\mathbf{x}) - f(\mathbf{x})\nabla g(\mathbf{x})}{g^2(\mathbf{x})}$

5. $\frac{\partial}{\partial x}[f(r)] = \frac{d}{dr}[f(r)]\frac{\partial r}{\partial x} = f'(r)\frac{\partial r}{\partial x} = f'(r)\frac{x}{r};$
   similarly.
   $\frac{\partial}{\partial y}[f(r)] = f'(r)\frac{y}{r} \quad \text{and} \quad \frac{\partial}{\partial z}[f(r)] = f'(r)\frac{z}{r}.$
   Therefore
   $\nabla f(r) = f'(r)\frac{x}{r}\mathbf{i} + f'(r)\frac{y}{r}\mathbf{j} + f'(r)\frac{z}{r}\mathbf{k}$
   $= f'(r)\frac{\mathbf{r}}{r}.$

6. (a) Use $\frac{d\mathbf{u}}{dt} = \frac{du_1}{dt}\mathbf{i} + \frac{du_2}{dt}\mathbf{j}$ and apply the chain rule to u_1, u_2.

   (b) (i) $\frac{d\mathbf{u}}{dt} = t(e^x \cos y\,\mathbf{i} + e^y \sin y\,\mathbf{j})$
   $+ \pi(-e^x \sin y\,\mathbf{i} + e^x \cos y\,\mathbf{j})$
   $= te^{t^2/2}(\cos \pi t\,\mathbf{i} + \sin \pi t\,\mathbf{j})$
   $+ \pi e^{t^2/2}(-\sin \pi t\,\mathbf{i} + \cos \pi t\,\mathbf{j})$

   (ii) $\mathbf{u}(t) = e^{t^2/2}\cos \pi t\,\mathbf{i} + e^{t^2/2}\sin \pi t\,\mathbf{j}$
   $\frac{d\mathbf{u}}{dt} = (-\pi e^{t^2/2}\sin \pi t + te^{t^2/2}\cos \pi t)\mathbf{i}$
   $+ (\pi e^{t^2/2}\cos \pi t + te^{t^2/2}\sin \pi t)\mathbf{j}$

7. All the tangent planes pass through the origin. To see this, write the equation of the surface as $xf(x/y) - z = 0$. The tangent plane at (x_0, y_0, z_0) has equation

   $(x - x_0)\left[\frac{x_0}{y_0}f'\left(\frac{x_0}{y_0}\right) + f\left(\frac{x_0}{y_0}\right)\right]$
   $-(y - y_0)\left[\frac{x_0^2}{y_0^2}f'\left(\frac{x_0}{y_0}\right)\right] - (z - z_0) = 0.$

   The plane passes through the origin:
   $-\frac{x_0^2}{y_0}f'\left(\frac{x_0}{y_0}\right) - x_0 f\left(\frac{x_0}{y_0}\right) + \frac{x_0^2}{y_0}f'\left(\frac{x_0}{y_0}\right) + z_0$
   $= z_0 - x_0 f\left(\frac{x_0}{y_0}\right) = 0.$

8. The equation of the tangent plane at (x_0, y_0, z_0) can be written
   $x_0^{-1/3}(x - x_0) + y_0^{-1/3}(y - y_0) + z_0^{-1/3}(z - z_0)$
   $= 0.$
   Setting $y = z = 0$, we get the x-intercept
   $x = x_0 + x_0^{1/3}(y_0^{2/3} + z_0^{2/3})$
   $= x_0 + x_0^{1/3}(a^{2/3} - x_0^{2/3}) \Longrightarrow x = x_0^{1/3}a^{2/3}$
   Similarly, the y-intercept is $y_0^{1/3}a^{2/3}$ and the z-intercept is $z_0^{1/3}a^{2/3}$.
   The sum of the squares of the intercepts is
   $(x_0^{2/3} + y_0^{2/3} + z_0^{2/3})a^{4/3} = a^{2/3}a^{4/3} = a^2.$

9. $\nabla f = \frac{y(ay - x^2)}{(a+x)^2(x+y)^2(b+y)}\mathbf{i}$
   $+ \frac{x(bx - y^2)}{(a+x)(x+y)^2(b+y)^2}\mathbf{j} = 0$
   at $(a^{2/3}b^{1/3}, a^{1/3}b^{2/3})$.
   Maximum is
   $f(a^{2/3}b^{1/3}, a^{1/3}b^{2/3}) = (a^{1/3} + b^{1/3})^{-3}.$

10. $\nabla f = \frac{2(ax + by + c)(ay^2 + a - bxy - cx)}{(x^2 + y^2 + 1)^2}\mathbf{i}$
    $+ \frac{2(ax + by + c)(bx^2 + b - axy - cy)}{(x^2 + y^2 + 1)^2}\mathbf{j}.$
    Since we want a maximum and $f(x, y) \geq 0$ for all (x, y), we may assume that $ax + by + c \neq 0$. Then
    $\nabla f = 0 \Longrightarrow x = \frac{a}{c}, y = \frac{b}{c}.$
    Maximum value:
    $f(a/c, b/c) = a^2 + b^2 + c^2.$

11. (a) $\nabla f = (2ax + by)\mathbf{i} + (bx + 2cy)\mathbf{j}$
    $\frac{\partial^2 f}{\partial x^2} = 2a, \quad \frac{\partial^2 f}{\partial y \partial x} = b, \quad \frac{\partial^2 f}{\partial y^2} = 2c;$
    $D = b^2 - 4ac.$

(b) The point (0, 0) is the only stationary point. If $D > 0$, $(0, 0)$ is a saddle point; if $D < 0$, $(0, 0)$ is a local minimum if $a > 0$ and a local maximum if $a < 0$.

(c) (i) if $b > 0$, $f(x, y) = (\sqrt{a}x + \sqrt{c}y)^2$; every point on the line $\sqrt{a}x + \sqrt{c}y = 0$ is a stationary point and at each such point f takes on a local and absolute min of 0 if $b < 0$, $f(x, y) = (\sqrt{a}x - \sqrt{c}y)^2$; every point on the line $\sqrt{a}x - \sqrt{c}y = 0$ is a stationary point and at each such point f takes on a local and absolute min of 0

(ii) if $b > 0$, $f(x, y) = -(\sqrt{|a|}x - \sqrt{|c|}y)^2$; every point on the line $\sqrt{|a|}x - \sqrt{|c|}y = 0$ is a stationary point and at each such point f takes on a local and absolute max of 0 if $b < 0$, $f(x, y) = -(\sqrt{|a|}x + \sqrt{|c|}y)^2$; every point on the line $\sqrt{|a|}x + \sqrt{|c|}y = 0$ is a stationary point and at each such point f takes on a local and absolute max of 0

12. (a) The maximum occurs when $x = y = z = \dfrac{k}{3}$, where $(xyz)^{1/3} = \dfrac{k}{3}$.

    (b) If $x + y + z = k$, then, by (a),
    $$(xyz)^{1/3} \leq \dfrac{k}{3} = \dfrac{x+y+z}{3}$$

13. $f(x, y, z) = 8xyz$,
    $g(x, y, z) = 4x^2 + 9y^2 + 36z^2 - 36$.
    $$\nabla f(x, y, z) = 8yz\,\mathbf{i} + 8xz\,\mathbf{j} + 8xy\,\mathbf{k}.$$
    $$\nabla g(x, y, z) = 8\lambda x\,\mathbf{i} + 18\lambda y\,\mathbf{j} + 72\lambda z\,\mathbf{k}.$$
    $$\nabla f = \lambda \nabla g,$$
    gives $yz = \lambda x$, $4xz = 9\lambda y$, $xy = 9\lambda z$.
    $$4\dfrac{xyz}{\lambda} = 4x^2, \quad 4\dfrac{xyz}{\lambda} = 9y^2, \quad 4\dfrac{xyz}{\lambda} = 36z^2.$$

Also notice
$$4x^2 + 9y^2 + 36z^2 - 36 = 0$$
We have
$$12\dfrac{xyz}{\lambda} = 36 \Longrightarrow x = \sqrt{3},\ y = \dfrac{2}{\sqrt{3}},\ z = \dfrac{1}{\sqrt{3}}.$$
Thus,
$$V = 8xyz = 8 \cdot \sqrt{3} \cdot \dfrac{2}{\sqrt{3}} \cdot \dfrac{1}{\sqrt{3}} = \dfrac{16}{\sqrt{3}}.$$

14. $f(r, h) = \pi r^2 h$, $r = 8$, $h = 12$,
    $\Delta r = -0.3$, $\Delta h = 0.2$
    $df = 2\pi rh\,\Delta r + \pi r^2\,\Delta h$
    $= 192\pi(-0.3) + 64\pi(0.2) = -44.8\pi$
    decreases by approximately 44.8π cubic inches.

15. Amount of paint is the increase in volume.
    $f(x, y, z) = xyz$, $x = 48$ in, $y = 24$ in, $z = 36$ in.
    $$\Delta x = \Delta y = \Delta z = \dfrac{2}{16}\ \text{in.}$$
    $$\Delta f \cong df = yz\,\Delta x + xz\,\Delta y + xy\,\Delta z = 3774\left(\dfrac{2}{16}\right)$$
    $= 468$
    Use about 468 cubic inches of paint.

16. $\dfrac{\partial f}{\partial x} = e^{g(x,y)} g_x(x, y) \Longrightarrow f(x, y) = e^{g(x,y)} + \phi(y);$
    $\dfrac{\partial f}{\partial x} = e^{g(x,y)} g_y(x, y) + \phi'(y) = e^{g(x,y)} g_y(x, y)$
    $\Longrightarrow f(x, y) = e^{g(x,y)} + C.$

17. $\mathbf{h}(\mathbf{r}) = \begin{cases} \nabla\left(\dfrac{k}{n+2} r^{n+2}\right), & n \neq 2 \\ \nabla(k \ln r), & n = -2 \end{cases}$

CHAPTER 16 Double and Triple Integrals
Part B

16.1 Multiple-Sigma Notation

1. Let $P_1 = \{x_0, x_1, \ldots, x_m\}$ be a partition of $[a_1, a_2]$
   Let $P_2 = \{y_0, y_1, \ldots, y_n\}$ be a partition of $[b_1, b_2]$
   Let $P_3 = \{z_0, z_1, \ldots, z_q\}$ be a partition of $[c_1, c_2]$
   Take $\Delta x_i = x_i - x_{i-1}$, $\Delta y_j = y_j - y_{j-1}$, $\Delta z_k = z_k - z_{k-1}$
   Evaluate the sum, $\sum_{i=1}^{m} \sum_{k=1}^{q} \Delta x_i \Delta z_k$

2. Let $P_1 = \{x_0, x_1, \ldots, x_m\}$ be a partition of $[a_1, a_2]$
   Let $P_2 = \{y_0, y_1, \ldots, y_n\}$ be a partition of $[b_1, b_2]$
   Let $P_3 = \{z_0, z_1, \ldots, z_q\}$ be a partition of $[c_1, c_2]$
   Take $\Delta x_i = x_i - x_{i-1}$, $\Delta y_j = y_j - y_{j-1}$, $\Delta z_k = z_k - z_{k-1}$
   Evaluate the sum, $\sum_{i=1}^{m} \sum_{j=1}^{n} \sum_{k=1}^{q} \Delta x_i \Delta y_j \Delta z_k$

3. Let $P_1 = \{x_0, x_1, \ldots, x_m\}$ be a partition of $[a_1, a_2]$
   Let $P_2 = \{y_0, y_1, \ldots, y_n\}$ be a partition of $[b_1, b_2]$
   Let $P_3 = \{z_0, z_1, \ldots, z_q\}$ be a partition of $[c_1, c_2]$
   Take $\Delta x_i = x_i - x_{i-1}$, $\Delta y_j = y_j - y_{j-1}$, $\Delta z_k = z_k - z_{k-1}$
   Evaluate the sum, $\sum_{j=1}^{n} \sum_{k=1}^{q} (z_k + z_{k-1}) \Delta y_j \Delta z_k$

4. Let $P_1 = \{x_0, x_1, \ldots, x_m\}$ be a partition of $[a_1, a_2]$
   Let $P_2 = \{y_0, y_1, \ldots, y_n\}$ be a partition of $[b_1, b_2]$
   Let $P_3 = \{z_0, z_1, \ldots, z_q\}$ be a partition of $[c_1, c_2]$
   Take $\Delta x_i = x_i - x_{i-1}$, $\Delta y_j = y_j - y_{j-1}$, $\Delta z_k = z_k - z_{k-1}$
   Evaluate the sum, $\sum_{i=1}^{m} \sum_{j=1}^{n} (4\Delta x_i - 3\Delta y_j)$

5. Let $P_1 = \{x_0, x_1, \ldots, x_m\}$ be a partition of $[a_1, a_2]$
   Let $P_2 = \{y_0, y_1, \ldots, y_n\}$ be a partition of $[b_1, b_2]$
   Let $P_3 = \{z_0, z_1, \ldots, z_q\}$ be a partition of $[c_1, c_2]$
   Take $\Delta x_i = x_i - x_{i-1}$, $\Delta y_j = y_j - y_{j-1}$, $\Delta z_k = z_k - z_{k-1}$
   Evaluate the sum, $\sum_{i=1}^{m} \sum_{j=1}^{n} \sum_{k=1}^{q} (y_j + y_{j-1}) \Delta x_i \Delta y_j \Delta z_k$

16.2 The Double Integral

6. Take $f(x, y) = 3x - 2y$ on $R : 0 \leq x \leq 1, 0 \leq y \leq 2$, and P as the partition $P = P_1 \times P_2$. Find $L_f(P)$ and $U_f(P)$ if $P_1 = \{0, 1/4, 1/2, 3/4, 1\}$ $P_2 = \{0, 1/2, 1, 3/2, 2\}$.

7. Take $f(x, y) = 3x - 2y$ on $R : 0 \leq x \leq 1, 0 \leq y \leq 2$, and P as the partition $P = P_1 \times P_2$. $P_1 = \{x_0, x_1, \ldots, x_m\}$ is an arbitrary partition of $[0, 1]$, and $P_2 = \{y_0, y_1, \ldots, y_n\}$ is an arbitrary partition of $[0, 2]$.

(a) Find $L_f(P)$ and $U_f(P)$

(b) Evaluate the double integral $\iint_R (3x - 2y)\,dx\,dy$ using part (a).

8. Take $f(x, y) = 2x(y - 1)$ on $R: 0 \le x \le 2, 0 \le y \le 1$, and P as the partition $P = P_1 \times P_2$. Find $L_f(P)$ and $U_f(P)$ if $P_1 = \{0, 1, 3/2, 2\}$ $P_2 = \{0, 1/2, 1\}$.

9. Take $f(x, y) = 2x(y - 1)$ on $R: 0 \le x \le 2, 0 \le y \le 1$, and P as the partition $P = P_1 \times P_2$. $P_1 = \{x_0, x_1, \ldots, x_m\}$ is an arbitrary partition of $[0, 2]$, and $P_2 = \{y_0, y_1, \ldots, y_n\}$ is an arbitrary partition of $[0, 1]$.

(a) Find $L_f(P)$ and $U_f(P)$

(b) Evaluate the double integral $\iint_R 2x(y - 1)\,dx\,dy$ using part (a).

10. Take $f(x, y) = 2x^2 - 3y^2$ on $R: 0 \le x \le 1, 0 \le y \le 1$, and P as the partition $P = P_1 \times P_2$. Find $L_f(P)$ and $U_f(P)$ if $P_1 = \{0, 1/2, 3/4, 1\}$ $P_2 = \{0, 1/2, 1\}$.

11. Take $f(x, y) = 2x^2 - 3y^2$ on $R: 0 \le x \le 1, 0 \le y \le 1$, and P as the partition $P = P_1 \times P_2$. $P_1 = \{x_0, x_1, \ldots, x_m\}$ is an arbitrary partition of $[0, 2]$, and $P_2 = \{y_0, y_1, \ldots, y_n\}$ is an arbitrary partition of $[0, 1]$.

(a) Find $L_f(P)$ and $U_f(P)$

(b) Evaluate the double integral $\iint_R (2x^2 - 3y^2))\,dx\,dy$ using part (a).

16.3 The Evaluation of a Double Integral by Repeated Integrals

12. Evaluate $\iint_\Omega (x^2 - y^2)\,dx\,dy$ taking $\Omega: 0 \le x \le 1, 0 \le y \le 1$.

13. Evaluate $\iint_\Omega y \cos xy\,dx\,dy$ taking $\Omega: 0 \le x \le 1, 0 \le y \le \pi$.

14. Evaluate $\iint_\Omega x^2 y^2\,dx\,dy$ taking $\Omega: -1 \le x \le 1, 0 \le y \le \pi/2$.

15. Evaluate $\iint_\Omega (3 - y)x^2\,dx\,dy$ taking $\Omega: 2 \le x \le 4, 0 \le y \le 3$.

16. Evaluate $\iint_\Omega (x - 1)\,dy\,dx$ taking $\Omega: 0 \le x \le 1, 0 \le y \le 2$.

17. Evaluate $\iint_\Omega e^{x+y}\,dy\,dx$ taking $\Omega: 0 \le x \le 1, 0 \le y \le 1$.

18. Evaluate $\iint_\Omega x\sqrt{x^2 + y}\,dx\,dy$ taking $\Omega: 0 \le x \le 1, 0 \le y \le 3$.

19. Evaluate $\iint_\Omega (x^2 - y)\,dx\,dy$ taking $\Omega: 0 \le x \le 3, 1 \le y \le 4$.

20. Evaluate $\iint_\Omega \dfrac{1}{\sqrt{1-x^2}}\, dy\, dx$ taking $\Omega: 0 \le x \le 1/2, 0 \le y \le 2$.

21. Evaluate $\iint_\Omega \dfrac{1}{1+y^2}\, dy\, dx$ taking $\Omega: 0 \le x \le 4, 0 \le y \le 1$.

22. Evaluate $\iint_\Omega \dfrac{1}{\sqrt{4-x^2}}\, dx\, dy$ taking $\Omega: 0 \le x \le 1, 0 \le y \le 2$.

23. Calculate the average value of $f(x, y) = x \cos y$ over the region $\Omega: 0 \le x \le 1, 0 \le y \le \pi/4$.

24. Calculate the average value of $f(x, y) = e^x$ over the region $\Omega: 0 \le x \le \ln y, 0 \le y \le e$.

25. Evaluate $\iint_\Omega (x^2 + 2)\, dx\, dy$ taking Ω: the bounded region between $y^2 = 2x$ and $y^2 = 8 - 2x$.

26. Evaluate $\iint_\Omega (2xy - x^2)\, dx\, dy$ taking Ω: the bounded region between $y = x^3$ and $y = x^2$.

27. Evaluate $\displaystyle\int_0^1 \int_y^1 e^{x^2}\, dx\, dy$ by first sketching the region of integration Ω then changing the order of integration.

28. Evaluate $\displaystyle\int_0^1 \int_{x^2}^{x} (x^2 + y^2)\, dx\, dy$ by first sketching the region of integration Ω then changing the order of integration.

29. Evaluate $\displaystyle\int_0^1 \int_0^{2\sqrt{1-y^2}} x\, dx\, dy$ by first sketching the region of integration Ω then changing the order of integration.

30. Evaluate $\displaystyle\int_1^2 \int_0^{\sqrt{x}} y \ln x^2\, dy\, dx$.

31. Evaluate $\displaystyle\int_0^1 \int_{2y}^2 \cos(x^2)\, dx\, dy$ by first expressing it as an equivalent double integral with order of integration reversed.

32. Evaluate $\displaystyle\int_0^1 \int_0^{x} y\sqrt{x^2 + y^2}\, dy\, dx$.

33. Sketch the region of integration Ω and express $\displaystyle\int_0^{\pi/4} \int_{\sin x}^{\cos x} f(x, y)\, dy\, dx$ as an equivalent double integral with order of integration reversed.

34. Sketch the region of integration Ω and express $\displaystyle\int_0^1 \int_{1-y}^{2-y} f(x, y)\, dx\, dy$ as an equivalent double integral with order of integration reversed.

35. Use a double integral to find the area bounded by $y = x^2$ and $y = \sqrt{x}$.

36. Use a double integral to find the area bounded by $x = y - y^2$ and $x + y = 0$.

37. Find the volume of the solid bounded by $y = x^2 - x$, $y = x$, $z = 0$, and $z = x + 1$.

38. Find the volume of the solid in the first octant bounded by $y = x^2/4$, $z = 0$, $y = 4$, $x = 0$, and $x - y + 2z = 2$.

39. Find the volume of the solid bounded by $x = 0$, $z = 0$, $z = 4 - x^2$, $y = 2x$, and $y = 4$.

40. Find the volume of the solid bounded by $y = x^2 - x + 1$, $y = x + 1$, $z = 0$, and $z = x + 1$.

41. Find the volume of the solid in the first octant bounded by $z = x^2 + y^2$, $z = 0$, and $x + y = 1$.

42. Find the volume of the solid in the first octant bounded by $z = 4 - y^2$, $z = 0$, $x = 0$, and $y = x$.

43. Find the volume of the solid in the first octant bounded by $x^2 + y^2 = 4$, $y = z$, and $z = 0$.

44. Find the volume bounded by $x^2 + y^2 = 1$ and $y^2 + z^2 = 1$.

16.4 Double Integrals in Polar Coordinates

45. Calculate $\displaystyle\int_{-2}^{2} \int_{0}^{\sqrt{4-x^2}} e^{-(x^2+y^2)} \, dy \, dx$ by changing to polar coordinates.

46. Calculate $\displaystyle\int_{0}^{3} \int_{-\sqrt{9-y^2}}^{\sqrt{9-y^2}} \frac{1}{\sqrt{x^2+y^2}} \, dx \, dy$ by changing to polar coordinates.

47. Calculate $\displaystyle\int_{-3}^{3} \int_{0}^{\sqrt{9-x^2}} y \, dy \, dx$ by changing to polar coordinates.

48. Calculate $\displaystyle\int_{0}^{2} \int_{0}^{\sqrt{4-x^2}} (x^2 + y^2) \, dy \, dx$ by changing to polar coordinates.

49. Integrate $f(x, y) = 2(x + y)$ over Ω, the region bounded by $x^2 + y^2 = 9$ and $x \geq 0$.

50. Find the volume in the first octant bounded by $x = 0$, $y = 0$, and $z = 0$, the plane $z + y = 3$, and the cylinder $x^2 + y^2 = 4$.

51. Use a double integral in polar coordinates to find the volume in the first octant of the solid bounded by $x^2 + y^2 = 4$, $y = z$, and $z = 0$.

52. Use a double integral in polar coordinates to find the volume of the solid bounded by $x^2 + y^2 = 5 - z$, and $z = 1$.

53. Use a double integral in polar coordinates to find the volume of the solid between the sphere $x^2 + y^2 + z^2 = 9$ and the cylinder $x^2 + y^2 = 1$.

54. Use a double integral in polar coordinates to find the volume of the solid bounded by the paraboloid $z = 4 - x^2 - y^2$ and $z = 0$.

55. Use a double integral in polar coordinates to find the volume of the solid in the first octant bounded by the ellipsoid $9x^2 + 9y^2 + 4z^2 = 36$ and the planes $x = \sqrt{3}y$, $x = 0$, and $z = 0$.

56. Use a double integral in polar coordinates to find the volume bounded by the sphere $x^2 + y^2 + z^2 = 16$ and the cylinder $(x - 2)^2 + y^2 = 4$.

57. Use a double integral in polar coordinates to find the volume bounded by $z = 0$, $x + 2y - z = -4$, and the cylinder $x^2 + y^2 = 1$.

58. Use a double integral in polar coordinates to find the volume that is inside the sphere $x^2 + y^2 + z^2 = 9$, outside the cylinder $x^2 + y^2 = 4$, and above $z = 0$.

59. Use a double integral in polar coordinates to find the area bounded by the limaçon $r = 4 + \sin\theta$.

60. Use a double integral in polar coordinates to find the area that is inside $r = 1 + \cos\theta$ and outside $r = 1$.

61. Use a double integral in polar coordinates to find the area that is inside $r = 3\sin 3\theta$. $0 \leq \theta \leq \pi$

16.5 Some Applications of Double Integration

62. Find the center of mass of a plate of mass M bounded by $x = 0$, $x = 4$, $y = 0$, and $y = 3$ if its density is given by $\lambda(x, y) = k(x + y^2)$.

63. Find the center of mass of a plate of mass M bounded by $y^2 = 4x$, $x = 4$, and $y = 0$ if its density is given by $\lambda(x, y) = ky$.

64. Find the center of mass of a homogeneous plate of mass M bounded by $x = 0$, $x = 4$, $y = 0$, and $y = 3$ if its density is given by $\lambda(x, y) = kx^2y$.

65. Find the center of mass of a plate of mass M bounded by $y = \sin x$, $y = 0$, and $0 \leq x \leq \pi$ if its density is proportional to the distance from the x-axis.

66. Find the center of mass of a plate of mass M bounded by $r = a\cos\theta$, $0 \leq \theta \leq \pi/2$ if its density is proportional to the distance from the origin.

67. Find the mass of a homogeneous plate of mass M in the first quadrant that is inside $r = 8\cos\theta$ and outside $r = 4$ if the density of the region is given by $\lambda(r, \theta) = \sin\theta$.

68. Find the mass of a homogeneous plate of mass M cut from the circle $x^2 + y^2 = 36$ by the line $x = 3$ if its density is given by $\lambda(x, y) = \dfrac{x^2}{x^2 + y^2}$.

69. Find the centroid of the region bounded by $x = 4y - y^2$ and the y-axis.

70. Find the centroid of the region bounded by $y = 4 - x$, $x = 0$, and $y = 0$.

71. Find the centroid of the region bounded by $y = x^2$ and the line $y = 4$.

72. Find the centroid of the region bounded by $y = x^3$, $x = 2$, and the line $y = 0$.

73. Find the centroid of the region bounded by $y = x^2 - 2x$ and $y = 0$.

74. Find the centroid of the region bounded by $x^2 = 8y$, $y = 0$, and $x = 4$.

75. Find the centroid of the region bounded by $y = \sqrt{4 - x^2}$ and $y = 0$.

76. Find the centroid of the region enclosed by the cardiod $r = 2 - 2\cos\theta$.

77. Find the moments of inertia I_x, I_y, I_z of the plate of mass density $\lambda(x, y) = 6x + 6y + 6$ occupying the region $\Omega: 0 \leq x \leq 1, 0 \leq y \leq 2x$.

78. Find the moments of inertia I_x, I_y, I_z of the plate of mass density $\lambda(x, y) = y + 1$ occupying the region Ω bounded by $y = x$, $y = -x$, $y = 1$.

16.6 Triple Integrals

79. Evaluate $\int_0^1 \int_0^z \int_0^{\sqrt{yz}} x \, dx \, dy \, dz$.

80. Evaluate $\iiint_T x \, dx \, dy \, dz$ where T is the solid in the first octant bounded by $x + y + z = 3$ and the coordinate planes.

16.7 Reduction to Repeated Integrals

81. Evaluate $\iiint_T yz \, dx \, dy \, dz$ where T is the solid in the first octant bounded by $y = 0$, $y = \sqrt{1 - x^2}$, and $z = x$.

82. Evaluate $\iiint_T y \, dx \, dy \, dz$ where T is the solid in the first octant bounded by $y = 1$, $y = x$, $z = x + 1$, and the coordinate planes.

83. Use a triple integral to find the volume of the solid bounded by $z = 0$, $y = 4 - x^2$, $y = 3x$, and $z = x + 4$.

84. Use a triple integral to find the volume of the solid whose base is the region in the xy-plane bounded by $y = x^2 - x + 1$ and $y = x + 1$, and whose height is given by $z = x + 1$.

85. Use a triple integral to find the volume of the solid bounded by $z = x^2 + y^2$, $y = x^2$, $z = 0$, and $y = x$.

86. Use a triple integral to find the volume of the solid bounded by $y = x^2$, $x = y^2$, $z = 0$, and $z = 3$.

87. Use a triple integral to find the volume of the solid bounded by $z = \dfrac{4}{y^2 + 1}$, $z = 0$, $y = x$, $y = 3$, and $x = 0$.

88. Use a triple integral to find the volume of the solid bounded by $z = 0$, $y = x^2 - x$, $y = x$, and $z = x + 1$.

89. Use a triple integral to find the volume of the solid bounded by $x^2 = 4y$, $y + z = 1$, and $z = 0$.

90. Use a triple integral to find the volume of the solid bounded by $y^2 = 4x$, $z = 0$, $z = x$, and $x = 4$.

91. Find the centroid of the tetrahedron bounded by $2x + 2y + z = 6$ and the coordinate planes.

92. Use a triple integral to find the volume of the solid in the first octant bounded by $z = y$, $y^2 = x$, and $x = 1$.

93. Use a triple integral to find the volume of the solid in the first octant bounded by the cylinder $x = 4 - y^2$, and the planes $z = y$, $x = 0$, and $z = 0$.

94. Use a triple integral to find the volume of the solid in the first octant bounded by $z = x^2 + y^2$, $y = x$, and $x = 1$.

95. Use a triple integral to find the volume of the solid in the first octant bounded by the cylinder $x = 4 - y^2$, and the planes $y = x$, $z = 0$, and $x = 0$.

96. Use a triple integral to find the volume of the tetrahedron bounded by the plane $3x + 6y + 4z = 12$ and the coordinate planes.

97. Find the centroid of the solid bounded below by the paraboloid $z = x^2 + y^2$ and above by the plane $z = 4$.

98. Find the centroid of the solid bounded by $z = 4y^2$, $z = 4$, $x = -1$, and $x = 1$.

99. Find the center of mass of the tetrahedron with vertices $(0, 0, 0)$, $(1, 0, 0)$, $(0, 1, 0)$, $(0, 0, 1)$ if the mass density is proportional to the distance from the yz-plane.

100. Find the moments of inertia about its three edges of a homogeneous box of mass M with edges of lengths a, b, and c.

101. Find the moments of inertia I_x, I_y, I_z of the homogeneous tetrahedron bounded by the coordinate planes and the plane $x + y + z = 1$.

102. Find the moment of inertia about the y-axis of the homogeneous solid bounded by $z = 1 - x^2$, $z = 0$, $y = -1$, and $y = 1$.

103. Find the moment of inertia about the z-axis of the cube $0 \le x \le 1$, $0 \le y \le 1$, $0 \le z \le 1$ if the mass density is $\lambda(x, y, z) = kz$.

16.8 Triple Integrals in Cylindrical Coordinates

104. Find the cylindrical coordinates (r, θ, z) of the point with rectangular coordinates $(2, 1, 2)$.

105. Find the rectangular coordinates of the point with cylindrical coordinates $(2, \pi/4, 2)$.

106. Evaluate $\int_0^{\pi/4} \int_1^{\cos\theta} \int_1^r \frac{1}{r^2 z^2} \, dz \, dr \, d\theta$.

107. Evaluate $\int_0^{2\pi} \int_1^2 \int_0^5 e^z r \, dz \, dr \, d\theta$.

108. Use cylindrical coordinates to find the volume of the solid in the first octant bounded by the coordinate planes, the cylinder $x^2 + y^2 = 4$, and the plane $z + y = 3$.

109. Use cylindrical coordinates to find the volume and centroid of the cylinder bounded by $x^2 + y^2 = 4$, $z = 0$, and $z = 4$.

110. Use cylindrical coordinates to find the volume inside $x^2 + y^2 = 4x$, above $z = 0$, and below $x^2 + y^2 = 4z$.

111. Use cylindrical coordinates to find the volume of the solid cut from the sphere $x^2 + y^2 + z^2 = 4$, bounded below $z = 0$, and on the sides by the cylinder $x^2 + y^2 = 1$.

112. Use cylindrical coordinates to evaluate $\iiint_T \sqrt{x^2 + y^2} \, dx \, dy \, dz$ where T is the solid bounded by $z = x^2 + y^2$ and $z = 8 - x^2 - y^2$.

113. Use cylindrical coordinates to find the volume and centroid of the solid bounded by the paraboloid $z = x^2 + y^2$ and the plane $z = 4$.

114. Use cylindrical coordinates to find the volume and centroid of the solid bounded by $z = \sqrt{x^2 + y^2}$, and the plane $z = 1$.

16.9 The Triple Integral as the Limit of Riemann Sums; Spherical Coordinates

115. Find the spherical coordinates (ρ, θ, ϕ) of the point with rectangular coordinates $(2, -1, 1)$.

116. Find the rectangular coordinates of the point with spherical coordinates $(2, 2\pi/3, \pi/4)$.

Calculus: One and Several Variables

117. Find the spherical coordinates of the point with cylindrical coordinates $(1, \pi/6, 2)$.

118. Find the cylindrical coordinates of the point with spherical coordinates $(3, \pi/3, \pi/6)$.

119. Evaluate $\displaystyle\int_0^{\pi/2} \int_0^{\theta} \int_{2\sin\theta}^{2} \rho \sin^2\phi \cos\phi \cos\theta \, d\rho \, d\phi \, d\theta$.

120. Evaluate $\displaystyle\int_0^{2\pi} \int_0^{\pi} \int_1^{2} \rho^4 \cos^2\phi \sin\phi \, d\rho \, d\phi \, d\theta$.

121. Use spherical coordinates to find the mass of the ball bounded by $x^2 + y^2 + z^2 \leq 9$ if its density is given by
$$\lambda(x, y, z) = \frac{z^2}{x^2 + y^2 + z^2}.$$

122. Use spherical coordinates to find the mass of the ball bounded by $x^2 + y^2 + z^2 = 2z$ if its density is given by $\lambda(x, y, z) = \sqrt{x^2 + y^2 + z^2}$.

123. Use spherical coordinates to find the mass and center of mass of the ball bounded by $x^2 + y^2 + z^2 \leq 4$ if its density is given by $\lambda(x, y, z) = x^2 + y^2$.

124. Use spherical coordinates to find the mass of a ball of radius 4 if its density is proportional to the distance from its center. Take k as the constant of proportionality.

16.10 Jacobians; Changing Variables in Multiple Integration

125. Find the Jacobian of the transformation $x = 2u + 3v$, $y = -u + 4v$.

126. Find the Jacobian of the transformation $x = u^2 v$, $y = u^2 + v^2$.

127. Find the Jacobian of the transformation $x = u \ln v$, $y = \ln u + v$.

128. Find the Jacobian of the transformation $x = u^2 - v^2$, $y = uv$.

129. Take Ω as the parallelogram bounded by $x + y = 0$, $x + y = 1$, $x - y = 0$, $x - y = 2$. Evaluate
$$\iint_\Omega (x + y) \, dx \, dy.$$

130. Take Ω as the parallelogram bounded by $x - y = 0$, $x - y = \pi$, $x + 2y = 0$, $x + 2y = {}^1/_2\pi$. Evaluate
$$\iint_\Omega (x^2 - y^2) \, dx \, dy.$$

131. Take Ω as the parallelogram bounded by $x - y = 0$, $x - y = \pi$, $x + 2y = 0$, $x + 2y = {}^1/_2\pi$. Evaluate
$$\iint_\Omega 2x^2 y \, dx \, dy.$$

132. Take Ω as the parallelogram bounded by $x + y = 0$, $x + y = 1$, $x - y = 0$, $x - y = 2$. Evaluate
$$\iint_\Omega \sin 2x \, dx \, dy.$$

Answers to Chapter 16, Part B Questions

1. $(a_2 - a_1)(c_2 - c_1)$

2. $(a_2 - a_1)(b_2 - b_1)(c_2 - c_1)$

3. $(b_2 - b_1)(c_2^2 - c_1^2)$

4. $4n(a_2 - a_1) - 3m(b_2 - b_1)$

5. $(a_2 - a_1)(b_2^2 - b_1^2)(c_2 - c_1)$

6. $L_f(P) = -11/4; U_f(P) = 3/4$

7. (a) $L_f(P) = \frac{3}{2}(a_2 - a_1)^2(b_2 - b_1)$
   $\qquad - (a_2 - a_1)(b_2 - b_1)^2$
   $\qquad - \frac{3}{2}(a_2 - a_1)(b_2 - b_1)\Delta x$
   $\qquad - (a_2 - a_1)(b_2 - b_1)\Delta y$
   (b) -1

8. $L_f(P) = -33/8; U_f(P) = -5/8$

9. (a) $L_f(P) = 2\sum_{i=1}^{m} x_i \Delta x_i \sum_{j=1}^{n}(y_{j-1} - 1)\Delta y_j$
   $U_f(P) = 2\sum_{i=1}^{m} x_{i-1}\Delta x_i \sum_{j=1}^{n}(y_j - 1)\Delta y_j$
   (b) -2

10. $L_f(P) = -47/32; U_f(P) = 21/32$

11. (a) $L_f(P) = 2\sum_{i=1}^{m} x_{i-1}^2 \Delta x_i \sum_{j=1}^{n} \Delta y_j - 3\sum_{i=1}^{m}\Delta x_i$
    $\qquad \sum_{j=1}^{n} y_j^2 \Delta y_j$
    $U_f(P) = 2\sum_{i=1}^{m} x_i^2 \Delta x_i \sum_{j=1}^{n} \Delta y_j - 3\sum_{i=1}^{m}\Delta x_i$
    $\qquad \sum_{j=1}^{n} y_{j-1}^2 \Delta y_j$
    (b) $-1/3$

12. 0

13. 2

14. $\pi^3/36$

15. 84

16. -1

17. $e^2 - 2e + 1$

18. $\dfrac{62}{15} - \dfrac{6\sqrt{3}}{5}$

19. $9/2$

20. $\pi/3$

21. π

22. $\pi/3$

23. $\sqrt{2}\pi$

24. $e^2/2 - e + 1/2$

25. $7744/105$

26. $1/120$

27. $\dfrac{1}{2}(e - 1)$

28. $3/35$

29. $4/3$

30. $2\ln 2 - 3/4$

31. $\dfrac{1}{4}\sin 4$

32. $\dfrac{2\sqrt{2} - 1}{12}$

33. $\displaystyle\int_0^{\frac{\sqrt{2}}{2}} \int_0^{\sin^{-1}} f(x, y)\,dx\,dy$
    $\displaystyle + \int_{\frac{\sqrt{2}}{2}}^{1} \int_0^{\cos^{-1} y} f(x, y)\,dx\,dy$

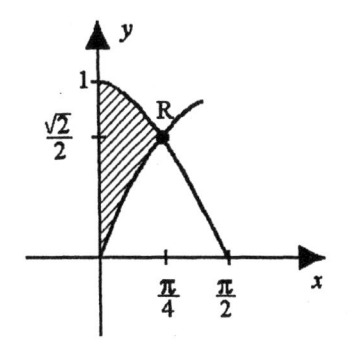

34. $\int_0^1 \int_{1-x}^1 f(x,y)\,dy\,dx + \int_1^2 \int_0^{2-x} f(x,y)\,dy\,dx$

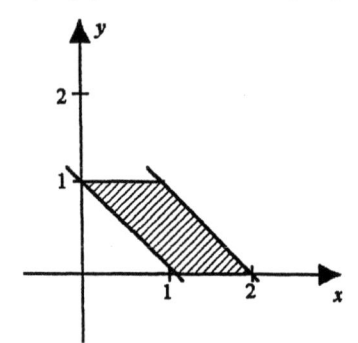

35. 1/3

36. 4/3

37. 8/3

38. 232/15

39. 40/3

40. 8/3

41. 1/6

42. 4

43. 8/3

44. 16/3

45. $\dfrac{\pi}{2}(1 - e^{-4})$

46. 3π

47. 18

48. 2π

49. 36

50. $3\pi - \dfrac{8}{3}$

51. 8/3

52. 8π

53. $\dfrac{64\pi}{3}\sqrt{2}$

54. 8π

55. $4\pi/3$

56. $\dfrac{128}{9}(3\pi - 4)$

57. 4π

58. $\dfrac{10\sqrt{5}}{3}\pi$

59. $33\pi/2$

60. $2 + \pi/4$

61. $9\pi/4$

62. (34/15, 39/20)

63. (8/3, 32/75)

64. (3, 2)

65. $(\pi/2, 16/9\pi)$

66. $(3a5, 9a/40)$

67. 16/3

68. $3\pi + \dfrac{9\sqrt{3}}{2}$

69. (8/5, 2)

70. (4/3, 4/3)

71. (0, 12/5)

72. (8/5, 16/7)

73. (1, −2/5)

74. (3, 3/5)

75. $(0, 8/3\pi)$

76. $(\bar{x}, \bar{y}) = \left(-\dfrac{5}{3}, 0\right)$

77. $I_x = 12,\ I_y = 39/5,\ Iz = 99/5$

78. $I_x = 9/10,\ I_y = 3/10,\ I_z = 6/5$

79. 1/16

80. 27/8

81. 1/30

82. 11/24

83. 625/12

84. 8/3

85. 3/35

86. 1

87. 2 ln 10

88. 8/3

89. 16/15

90. 256/5

91. (3/4, 3/4, 3/2)

92. 1/4

93. 4

94. 1/3

95. 4

96. 4

97. (0, 0, 8/3)

98. (0, 0, 12/5)

99. (2/5, 1/5, 1/5)

100. $I_a = \dfrac{M}{3}(b^2 + c^2)$, $I_b = \dfrac{M}{3}(a^2 + c^2)$, $I_c = \dfrac{M}{3}(a^2 + b^2)$

101. $I_x = I_y = I_z = 1/30$

102. $I_y = 8/7$

103. k/3

104. $\left(\sqrt{5}, \tan^{-1}\dfrac{1}{2}, 2\right)$

105. $\left(\sqrt{2}, \sqrt{2}, 2\right)$

106. $\dfrac{1}{2} + \dfrac{\pi}{8} - \ln\left(\sqrt{2}+1\right)$

107. $3\pi(e^5 - 1)$

108. $3\pi - 8/3$

109. 16π; (0, 0, 2)

110. 6π

111. $\dfrac{2\pi}{3}(8 - 3\sqrt{3})$

112. 16π

113. 8π; (0, 0, 8/3)

114. $\pi/3$; (0, 0, 3/4)

115. $\left(\sqrt{6}, \tan^{-1}\left(-\dfrac{1}{2}\right), \cos^{-1}\left(\dfrac{1}{\sqrt{6}}\right)\right)$

116. $\left(-\dfrac{\sqrt{2}}{2}, \dfrac{\sqrt{6}}{2}, \sqrt{2}\right)$

117. $\left(\sqrt{5}, \dfrac{\pi}{6}, \cos^{-1}\dfrac{2}{\sqrt{5}}\right)$

118. $\left(\dfrac{3}{2}, \dfrac{\pi}{3}, \dfrac{3\sqrt{3}}{2}\right)$

119. 1/10

120. $124\pi/15$

121. 12π

122. $8\pi/5$, (0, 0, 8/7)

123. $256\pi/15$

124. $256\pi k$

125. 11

126. $4uv^2 - 2u^3$

127. $\ln v - 1/v$

128. $2u^2 + 2v^2$

129. $\dfrac{1}{2}$

130. $7\pi^4/216$

131. $\dfrac{-1}{96}\pi^5$

132. $\dfrac{-1}{2}(\sin 3 - \sin 2 - \sin 1)$

Part C

1. Verify property (16.1.4). (16.1)

2. Let $f = f(x, y)$ be continuous on the rectangle $R : a \le x \le b, c \le y \le d$. Suppose that $L_f(P) = U_f(P)$ for some partition P of R. What can you conclude about f? What is (16.2)
$$\iint_R f(x, y)\, dx\, dy?$$

3. Let $f(x, y) = \sin(x + y)$ on $R : 0 \le x \le 1, 0 \le y \le 1$. Show that (16.2)
$$0 \le \iint_R \sin(x + y)\, dx\, dy \le 1.$$

4. Find the volume of the tetrahedron bounded by the coordinate planes and the plane (16.3)
$$\frac{x}{a} + \frac{y}{b} + \frac{z}{c} = 1, \quad a, b, c > 0.$$

5. Let R be a rectangle symmetric about the x-axis, sides parallel to the coordinate axes. Show that, if f is odd with respect to y, then the double integral of f over R is 0. (16.3)

6. (*Differentiation under the integral sign*) If f and $\partial f / \partial x$ are continuous, then the function (16.3)
$$H(t) = \int_a^b \frac{\partial f}{\partial x}(t, y)\, dy$$
can be shown to be continuous. Use the identity
$$\int_0^x \int_a^b \frac{\partial f}{\partial x}(t, y)\, dy\, dt = \int_a^b \int_0^x \frac{\partial f}{\partial x}(t, y)\, dt\, dy$$
to verify that
$$\frac{d}{dx}\left[\int_a^b f(x, y)\, dy\right] = \int_a^b \frac{\partial f}{\partial x}(x, y)\, dy.$$

7. Find the volume of the solid bounded below by the xy-plane and above by the paraboloid $z = 1 - (x^2 + y^2)$. (16.4)

8. Find the volume of the solid bounded above by the cone $z^2 = x^2 + y^2$ and below by the region Ω which lies inside the curve $x^2 + y^2 = 2ax$. (16.4)

9. A plate in the xy-plane undergoes a rotation in that plane about its center of mass. Show that I_z remains unchanged. (16.5)

10. A plate of mass M is in the form of a right triangle of base b and height h. Given that the mass density of the plate varies directly as the square of the distance from the vertical of the right angle, locate the center of mass of the plane. (16.5)

11. Show that, if f is continuous and nonnegative on a basic solid T, then the triple integral of f over T is nonnegative. (16.6)

12. Express the indicated quantity by repeated integrals. Do not solve! The volume of the solid bounded above by the paraboloid $z = 4 - x^2 - y^2$ and bounded below by the parabolic cylinder $z = 2 + y^2$. (16.7)

13. Show that, if $(\bar{x}, \bar{y}, \bar{z})$ is the centroid of a solid T, then (16.7)
$$\iiint_T (x - \bar{x})\, dx\, dy\, dz = 0,$$

$$\iiint_T (y - \bar{y})\, dx\, dy\, dz = 0,$$

$$\iiint_T (z - \bar{z})\, dx\, dy\, dz = 0.$$

14. Use triple integrals to find the volume enclosed by the ellipsoid (16.7)
$$\frac{x^2}{a^2} + \frac{y^2}{b^2} + \frac{z^2}{c^2} = 1.$$

15. Let T be a homogeneous right circular cylinder of mass M, base radius R, and height h. Find the moment of inertia of the cylinder about: (a) the central axis; (b) a line that lies in the plane of one of the bases and passes through the center of that base; (c) a line that passes through the center of the cylinder and is parallel to the bases. (16.8)

16. Find the volume of the solid bounded above by the plane $2z = 4 + x$, below by the xy-plane, and on the sides by the cylinder $x^2 + y^2 = 2x$. (16.8)

17. $$\int_0^{\pi/4} \int_0^{2\pi} \int_0^{\sec\phi} \rho^3 \sin\phi \cos\phi \, d\rho\, d\theta\, d\phi.$$ (16.9)

18. Let T be the solid bounded below by the half-cone $z = \sqrt{x^2 + y^2}$ and above by the spherical surface $x^2 + y^2 + z^2 = 1$. Use spherical coordinates to evaluate (16.9)
$$\iiint_T e^{(x^2+y^2+z^2)^{3/2}}\, dx\, dy\, dz.$$

19. Show that the ellipse $b^2 x^2 + a^2 y^2 = a^2 b^2$ has area πab by setting $x = ar\cos\theta$, $y = br\sin\theta$. (16.10)

20. Evaluate (16.10)
$$\int_{-\infty}^{\infty} \int_{-\infty}^{\infty} \frac{e^{-(x-y)^2}}{1 + (x+y)^2}\, dx\, dy$$
by integrating over the square $S_a: -a \leq x \leq a, -a \leq y \leq a$ and taking the limit as $a \to \infty$.

21. Evaluate (16.10)
$$\iiint_T \left(\frac{x^2}{a^2} + \frac{y^2}{b^2} + \frac{z^2}{c^2}\right) dx\, dy\, dz.$$

Answers to Chapter 16, Part C Questions

1. Follows from the additive property of single-sigma summations.

2. On each subrectangle, the minimum and the maximum of f are equal, so f is constant on each subrectangle and therefore (since f is continuous) on the entire rectangle R. Then
$$\iint_R f(x,y)\,dx\,dy = f(a,c)(b-a)(d-c).$$

3. $0 \le \sin(x+y) \le 1$ for all $(x,y) \in R$. Thus,
$$0 \le \iint_R \sin(x+y)\,dx\,dy \le \iint_R dx\,dy = 1$$

4. $\int_0^a \int_0^{b(1-x/a)} c\left(1 - \frac{x}{a} - \frac{y}{b}\right) dy\,dx$
$$= \int_0^a \frac{bc}{2}\left(1 - \frac{x}{a}\right)^2 dx = \frac{abc}{6}$$

5. We have $R : a \le x \le b,\ -c \le y \le c$. Set $g_x(y) = f(x,y)$. The function g_x is odd and thus
$$\int_{-c}^c g_x(y)\,dy = 0.$$

It follows that
$$\iint_R f(x,y)\,dx\,dy = \int_a^b \left(\int_{-c}^c f(x,y)\,dy\right) dx$$
$$= \int_a^b \left(\int_{-c}^c g_x(y)\,dy\right) dx$$
$$= \int_a^b 0\,dx = 0$$

6. $\int_0^x \int_a^b \frac{\partial f}{\partial x}(t,y)\,dy\,dt = \int_a^b \int_0^x \frac{\partial f}{\partial x}(t,y)\,dt\,dy$
$$= \int_a^b [f(x,y) - f(0,y)]\,dy$$
$$= \int_a^b f(x,y)\,dy$$
$$- \int_a^b f(0,y)\,dy.$$

Thus
$$\int_a^b f(x,y)\,dy = \int_0^x \int_a^b \frac{\partial f}{\partial x}(t,y)\,dy\,dt$$
$$+ \int_a^b f(0,y)\,dy$$

and
$$\frac{d}{dx}\left[\int_a^b f(x,y)\,dy\right]$$
$$= \frac{d}{dx}\left[\int_0^x \int_a^b \frac{\partial f}{\partial x}(t,y)\,dy\,dt\right]$$
$$+ \frac{d}{dx}\left[\int_a^b f(0,y)\,dy\right]$$
$$= \frac{d}{dx}\left[\int_0^x H(t)\,dt\right] + 0 = H(x)$$
$$= \int_a^b \frac{\partial f}{\partial x}(x,y)\,dy,$$
by Theorem 5.3.5.

7. $V = \int_0^{2\pi} \int_0^1 (1 - r^2)\,r\,dr\,d\theta$
$$= 2\pi \int_0^1 (r - r^3)\,dr = \frac{\pi}{2}$$

8. $\int_{-\pi/2}^{\pi/2} \int_0^{2a\cos\theta} r^2\,dr\,d\theta = \int_{-\pi/2}^{\pi/2} \frac{8a^3}{3}\cos^3\theta\,d\theta$
$$= \frac{32}{9}a^3$$

9. $I_z = I_M + d^2 M$. Rotation doesn't change d, doesn't change M, and doesn't change I_M.

10. Putting the right angle at the origin, we have $\lambda(x,y) = k(x^2 + y^2)$.
$$M = \int_0^b \int_0^{h - \frac{h}{b}x} k(x^2 + y^2)\,dy\,dx$$
$$= \frac{1}{12}kbh(b^2 + h^2)$$
$$x_M M = \int_0^b \int_0^{h - \frac{h}{b}x} kx(x^2 + y^2)\,dy\,dx$$
$$= \frac{kb^2 h(3b^2 + h^2)}{60} \Rightarrow x_M = \frac{b(3b^2 + h^2)}{5(b^2 + h^2)}$$
$$y_M M = \int_0^b \int_0^{h - \frac{h}{b}x} ky(x^2 + y^2)\,dy\,dx$$
$$= \frac{kbh^2(b^2 + 3h^2)}{60} \Rightarrow y_M = \frac{h(b^2 + 3h^2)}{5(b^2 + h^2)}$$

11. Encase T in a box Π. A partition P of Π breaks up Π into little boxes Π_{ijk}. Since f is nonnegative on Π, all the m_{ijk} are nonnegative. Therefore
$$0 \le L_f(P) \le \iiint_T f(x,y,z)\,dx\,dy\,dz.$$

12. $\int_{-\sqrt{2}}^{\sqrt{2}} \int_{-\sqrt{1-x^2/2}}^{\sqrt{1-x^2/2}} \int_{2+y^2}^{4-x^2-y^2} dz\, dy\, dx$

13. $\iiint_T (x - \bar{x})\, dx\, dy\, dz$
$= \iiint_T x\, dx\, dy\, dz - \bar{x} \iiint_T dx\, dy\, dz$
$= \bar{x}V - \bar{x}V = 0$
similarly the other two integrals are zero

14. $8\int_0^a \int_0^{b\sqrt{1-x^2/a^2}} \int_0^{c\sqrt{1-x^2/a^2-y^2/b^2}} dz\, dy\, dx$
$= \frac{4}{3}\pi abc.$

15. (a) $I = \frac{M}{\pi r^2 h}\int_0^{2\pi}\int_0^R \int_0^h r^3 dz\, dr\, d\theta = \frac{1}{2}MR^2$
(b) $I = \frac{M}{\pi r^2 h}\int_0^{2\pi}\int_0^R \int_0^h (r^2 \sin^2\theta + z^2)\, r\, dz\, dr\, d\theta$
$= \frac{1}{4}MR^2 + \frac{1}{3}Mh^2$
(c) $I = \frac{1}{4}MR^2 + \frac{1}{3}Mh^2 - M\left(\frac{1}{2}h\right)^2$
$= \frac{1}{4}MR^2 + \frac{1}{12}Mh^2$

16. $V = \int_{-\pi/2}^{\pi/2} \int_0^{2\cos\theta} \int_0^{2+\frac{1}{2}r\cos\theta} r\, dz\, dr\, d\theta = \frac{5}{2}\pi$

17. $\int_0^{\pi/4}\int_0^{2\pi}\int_0^{\sec\phi} \rho^3 \cos\phi \sin\phi\, d\rho\, d\theta\, d\phi$
$= \frac{\pi}{2}\int_0^{\pi/4} \sec^4\phi \cos\phi \sin\phi\, d\phi = \frac{3}{4}\pi$

18. $\int_0^{2\pi}\int_0^{\pi/4}\int_0^1 e^{\rho^3} \rho^2 \sin\phi\, d\rho\, d\phi\, d\theta$
$= \frac{1}{3}\pi(e-1)(2-\sqrt{2})$

19. $J(r,\theta) = abr,\ \Gamma: 0 \le r \le 1,\ 0 \le \theta \le 2\pi$
$A = \iint_\Gamma abr\, dr\, d\theta = ab\int_0^{2\pi}\int_0^1 r\, dr\, d\theta = \pi ab$

20. $\iint_{S_a} \frac{e^{-(x-y)^2}}{1+(x+y)^2} dx\, dy = \frac{1}{2}\iint_\Gamma \frac{e^{-u^2}}{1+v^2} du\, dv$

where Γ is the square in the uv-plane with vertices $(-2a, 0), (0, -2a), (2a, 0), (0, 2a)$. Γ contains the square $-a \le u \le a,\ -a \le v \le a$ and is contained in the square $-2a \le u \le 2a,\ -2a \le v \le 2a$. Therefore

$\frac{1}{2}\int_{-a}^a \int_{-a}^a \frac{e^{-u^2}}{1+v^2} du\, dv$
$\le \frac{1}{2}\iint_\Gamma \frac{e^{-u^2}}{1+v^2} du\, dv$
$\le \frac{1}{2}\int_{-2a}^{2a}\int_{-2a}^{2a} \frac{e^{-u^2}}{1+v^2} du\, dv.$

The two extremes can be written

$\frac{1}{2}\left(\int_{-a}^a e^{-u^2} du\right)\left(\int_{-a}^a \frac{1}{1+v^2} dv\right)$

and

$\frac{1}{2}\left(\int_{-2a}^{2a} e^{-u^2} du\right)\left(\int_{-2a}^{2a} \frac{1}{1+v^2} dv\right).$

As $a \to \infty$ both expressions tend to

$\frac{1}{2}(\sqrt{\pi})(\pi) = \frac{1}{2}\pi^{3/2}.$ It follows that

$\int_{-\infty}^\infty \int_{-\infty}^\infty \frac{e^{-(x-y)^2}}{1+(x+y)^2} dx\, dy = \frac{1}{2}\pi^{3/2}.$

21. $I = \int_0^{2\pi}\int_0^1 \int_0^\pi \rho^2(abc\rho^2 \sin\phi)\, d\phi\, d\rho\, d\theta$
$= \frac{4}{5}\pi abc$

CHAPTER 17 Line Integrals and Surface Integrals
Part B

17.1 Line Integrals

1. Integrate $\mathbf{h}(x, y) = x^2 \mathbf{i} + y \mathbf{j}$ over
   (a) $\mathbf{r}(u) = u^2 \mathbf{i} - 2u \mathbf{j}, u \in [0, 1]$
   (b) the line segment from $(2, 3)$ to $(1, 2)$.

2. Integrate $\mathbf{h}(x, y) = 2xy \mathbf{i} + 3y^2 \mathbf{j}$ over
   (a) $\mathbf{r}(u) = e^u \mathbf{i} + e^{-u} \mathbf{j}, u \in [0, 2]$
   (b) the line segment from $(1, 2)$ to $(2, 3)$.

3. Integrate $\mathbf{h}(x, y) = (2xy - y) \mathbf{i} + 3xy \mathbf{j}$ over
   (a) $\mathbf{r}(u) = (1 - u) \mathbf{i} + 2u \mathbf{j}, u \in [0, 1]$
   (b) the line segment from $(1, 1)$ to $(2, 2)$.

4. Integrate $\mathbf{h}(x, y) = x^{-2} y^{-2} \mathbf{i} - x^{-1} y^{-1} \mathbf{j}$ over
   (a) $\mathbf{r}(u) = \sqrt{2u - 1} \mathbf{i} + \sqrt{2 + u} \mathbf{j}, u \in [1, 3]$
   (b) the line segment from $(2, 3)$ to $(4, 5)$.

5. Integrate $\mathbf{h}(x, y) = 2y \mathbf{i} - 3x \mathbf{j}$ over the triangle with vertices $(-2, 0), (2, 0), (0, 2)$ traversed counterclockwise.

6. Integrate $\mathbf{h}(x, y) = e^{x-2y} \mathbf{i} - e^{2x+y} \mathbf{j}$ over the line segment from $(-1, 1)$ to $(1, 2)$.

7. Integrate $\mathbf{h}(x, y) = (x^2 + y) \mathbf{i} + (2y^2 - xy) \mathbf{j}$ over the closed curve that begins at $(-2, 0)$, goes along the x-axis to $(2, 0)$, and returns to $(-2, 0)$ by the upper part of the circle.

8. Integrate $\mathbf{h}(x, y) = 2xy^2 \mathbf{i} + (xy^2 - 2x^3) \mathbf{j}$ over the square with vertices $(0, 0), (1, 0), (1, 1), (0, 1)$ traversed counterclockwise.

9. Integrate $\mathbf{h}(x, y, z) = xz^2 \mathbf{i} + y^2 z \mathbf{j} + xy \mathbf{k}$ over
   (a) $\mathbf{r}(u) = 2u \mathbf{i} - u^2 \mathbf{j} + u^3 \mathbf{k}, u \in [0, 1]$
   (b) the line segment from $(0, 0, 0)$ to $(1, 1, 1)$.

10. Integrate $\mathbf{h}(x, y, z) = xe^x \mathbf{i} + e^z \mathbf{j} + e^{-y} \mathbf{k}$ over
    (a) $\mathbf{r}(u) = 2u \mathbf{i} - 3u \mathbf{j} + 2u \mathbf{k}, u \in [0, 1]$
    (b) the line segment from $(0, 0, 0)$ to $(1, 1, 1)$.

11. Calculate the work done by the force $\mathbf{F}(x, y, z) = xy \mathbf{i} - y^2 \mathbf{j} - xyz \mathbf{k}$ applied to an object that moves in a straight line from $(0, 2, -1)$ to $(2, 1, 1)$.

12. Calculate the work done by the force $\mathbf{F}(x, y, z) = x^2 \mathbf{i} + xy \mathbf{j} + z \mathbf{k}$ applied to an object that moves in a straight line from $(-1, 1, 2)$ to $(2, -1, -1)$.

13. An object of mass m moves from time $t = 0$ to $t = 1$ so that its position at time t is given by the vector function $\mathbf{r}(t) = 2t \mathbf{i} - t^2 \mathbf{j}$. Find the total force acting on the object at time t and calculate the work done by that force during the time interval $[0, 1]$.

17.2 The Fundamental Theorem for Line Integrals

14. Calculate the line integral of $\mathbf{h}(x, y) = 2xy\,\mathbf{i} + (x^2 + y)\,\mathbf{j}$ over the curve $\mathbf{r}(u) = 2\cos u\,\mathbf{i} + \sin u\,\mathbf{j}$, $u \in [0, 2\pi]$.

15. Calculate the line integral of $\mathbf{h}(x, y) = (y^2 + 2xy)\,\mathbf{i} + (x^2 + 2xy)\,\mathbf{j}$ over the curve $\mathbf{r}(u) = \sin u\,\mathbf{i} + (2 - 2\cos u)\,\mathbf{j}$, $u \in [0, \pi/2]$.

16. Calculate the line integral of $\mathbf{h}(x, y) = (\sin y + y\cos x)\,\mathbf{i} + (\sin x + x\cos y)\,\mathbf{j}$ over the straight-line segment from $(\pi/2, \pi/2)$ to (π, π).

17. Calculate the line integral of $\mathbf{h}(x, y, z) = 8xz\,\mathbf{i} - 2yz\,\mathbf{j} + (4x^2 - y^2)\,\mathbf{k}$ over the curve $\mathbf{r}(u) = (1 + 3u)\,\mathbf{i} + 2u^{5/2}\,\mathbf{j} + (2 + u)\,\mathbf{k}$, $u \in [0, 1]$.

18. Calculate the line integral of $\mathbf{h}(x, y, z) = (y + z)\,\mathbf{i} + (x + z)\,\mathbf{j} + (x + y)\,\mathbf{k}$ over the curve $\mathbf{r}(u) = u^4\,\mathbf{i} + 2\sin\frac{\pi}{2}u\,\mathbf{j} + 3u^2\,\mathbf{k}$, $u \in [0, 1]$.

19. Calculate the work done by the force $F(x, y, z) = z\,\mathbf{i} + y\,\mathbf{j} + x\,\mathbf{k}$ applied to a particle that moves along the curve $\mathbf{r}(u) = \mathbf{i} + \sin u\,\mathbf{j} + \cos u\,\mathbf{k}$ for $0 \le u \le \pi/3$.

20. Calculate the work done by the force $F(x, y, z) = 3x^2\,\mathbf{i} + \dfrac{z^2}{y}\,\mathbf{j} + 2z\ln y\,\mathbf{k}$ applied to a particle that moves from the point $(0, 1, 1)$ to the point $(2, 2, 1)$.

17.4 Line Integrals with Respect to Arc Length

21. Evaluate $\displaystyle\int_C 2xy\,dx + (e^x + x^2)\,dy$, where C is the line segment from $(0, 0)$ to $(1, 1)$.

22. Evaluate $\displaystyle\int_C y^2\,dx - x^2\,dy$, where C is the line segment from $(0, 1)$ to $(1, 0)$.

23. Evaluate $\displaystyle\int_C xy\,dx - y^2\,dy$, where C is the line segment from $(0, 0)$ to $(2, 1)$.

24. Evaluate $\displaystyle\int_C (x^2 - y^2)\,dx - 2xy\,dy$, where C is the parabola $y = 2x^2$ from $(0, 0)$ to $(1, 2)$.

25. Evaluate $\displaystyle\int_C (3x^2 + y)\,dx + 4xy\,dy$, where C is the broken line path from $(0, 0)$ to $(2, 0)$ to $(0, 4)$ to $(0, 0)$.

26. Evaluate $\displaystyle\int_C (e^x - 3y)\,dx + (e^y + 6x)\,dy$, where C is the broken line path from $(0, 0)$ to $(1, 0)$ to $(0, 2)$ to $(0, 0)$.

27. Evaluate $\displaystyle\int_C -yz\,dx - xz\,dy + (1 + xy)\,dz$, where C is the circular helix $\mathbf{r}(t) = 2\cos t\,\mathbf{i} + 2\sin t\,\mathbf{j} + 3t\,\mathbf{k}$ from $(2, 0, 0)$ to $(2, 0, 6\pi)$.

28. Evaluate $\displaystyle\int_C x^2 y\,dx + 4\,dy$, where C is the curve $\mathbf{r}(t) = e^t\,\mathbf{i} + e^{-t}\,\mathbf{j}$ for $0 \le t \le 1$.

29. Evaluate $\displaystyle\int_C z\,dx + x\,dy + y\,dz$, where C is the helix $\mathbf{r}(t) = \sin t\,\mathbf{i} + 3\sin t\,\mathbf{j} + \sin^2 t\,\mathbf{k}$ for $0 \le t \le \pi/2$.

30. Evaluate $\displaystyle\int_C -y\,dx - x\,dy + z\,dz$, where C is the circle $x = \cos t$, $y = \sin t$ for $0 \le t \le 2\pi$.

31. Evaluate $\displaystyle\int_C 4xy\,dx - 8y\,dy + 3\,dz$, where C is the curve given by $y = 2x$, $z = 3$ from $(0, 0, 3)$ to $(3, 6, 3)$.

32. Evaluate $\int_C x \sin y \, dx + \cos y \, dy + (x+y) \, dz$, where C is the straight line $x = y = z$ from $(0, 0, 0)$ to $(1, 1, 1)$.

33. Find the length and centroid of a wire shaped like the helix $x = 3\cos u$, $y = 3\sin u$, $z = 4u$, $u \in [0, 2\pi]$.

34. Find the mass, center of mass, and moments of inertia I_x, I_y of a wire shaped like the first-quadrant portion of the circle $x^2 + y^2 = a^2$ with mass density $\lambda(x, y) = kxy$.

35. Find the moment of inertia about the z-axis of a thin homogeneous rod of mass M that lies along the interval $0 \le x \le L$ of the x-axis.

36. Find the center of mass and moments of inertia I_x, I_y, I_z of a wire shaped like the curve $\mathbf{r}(u) = (u^2 - 1)\mathbf{j} + 2u\mathbf{k}$, $u \in [0, 1]$ if the mass density is $\lambda(x, y, z) = \sqrt{y+2}$.

17.5 Green's Theorem

37. Use Green's Theorem to evaluate $\oint_C (3x^2 + y)dx + 4xy \, dy$, where C is the triangular region with vertices $(0, 0)$, $(2, 0)$, and $(0, 4)$. Assume that the curve is traversed in a counterclockwise manner.

38. Use Green's Theorem to evaluate $\oint_C (2xy - y^2)dx + (x^2 - y^2)dy$, where C is the boundary of the region enclosed by $y = x$ and $y = x^2$. Assume that the curve C is traversed in a counterclockwise manner.

39. Use Green's Theorem to evaluate $\oint_C (3x^2 + y)dx + 4y^2 dy$, where C is the boundary of the region enclosed by $x = y^2$ and $y = x/2$ traversed in a counterclockwise manner.

40. Use Green's Theorem to evaluate $\oint_C (y - \sin x)dx + \cos x \, dy$, where C is the boundary of the region with vertices $(0, 0)$, $(\pi/2, 0)$, and $(\pi/2, 1)$ traversed in a counterclockwise manner.

41. Use Green's Theorem to evaluate $\oint_C (e^x - 3y)dx + (e^y + 6x)dy$, where C is the boundary of the triangular region with vertices $(0, 0)$, $(1, 0)$, and $(0, 2)$ traversed in a counterclockwise manner.

42. Use Green's Theorem to evaluate $\oint_C (2xy - y^2)dx + x^2 \, dy$, where C is the boundary of the region enclosed by $y = x + 1$ and $y = x^2 + 1$ traversed in a counterclockwise manner.

43. Use Green's Theorem to evaluate $\oint_C (x^3 - 3y)dx + (x + \sin y)dy$, where C is the boundary of the triangular region with vertices $(0, 0)$, $(1, 0)$, and $(0, 2)$ traversed in a counterclockwise manner.

44. Use Green's Theorem to evaluate $\oint_C (x^2 - \cosh y)dx + (y + \sin x)dy$, where C is the boundary of the region enclosed by $0 \le x \le \pi$ and $0 \le y \le 1$ traversed in a counterclockwise manner.

45. Use Green's Theorem to evaluate $\oint_C (-xy^2 \, dx + x^2 y)dy$, where C is the boundary of the region in the first quadrant enclosed by $y = 1 - x^2$ traversed in a counterclockwise manner.

46. Use Green's Theorem to evaluate $\oint_C [y^3 dx + (x^3 + 3xy^2)dy]$, where C is the boundary of the region enclosed by $y = x^2$ and $y = x$ traversed in a counterclockwise manner.

47. Use Green's Theorem to evaluate $\oint_C (-x^2 y \, dx + xy^2 \, dy)$, where C is the circle $x^2 + y^2 = 16$ traversed in a counterclockwise manner.

48. Use Green's Theorem to evaluate $\oint_C [2xy \, dx + (e^x + x^2) dy]$, where C is the boundary of the triangular region with vertices $(0, 0)$, $(1, 0)$, and $(1, 1)$ traversed in a counterclockwise manner.

49. Use a line integral to find the area of the region in the first quadrant enclosed by $y = x$ and $y = x^3$.

50. Use a line integral to find the area of the region enclosed by $y = 1 - x^4$ and $y = 0$.

51. Use Green's Theorem to evaluate $\oint_C [2 \tan^{-1} \frac{y}{x} dx + \ln(x^2 + y^2) dy]$, where C is the boundary of the circle $(x - 2)^2 + y^2 = 1$ traversed in a counterclockwise manner.

52. Use a line integral to find the area of the region enclosed by $x^2 + 4y^2 = 4$.

53. Use a line integral to find the area of the region enclosed by $y = x$ and $y = x^2$.

54. Use a line integral to find the area of the region enclosed by $y = \sin x$, $y = \cos x$, and $x = 0$.

17.6 Parameterized Surfaces; Surface Area

55. Find the surface area cut from the plane $z = 4x + 3$ by the cylinder $x^2 + y^2 = 25$.

56. Find the surface area of that portion of the paraboloid $z = x^2 + y^2$ that lies below the plane $z = 1$.

57. Find the surface area cut from the plane $2x - y - z = 0$ by the cylinder $x^2 + y^2 = 4$.

58. Find the surface area of that portion of the plane $3x + 4y + 6z = 12$ that lies in the first octant.

59. Find the surface area of that portion of the paraboloid $z = 25 - x^2 - y^2$ for which $z \geq 0$.

60. Find the surface area of that portion of the sphere $x^2 + y^2 + z^2 = 4$ that lies inside the cylinder $x^2 + y^2 = 2x$ and above the xy-plane.

61. Find the surface area of that portion of the paraboloid $z = 25 - x^2 - y^2$ that lies inside the cylinder $x^2 + y^2 = 9$ and above the xy-plane.

62. Find the surface area of the surface $z = \frac{1}{a}(y^2 - x^2)$ cut by the cylinder $x^2 + y^2 = a^2$ that lies above the xy-plane.

63. Find the surface area of that portion of the cylinder $y^2 + z^2 = 4$ in the first octant cut out by the planes $x = 0$ and $y = x$.

64. Find the surface area of that portion of the cylinder $x^2 + z^2 = 25$ that lies inside the cylinder $x^2 + y^2 = 25$.

65. Find the surface area of the surface $z = 2x + y^2$ that lies above the triangular region with vertices at $(0, 0, 0)$, $(0, 1, 0)$, and $(1, 1, 1)$.

66. Find the surface area of that portion of the plane $z = x + y$ in the first octant that lies inside the cylinder $4x^2 + 9y^2 = 36$.

67. Find the surface area of that portion of the cylinder $y^2 + z^2 = 4$ that lies above the region in the xy-plane enclosed by the lines $x + y = 1$, $x = 0$, and $y = 0$.

68. Find the surface area of that portion of the cylinder $z = y^2$ that lies above the triangular region with vertices at $(0, 0, 0)$, $(0, 1, 0)$, and $(1, 1, 0)$.

69. Find the surface area of that portion of the cylinder $x^2 = 1 - z$ that lies above the triangular region with vertices at $(0, 0, 0)$, $(1, 0, 0)$, and $(1, 1, 0)$.

70. Find the surface area of that portion of the sphere $x^2 + y^2 + z^2 = 4$ that lies inside the cylinder $x^2 + y^2 = 2y$.

71. Find the surface area of that portion of the sphere $x^2 + y^2 + z^2 = 4$ that lies in the first octant between the planes $y = 0$, and $y = x$.

72. Find the surface area of that portion of the sphere $x^2 + y^2 + z^2 = 4$ that lies in the first octant between the planes $y = 0$, and $y = \sqrt{3}x$.

17.7 Surface Integrals

73. Evaluate the surface integral $\iint_S (x^2 + y^2) \, d\sigma$ where S is the portion of the cone $z = \sqrt{3(x^2 + y^2)}$ for $0 \leq z \leq 3$.

74. Evaluate the surface integral $\iint_S 8x \, d\sigma$ where S is the surface enclosed by $z = x^2$, $0 \leq x \leq 2$, and $-1 \leq y \leq 2$.

75. Evaluate the surface integral $\iint_S 3x^3 \sin y \, d\sigma$ where S is the surface enclosed by $z = x^3$, $0 \leq x \leq 2$, and $0 \leq y \leq \pi$.

76. Evaluate the surface integral $\iint_S (\cos x + \sin y) \, d\sigma$ where S is that portion of the plane $x + y + z = 1$ that lies in the first octant.

77. Evaluate the surface integral $\iint_S \tan^{-1} \frac{y}{x} \, d\sigma$ where S is that portion of the paraboloid $z = x^2 + y^2$ enclosed by $1 \leq z \leq 9$.

78. Evaluate the surface integral $\iint_S x \, d\sigma$ where S is that portion of the plane $x + 2y + 3z = 6$ that lies in the first octant.

79. Evaluate the surface integral $\iint_S (x^2 + y^2) \, d\sigma$ where S is that portion of the plane $z = 4x + 20$ intercepted by the cylinder $x^2 + y^2 = 9$.

80. Evaluate the surface integral $\iint_S y \, d\sigma$ where S is that portion of the plane $z = x + y$ inside the elliptic cylinder $4x^2 + 9y^2 = 36$ that lies in the first octant.

81. Evaluate the surface integral $\iint_S y \, d\sigma$ where S is that portion of the cylinder $y^2 + z^2 = 4$ that lies above the region in the xy-plane enclosed by the lines $x + y = 1$, $x = 0$, and $y = 0$.

82. Evaluate the surface integral $\iint_S y^4 \, d\sigma$ where S is that portion of the surface $z = y^4$ that lies above the triangle in the xy-plane with vertices $(0, 0)$, $(0, 1)$, and $(1, 1)$.

83. Evaluate the surface integral $\iint_S x^2 \, d\sigma$ where S is that portion of the surface $z = x^3$ that lies above the triangle in the xy-plane with vertices $(0, 0)$, $(1, 0)$, and $(1, 1)$.

84. Evaluate the surface integral $\iint_S x^2 \, d\sigma$ where S is that portion of the plane $x + y + z = 1$ that lies inside the cylinder $x^2 + y^2 = 1$.

85. Evaluate the surface integral $\iint_S y^2 \, d\sigma$ where S is that portion of the plane $x + y + z = 1$ that lies in the first octant.

86. Evaluate the surface integral $\iint_S y^2 \, d\sigma$ where S is that portion of the cylinder $y^2 + z^2 = 1$ that lies above the xy-plane between $x = 0$ and $x = 5$.

87. Evaluate the surface integral $\iint_S (x^2 + y^2) \, d\sigma$ where S is that portion of the cylinder $x^2 + z^2 = 1$ that lies above the xy-plane enclosed by $0 \le y \le 5$.

88. Evaluate the surface integral $\iint_S (y^2 + z^2) \, d\sigma$ where S is the portion of the cone $x = \sqrt{3(y^2 + z^2)}$ for $0 \le x \le 3$.

89. Evaluate the surface integral $\iint_S 8x \, d\sigma$ where S is the surface enclosed by $y = x^2$, $0 \le x \le 2$, and $-1 \le z \le 2$.

90. Evaluate the surface integral $\iint_S (\sin y + \cos z) \, d\sigma$ where S is that portion of the plane $x + y + z = 1$ that lies in the first octant.

91. Evaluate $\iint_S (\mathbf{v} \cdot \mathbf{n}) \, d\sigma$ where $\mathbf{v} = y\mathbf{i} - x\mathbf{j} + 8\mathbf{k}$ and S is that portion of the paraboloid $z = x^2 + y^2$ that lies below the plane $z = 4$. Take $\mathbf{n}$ as the downward unit normal.

92. Evaluate $\iint_S (\mathbf{v} \cdot \mathbf{n}) \, d\sigma$ where $\mathbf{v} = y\mathbf{i} - x\mathbf{j} + 9\mathbf{k}$ and S is that portion of the paraboloid $z = 4 - x^2 - y^2$ that lies above $z = 0$. Take $\mathbf{n}$ as the upward unit normal.

93. Evaluate $\iint_S (\mathbf{v} \cdot \mathbf{n}) \, d\sigma$ where $\mathbf{v} = x\mathbf{i} - y\mathbf{j} + z\mathbf{k}$ and S is that portion of the plane $2x + 3y + 4z = 12$ that lies in the first octant. Take $\mathbf{n}$ as the upward unit normal.

94. Evaluate $\iint_S (\mathbf{v} \cdot \mathbf{n}) \, d\sigma$ where $\mathbf{v} = yz\mathbf{i} - xz\mathbf{j} + xy\mathbf{k}$ and S is that portion of the hemisphere $z = \sqrt{4 - x^2 - y^2}$ that lies above the xy-plane. Take $\mathbf{n}$ as the upward unit normal.

95. Evaluate $\iint_S (\mathbf{v} \cdot \mathbf{n}) \, d\sigma$ where $\mathbf{v} = y\mathbf{i} - x\mathbf{j} - 4z^2 \mathbf{k}$ and S is that portion of the cone $z = \sqrt{x^2 + y^2}$ that lies above the square in the xy-plane with vertices $(0, 0)$, $(1, 0)$, $(1, 1)$, and $(0, 1)$. Take $\mathbf{n}$ as the downward unit normal.

96. Evaluate $\iint_S (\mathbf{v} \cdot \mathbf{n}) d\sigma$ where $\mathbf{v} = y\mathbf{i} - x\mathbf{j} - \mathbf{k}$ and S is that portion of the hemisphere $z = -\sqrt{4 - x^2 - y^2}$ that lies above the plane $z = 0$. Take $\mathbf{n}$ as the downward unit normal.

97. Evaluate $\iint_S (\mathbf{v} \cdot \mathbf{n}) d\sigma$ where $\mathbf{v} = z\mathbf{i} + x\mathbf{j} + y\mathbf{k}$ and S is that portion of the cylinder $x^2 + y^2 = 4$ in the first octant between $z = 0$ and $z = 4$. Take $\mathbf{n}$ as the outward unit normal.

98. Evaluate $\iint_S (\mathbf{v} \cdot \mathbf{n}) d\sigma$ where $\mathbf{v} = x\mathbf{i} + y\mathbf{j} + z\mathbf{k}$ and S is that portion of the cone $z = \sqrt{x^2 + y^2}$ that lies in the first octant between $z = 1$ and $z = 2$. Take $\mathbf{n}$ as the downward unit normal.

99. Evaluate $\iint_S (\mathbf{v} \cdot \mathbf{n}) d\sigma$ where $\mathbf{v} = -xy^2\mathbf{i} + z\mathbf{j} + xz\mathbf{k}$ and S is that portion of the surface $z = xy$ bounded by $0 \leq x \leq 3$ and $0 \leq y \leq 2$. Take $\mathbf{n}$ as the upward unit normal.

100. Evaluate $\iint_S (\mathbf{v} \cdot \mathbf{n}) d\sigma$ where $\mathbf{v} = y\mathbf{i} + 2x\mathbf{j} + xy\mathbf{k}$ and S is that portion of the cylinder $x^2 + y^2 = 9$ in the first octant between $z = 1$ and $z = 4$.

101. Evaluate $\iint_S (\mathbf{v} \cdot \mathbf{n}) d\sigma$ where $\vec{V} = \vec{x}\mathbf{i} + \vec{y}\mathbf{j} + \vec{z}\mathbf{k}$ and S is that portion of the cone $x = \sqrt{y^2 + z^2}$ that lies in the first octant between $x = 1$ and $x = 3$. Take $\mathbf{n}$ as the unit normal that points away from the yz-plane.

102. Calculate the flux of $\mathbf{v} = x\mathbf{i} + y\mathbf{j} - 2z\mathbf{k}$ across the portion of the sphere $x^2 + y^2 + z^2 = 9$ that lies above the xy-plane, with upward unit normal.

103. Evaluate $\iint_S (\mathbf{v} \cdot \mathbf{n}) d\sigma$ where $\mathbf{v} = x\mathbf{i} + 4\mathbf{j} + 2x^2\mathbf{k}$ and S is that portion of the paraboloid $z = x^2 + y^2$ that lies above the xy-plane enclosed by the parabolas $y = 1 - x^2$ and $y = x^2 - 1$. Take $\mathbf{n}$ as the downward unit normal.

104. Evaluate $\iint_S (\mathbf{v} \cdot \mathbf{n}) d\sigma$ where $\mathbf{v} = 2\mathbf{i} - z\mathbf{j} + y\mathbf{k}$ and S is that portion of the paraboloid $x = y^2 + z^2$ between $x = 0$ and $x = 4$. Take $\mathbf{n}$ as the unit normal that points away from the yz-plane.

105. Calculate the flux of $\mathbf{v} = 9\mathbf{i} - z\mathbf{j} + y\mathbf{k}$ across the portion of the paraboloid $x = 4 - y^2 - z^2$ for which $x \geq 0$, in the direction pointing away from the xy-plane.

106. Evaluate $\iint_S (\mathbf{v} \cdot \mathbf{n}) d\sigma$ where $\mathbf{v} = -x\mathbf{i} - 2x\mathbf{j} + (z - 1)\mathbf{k}$ and S is the surface enclosed by that portion of the paraboloid $z = 4 - y^2$ that lies in the first octant and is bounded by the coordinate planes and the plane $y = x$. Take $\mathbf{n}$ as the upward unit normal.

17.8 The Vector Differential Operator ∇

107. Given that $\mathbf{v}(x, y) = x^2\mathbf{i} + 2y\mathbf{j}$, find $\nabla \cdot \mathbf{v}$ and $\nabla \times \mathbf{v}$.

108. Given that $\mathbf{v}(x, y) = 3y\mathbf{i} - 2x^2\mathbf{j}$, find $\nabla \cdot \mathbf{v}$ and $\nabla \times \mathbf{v}$.

109. Given that $\mathbf{v}(x, y, z) = yz\mathbf{i} - xz\mathbf{j} + 3xy\mathbf{k}$, find $\nabla \cdot \mathbf{v}$ and $\nabla \times \mathbf{v}$.

110. Given that $\mathbf{v}(x, y, z) = -2xy^2\mathbf{i} + z\mathbf{j} + xz\mathbf{k}$, find $\nabla \cdot \mathbf{v}$ and $\nabla \times \mathbf{v}$.

111. Given that $v(x, y, z) = -x\mathbf{i} + \mathbf{j} - 2x^2\mathbf{k}$, find $\nabla \cdot \mathbf{v}$ and $\nabla \times \mathbf{v}$.

112. Given that $v(x, y, z) = -x\mathbf{i} - 2x\mathbf{j} + (z-1)\mathbf{k}$, find $\nabla \cdot \mathbf{v}$ and $\nabla \times \mathbf{v}$.

113. Given that $v(x, y, z) = (2x + \cos z)\mathbf{i} + (y - e^x)\mathbf{j} - (2z - \ln y)\mathbf{k}$, find $\nabla \cdot \mathbf{v}$ and $\nabla \times \mathbf{v}$.

114. Given that $v(x, y, z) = e^x\mathbf{i} - ye^x\mathbf{j} + 3yz\mathbf{k}$, find $\nabla \cdot \mathbf{v}$ and $\nabla \times \mathbf{v}$.

115. Given that $v(x, y, z) = (x^3 + 3xy^2)\mathbf{i} + z^3\mathbf{k}$, find $\nabla \cdot \mathbf{v}$ and $\nabla \times \mathbf{v}$.

116. Given that $v(x, y, z) = -y^3\mathbf{i} + x^3\mathbf{j} - (x + z)\mathbf{k}$, find $\nabla \cdot \mathbf{v}$ and $\nabla \times \mathbf{v}$.

117. Given that $f(x, y, z) = x^3 + y^3 + z^3$, calculate the Laplacian $\nabla^2 f$.

118. Given that $f(x, y, z) = 2x^2y^3z$, calculate the Laplacian $\nabla^2 f$.

119. Given that $f(x, y, z) = 2(x^2 + y^2)$, calculate the Laplacian $\nabla^2 f$.

120. Given that $f(r) = \sin r$, calculate the Laplacian $\nabla^2 f$.

17.9 The Divergence Theorem

121. Find the divergence of $v(x, y, z) = x^2y\mathbf{i} + xy^2\mathbf{j} + xyz\mathbf{k}$.

122. Find the divergence of $v(x, y, z) = \cosh x\,\mathbf{i} + \sinh y\,\mathbf{j} + \ln(xy)\mathbf{k}$.

123. Find the divergence of $v(x, y, z) = e^x \cos y\,\mathbf{i} + e^x \sin y\,\mathbf{j} + z\mathbf{k}$.

124. Use the divergence theorem to evaluate $\iint_S (\mathbf{v} \cdot \mathbf{n})\,d\sigma$ where $v(x, y, z) = x\mathbf{i} + y\mathbf{j} + z\mathbf{k}$, $\mathbf{n}$ is the outer unit normal to S, and S is the surface of the paraboloid $z = x^2 + y^2$ that is inside the cylinder $x^2 + y^2 = 1$.

125. Use the divergence theorem to evaluate $\iint_S (\mathbf{v} \cdot \mathbf{n})\,d\sigma$ where $v(x, y, z) = x\mathbf{i} + y\mathbf{j} + z\mathbf{k}$, $\mathbf{n}$ is the outer unit normal to S, and S is the surface of the cube $-1 \le x \le 1, -1 \le y \le 1, -1 \le z \le 1$.

126. Use the divergence theorem to evaluate $\iint_S (\mathbf{v} \cdot \mathbf{n})\,d\sigma$ where $v(x, y, z) = x\mathbf{i} + y\mathbf{j} + z\mathbf{k}$, $\mathbf{n}$ is the outer unit normal to S, and S is the surface formed by the intersection of two paraboloids $z = x^2 + y^2$ and $z = 4 - (x^2 + y^2)$.

127. Use the divergence theorem to evaluate $\iint_S (\mathbf{v} \cdot \mathbf{n})\,d\sigma$ where $v(x, y, z) = (2x + z)\mathbf{i} + y\mathbf{j} - (2z + \sin x)\mathbf{k}$, $\mathbf{n}$ is the outer unit normal to S, and S is the surface of the cylinder $x^2 + y^2 = 4$ enclosed between the planes $z = 0$ and $z = 4$.

128. Use the divergence theorem to evaluate $\iint_S (\mathbf{v} \cdot \mathbf{n})\,d\sigma$ where $v(x, y, z) = \frac{x^3}{3}\mathbf{i} + \frac{y^3}{3}\mathbf{j} - \frac{z^3}{3}\mathbf{k}$, $\mathbf{n}$ is the outer unit normal to S, and S is the surface of the cylinder $x^2 + y^2 = 1$ enclosed between the planes $z = 0$ and $z = 1$.

129. Use the divergence theorem to evaluate $\iint_S (\mathbf{v} \cdot \mathbf{n})\,d\sigma$ where $v(x, y, z) = x\mathbf{i} + y\mathbf{j} - z\mathbf{k}$, $\mathbf{n}$ is the outer unit normal to S, and S is the surface of the solid bounded by $x + y + z = 1, x = 0, y = 0,$ and $z = 0$.

130. Use the divergence theorem to evaluate $\iint_S (\mathbf{v} \cdot \mathbf{n})\, d\sigma$ where $\mathbf{v}(x, y, z) = (x^3 + 3xy^2)\mathbf{i} + z^3\mathbf{k}$, $\mathbf{n}$ is the outer unit normal to S, and S is the surface of the sphere of radius a centered at the origin.

131. Use the divergence theorem to evaluate $\iint_S (\mathbf{v} \cdot \mathbf{n})\, d\sigma$ where $\mathbf{v}(x, y, z) = e^x\mathbf{i} + ye^x\mathbf{j} + 4x^2z\mathbf{k}$, $\mathbf{n}$ is the outer unit normal to S, and S is the surface of the solid enclosed by $x^2 + y^2 = 4$ and the planes $z = 0$ and $z = 9$.

132. Use the divergence theorem to calculate the total flux of $\mathbf{v}(x, y, z) = e^x\mathbf{i} - ye^x\mathbf{j} + 3z\mathbf{k}$, out of the sphere $x^2 + y^2 + z^2 = 9$.

133. Use the divergence theorem to calculate the total flux of $\mathbf{v}(x, y, z) = x^2\mathbf{i} - y^2\mathbf{j} + z^2\mathbf{k}$, out of the cube $0 \le x \le 1, 0 \le y \le 1, 0 \le z \le 1$.

134. Use the divergence theorem to evaluate $\iint_S (\mathbf{v} \cdot \mathbf{n})\, d\sigma$ where $\mathbf{v}(x, y, z) = x^3\mathbf{i} + x^2y\mathbf{j} + x^2z\mathbf{k}$, $\mathbf{n}$ is the outer unit normal to S, and S is the surface of the solid enclosed by the cylinder $x^2 + y^2 = 2$ and the planes $z = 0$ and $z = 2$.

135. Use the divergence theorem to evaluate $\iint_S (\mathbf{v} \cdot \mathbf{n})\, d\sigma$ where
$\mathbf{v}(x, y, z) = x(x^2 + y^2 + z^2)\mathbf{i} + y(x^2 + y^2 + z^2)\mathbf{j} + z(x^2 + y^2 + z^2)\mathbf{k}$, $\mathbf{n}$ is the outer unit normal to S, and S is the sphere $x^2 + y^2 + z^2 = 16$.

136. Use the divergence theorem to evaluate $\iint_S (\mathbf{v} \cdot \mathbf{n})\, d\sigma$ where $\mathbf{v}(x, y, z) = x^3\mathbf{i} + x^2y\mathbf{j} + x^2z\mathbf{k}$, $\mathbf{n}$ is the outer unit normal to S, and S is the surface of the solid enclosed by the hemisphere $z = \sqrt{4 - x^2 - y^2}$ and the xy-plane.

137. Use the divergence theorem to evaluate $\iint_S (\mathbf{v} \cdot \mathbf{n})\, d\sigma$ where $\mathbf{v}(x, y, z) = x^2\mathbf{i} + y^2\mathbf{j} + z^2\mathbf{k}$, $\mathbf{n}$ is the outer unit normal to S, and S is the surface of the solid enclosed by the cylinder $x^2 + y^2 = 4$ and the planes $z = 0$ and $z = 5$.

138. Use the divergence theorem to evaluate $\iint_S (\mathbf{v} \cdot \mathbf{n})\, d\sigma$ where $\mathbf{v}(x, y, z) = yz\mathbf{i} + xy\mathbf{j} + xz\mathbf{k}$, $\mathbf{n}$ is the outer unit normal to S, and S is the solid enclosed by the cylinder $x^2 + z^2 = 1$ and the planes $y = -1$ and $y = 1$.

139. Use the divergence theorem to evaluate $\iint_S (\mathbf{v} \cdot \mathbf{n})\, d\sigma$ where $\mathbf{v}(x, y, z) = y^2x\mathbf{i} + yz^2\mathbf{j} + x^2y^2\mathbf{k}$, $\mathbf{n}$ is the outer unit normal to S, and S is the sphere $x^2 + y^2 + z^2 = 4$.

17.10 Stokes's Theorem

140. Find the curl of $\mathbf{v}(x, y, z) = x^2y\mathbf{i} + y^2x\mathbf{j} + xyz\mathbf{k}$.

141. Find the curl of $\mathbf{v}(x, y, z) = \cosh x\mathbf{i} + \sinh y\mathbf{j} + \ln xy\mathbf{k}$.

142. Find the curl of $\mathbf{v}(x, y, z) = e^x \cos y\mathbf{i} + e^x \sin y\mathbf{j} + z\mathbf{k}$.

143. Verify Stokes's Theorem if S is the portion of the sphere $x^2 + y^2 + z^2 = 1$ for which $z \ge 0$ and $\mathbf{v}(x, y, z) = (2x - y)\mathbf{i} - yz^2\mathbf{j} - y^2z\mathbf{k}$.

144. Use Stokes's Theorem to evaluate $\oint_C (z-y)dx + (x-z)dy + (y-x)dz$ where C is the boundary, in the xy-plane, of the surface given by $z = 4 - (x^2 + y^2)$, $z \geq 0$.

145. Use Stokes's Theorem to evaluate $\oint_C y^2 dx + x^2 dy - (x+z)dz$ where C is the triangle in the xy-plane with vertices $(0, 0, 0)$, $(1, 0, 0)$, and $(1, 1, 0)$ with a counterclockwise orientation looking down the positive z-axis.

146. Use Stokes's Theorem to evaluate $\oint_C -3y\,dx + 3x\,dy + z\,dz$ over the circle $x^2 + y^2 = 1$, $z = 1$ traversed counterclockwise.

147. Use Stokes's Theorem to evaluate $\oint_C z\,dx + x\,dy + y\,dz$ over the triangle with vertices $(1, 0, 0)$, $(0, 1, 0)$, and $(0, 0, 1)$ traversed in a counterclockwise manner.

148. Use Stokes's Theorem to evaluate $\iint_S [(\nabla \times \mathbf{v}) \cdot \mathbf{n}]\,d\sigma$ where $\mathbf{v}(x, y, z) = x^2\mathbf{i} + z^2\mathbf{j} - y^2\mathbf{k}$ and S is that portion of the paraboloid $z = 4 - x^2 - y^2$ for which $z \geq 0$ and $\mathbf{n}$ is the upper unit normal.

149. Use Stokes's Theorem to evaluate $\iint_S [(\nabla \times \mathbf{v}) \cdot \mathbf{n}]\,d\sigma$ where

$\mathbf{v}(x, y, z) = (z - y)\mathbf{i} - (z^2 + x)\mathbf{j} + (x^2 - y^2)\mathbf{k}$ and S is that portion of the sphere $x^2 + y^2 + z^2 = 4$ for which $z \geq 0$ and $\mathbf{n}$ is the upper unit normal.

150. Use Stokes's Theorem to evaluate $\iint_S [(\nabla \times \mathbf{v}) \cdot \mathbf{n}]\,d\sigma$ where $\mathbf{v}(x, y, z) = y\mathbf{k}$ and S is that portion of the ellipsoid $4x^2 + 4y^2 + z^2 = 4$ for which $z \geq 0$ and $\mathbf{n}$ is the upper unit normal.

151. Use Stokes's Theorem to evaluate $\oint_C \sin z\,dx - \cos x\,dy + \sin y\,dz$ over the boundary of rectangle $0 \leq x \leq \pi$, $0 \leq y \leq 1$, $z = 2$, traversed in a counterclockwise manner.

152. Use Stokes's Theorem to evaluate $\oint_C (x+y)dx + (2x-3)dy + (y+z)dz$ over the boundary of the triangle with vertices $(2, 0, 0)$, $(0, 3, 0)$, and $(0, 0, 6)$ traversed in a counterclockwise manner.

153. Use Stokes's Theorem to evaluate $\oint_C 4z\,dx - 2x\,dy + 2x\,dz$ where C is the intersection of the cylinder $x^2 + y^2 = 1$ and the plane $z = y + 1$, traversed in a counterclockwise manner.

154. Use Stokes's Theorem to evaluate $\oint_C -yz\,dx + xz\,dy + xy\,dz$ over the circle $x^2 + y^2 = 2$, $z = 1$, traversed in a counterclockwise manner.

155. Use Stokes's Theorem to evaluate $\oint_C (4x - 2y)dx + yz^2\,dy + y^2z\,dz$ over the circle $x^2 + y^2 = 4$, $z = 2$, traversed in a counterclockwise manner.

156. Use Stokes's Theorem to evaluate $\oint_C (e^{-x^2} - yz)dx + (e^{-y^2} + xz + 2x)dy + e^{-z}dz$ over the circle $x^2 + y^2 = 1$, $z = 1$, traversed in a counterclockwise manner.

157. Use Stokes's Theorem to evaluate $\oint_C xz\,dx + y^2\,dy + x^2\,dz$ where C is the intersection of the plane $x + y + z = 5$ and the cylinder $x^2 + \dfrac{y^2}{4} = 1$, traversed in a counterclockwise manner.

Answers to Chapter 17, Part B Questions

1. (a) $7/3$ (b) $-29/6$
2. (a) $2e^2 - e^{-6} - 1$ (b) $-34/3$
3. (a) 2 (b) $17/2$
4. (a) $\dfrac{-2}{5\sqrt{5}} + \dfrac{2}{5} - \dfrac{7}{5\sqrt{5}}\dfrac{\pi}{4} + \dfrac{7}{5\sqrt{5}}\tan^{-1}\dfrac{1}{\sqrt{5}}$
   ≈ -0.007326
   (b) $23/60 + 3\ln 5/6$
5. -20
6. $2e^{-3} - \dfrac{1}{5}(e^4 - e^{-1})$
7. $-16/3 - 2\pi$
8. $-8/3$
9. (a) $-13/18$ (b) $5/6$
10. (a) $-\dfrac{1}{2}e^2 + \dfrac{2}{3}e^3 + \dfrac{11}{6}$
    (b) $1 + e - e^{-1}$
11. $13/3$
12. $1/2$
13. $\mathbf{F} = -2m\,\mathbf{j}$; work $= 2m$
14. 0
15. 6
16. $-\pi$
17. 172
18. 11
19. $-1/8$
20. $8 + \ln 2$
21. e
22. $2/3$
23. 1
24. $-11/3$
25. $52/3$
26. 9
27. 6π
28. $e^2/2 + 4/e - 9/2$
29. $23/6$
30. 2π
31. -72
32. $2\sin 1 - \cos 1 + 1$
33. length $= 10\pi$, centroid $(0, 0, 4\pi)$
34. mass $= \dfrac{1}{2}ka^3$, center of mass $(2a/3, 2a/3)$,
    $I_x = I_y = \dfrac{1}{4}ka^5$
35. $L^2 M/3$
36. center of mass $(0, -3/5, 9/8)$,
    $I_x = 192/35$, $I_y = 65/15$, $I_z = 128/105$
37. $52/3$
38. $2/15$
39. $4/3$
40. $-2/\pi - \pi/4$
41. 9
42. $7/15$
43. 4
44. $\pi(\cosh 1 - 1)$
45. $1/3$
46. $3/20$
47. 128π
48. 1
49. $1/4$
50. $8/5$
51. 0

52. 2π

53. $1/6$

54. $\sqrt{2} - 1$

55. $25\sqrt{17}\pi$

56. $\dfrac{\pi}{6}\left(5\sqrt{5} - 1\right)$

57. $4\sqrt{6}\pi$

58. $\sqrt{61}$

59. $\dfrac{\pi}{6}\left(101\sqrt{101} - 1\right)$

60. 4π

61. $\dfrac{\pi}{6}\left(37\sqrt{37} - 1\right)$

62. $\dfrac{\pi a^2}{6}\left(5\sqrt{5} - 1\right)$

63. 4

64. 200

65. $\dfrac{27 - 5\sqrt{5}}{12}$

66. $\dfrac{3\sqrt{3}\pi}{2}$

67. $\dfrac{\pi}{3} + 2\sqrt{3} - 4$

68. $\dfrac{1}{12}\left(5\sqrt{5} - 1\right)$

69. $\dfrac{1}{12}\left(5\sqrt{5} - 1\right)$

70. $8\pi - 16$

71. π

72. $4\pi/3$

73. 9π

74. $2\left(17\sqrt{17} - 1\right)$

75. $\dfrac{1}{9}\left(145\sqrt{145} - 1\right)$

76. $\sqrt{3}(2 - \cos 1 - \sin 1)$

77. $\dfrac{\pi^2}{6}\left(37\sqrt{37} - 5\sqrt{5}\right)$

78. $6\sqrt{14}$

79. $\dfrac{81\sqrt{17}}{2}\pi$

80. $4\sqrt{3}$

81. $4 - \sqrt{3} - \dfrac{2\pi}{3}$

82. $\dfrac{1}{144}\left(17\sqrt{17} - 1\right)$

83. $\dfrac{1}{54}\left(10\sqrt{10} - 1\right)$

84. $\dfrac{\sqrt{3}}{4}\pi$

85. $\dfrac{\sqrt{3}}{12}$

86. $5\pi/2$

87. $265\pi/6$

88. 9π

89. $2\left(17\sqrt{17} - 1\right)$

90. $\sqrt{3}(2 - \cos 1 - \sin 1)$

91. -32π

92. 36π

93. 36

94. 0

95. $8/3$

96. 4π

97. 24

98. $\dfrac{\pi}{3}\left(2\sqrt{2} - 1\right)$

99. 18

100. $81/2$

101. 0

Answers

102. 0
103. 0
104. 8π
105. 36π
106. -6
107. $2x+2; 0$
108. $0; -(4x+3)\mathbf{k}$
109. $0; 4x\mathbf{i} - 2y\mathbf{j} - 2z\mathbf{k}$
110. $-2y^2 + x; -\mathbf{i} - z\mathbf{j} + 4xy\mathbf{k}$
111. $-1; 4x\mathbf{j}$
112. $0; -2\mathbf{k}$
113. $1; \dfrac{1}{y}\mathbf{i} - \sin z\, \mathbf{j} - e^x\mathbf{k}$
114. $3y; 3z\mathbf{i} - ye^x\mathbf{k}$
115. $2x + 3y^2 + 3z^2; -6xy\mathbf{k}$
116. $-1; \mathbf{j} + (3x^2 + 3y^2)\mathbf{k}$
117. $6(x+y+z)$
118. $4y^3z + 12x^2yz$
119. 8
120. $-\sin x - \sin y - \sin z$
121. $5xy$
122. $\sinh x + \cosh y$
123. $2e^x \cos y + 1$
124. $\pi/2$
125. 24
126. 4π
127. 16π
128. $5\pi/6$
129. $\tfrac{1}{2}$
130. $12\pi a^5/5$
131. 144π
132. 108π
133. 1
134. 10π
135. 4096π
136. $64\pi/3$
137. 100π
138. 0
139. $256\pi/15$
140. $xz\mathbf{i} - yz\mathbf{j} + (y^2 - x^2)\mathbf{k}$
141. $1/y\,\mathbf{i} - 1/x\,\mathbf{j}$
142. $2e^x \sin y\, \mathbf{k}$
143. π
144. 8π
145. $1/3$
146. 3π
147. $3/2$
148. 0
149. 8π
150. 0
151. 2
152. 12
153. -4π
154. 4π
155. 8π
156. 4π
157. 0

Part C

1. A mass m, moving in a force field, traces out a circular arc at constant speed. Show that the force field does no work. Give a physical explanation for this. (17.1)

2. *(Linearity)* Show that, if **f** and **g** are continuous vector fields and C is piecewise smooth, then (17.1)

   (17.1.9)
   $$\boxed{\int_C [\alpha\, \mathbf{f}(\mathbf{r}) + \beta\, \mathbf{g}(\mathbf{r})] \cdot d\mathbf{r} = \alpha \int_C \mathbf{f}(\mathbf{r}) \cdot d\mathbf{r} + \beta \int_C \mathbf{g}(\mathbf{r}) \cdot d\mathbf{r}}$$

   for all real α, β.

3. *(Important)* The *circulation* of a vector field **v** around an oriented closed curve C is by definition the line integral (17.1)
   $$\int_C \mathbf{v}(\mathbf{r}) \cdot d\mathbf{r}.$$
   Let **v** be the velocity field of a fluid in counterclockwise circular motion about the z-axis with constant angular speed ω.
   (a) Verify that $\mathbf{v}(\mathbf{r}) = \omega\,\mathbf{k} \times \mathbf{r}$.
   (b) Show that the circulation of **v** around any circle C the xy-plane with center at the origin is $\pm 2\omega$ times the area of the circle.

4. Let $\mathbf{r} = x\,\mathbf{i} + y\,\mathbf{j} + z\,\mathbf{k}$ and set $r = \|\mathbf{r}\|$. The central force field (17.2)
   $$\mathbf{F}(\mathbf{r}) = \frac{K}{r^n}\mathbf{r}, \quad n \text{ a positive integer}$$
   is a gradient field. Find f such that $\nabla f(\mathbf{r}) = \mathbf{F}(\mathbf{r})$ if: (a) $n = 2$; (b) $n \neq 2$.

5. Set (17.2)
   $$P(x, y) = \frac{y}{x^2 + y^2} \quad \text{and} \quad Q(x, y) = -\frac{x}{x^2 + y^2}$$
   on *the punctured unit disc* $\Omega: 0 < x^2 + y^2 < 1$.
   (a) Verify that P and Q are continuously differentiable on Ω and that
   $$\frac{\partial P}{\partial y}(x, y) = \frac{\partial Q}{\partial x}(x, y) \quad \text{for all } (x, y) \in \Omega.$$
   (b) Verify that, in spite of (a), the vector field $\mathbf{h}(x, y) = P(x, y)\mathbf{i} + Q(x, y)\mathbf{j}$ is not a gradient on Ω.
   Hint: Integrate **h** over a circle of radius less than 1 centered at the origin.
   (c) Explain how (b) does not contradict Theorem 15.8.2.

6. Suppose a force field **F** is directed away from the origin with a magnitude that is inversely proportional to the distance from the origin. Show that **F** is a conservative field. (17.3)

7. Evaluate $\int_C y\,dx + yz\,dy + z(x-1)\,dz$, where C is the curve of intersection of the sphere $x^2 + y^2 + z^2 = 4$ and the cylinder $(x-1)^2 + y^2 = 1$ from $(2, 0, 0)$ to $(0, 0, 2)$. (17.4)

8. Given the vector field
   $$\mathbf{F}(x, y, z) = (2xy + z^2)\mathbf{i} + (x^2 - 2yz)\mathbf{j} + (2xz - y^2)\mathbf{k}. \tag{17.4}$$

(a) Show that **F** is a gradient field.
(b) Evaluate

$$\int_C (2xy + z^2)\,dx + (x^2 - 2yz)\,dy + (2xz - y^2)\,dz$$

along any piecewise smooth curve from $(1, 0, 1)$ to $(3, 2, -1)$.
(c) What is the value of

$$\int_{C'} (2xy + z^2)\,dx + (x^2 - 2yz)\,dy + (2xz - y^2)\,dz$$

where C' is any piecewise smooth curve from $(3, 2, -1)$ to $(1, 0, 1)$?

9. Find the Jordan curve C that maximizes the line integral (17.5)

$$\oint_C y^3\,dx + (3x - x^3)\,dy.$$

10. Show that, if $f = f(x)$ and $g = g(y)$ are continuously differentiable, then (17.5)

$$\int_C f(x)\,dx + g(y)\,dy = 0$$

for all piecewise-smooth Jordan curves C.

11. Let C be the line segment from the point (x_1, y_1) to the point (x_2, y_2). Show that (17.5)

$$\int_C -y\,dx + x\,dy = x_1 y_2 - x_2 y_1.$$

12. (a) Determine the fundamental vector product for the cylindrical surface (17.6)

$$\mathbf{r}(u, v) = a\cos u\,\mathbf{i} + a\sin u\,\mathbf{j} + v\mathbf{k}; \quad 0 \le u \le 2\pi, \quad 0 \le v \le l.$$

(b) Use your answer to (a) to find the area of the surface.

In Exercises 38–40, the mass density of a material cone $z = \sqrt{x^2 + y^2}$ with $0 \le z \le 1$ varies directly as the distance from the z-axis.

13. Determine the moments of inertia about the coordinate axes:
(a) I_x. (b) I_y. (c) I_z. (17.7)

14. Show that the scalar field (17.8)

$$f(x, y, z) = \frac{1}{\sqrt{x^2 + y^2 + z^2}}$$

is harmonic on every solid T that excludes the origin. Except for a constant multiplier, f is a potential function for the gravitational field.

15. Show that if $f = f(x, y, z)$ satisfies Laplace's equation, then its gradient field is both solenoidal and irrotational. (17.8)

16. Let T be a solid with a piecewise-smooth boundary S. Express the volume of T as a surface integral over S. (17.9)

17. The sphere $x^2 + y^2 + z^2 = a^2$ intersects the plane $x + 2y + z = 0$ in a curve C. Calculate the circulation of $\mathbf{v} = 2y\mathbf{i} - z\mathbf{j} + 2x\mathbf{k}$ about C by using Stokes's Theorem. (17.10)

18. Let S be a smooth surface with smooth bounding curve C. Show that, if ϕ and ψ are sufficiently differentiable scalar fields, then (17.10)

$$\iint_S [(\nabla\phi \times \nabla\psi) \cdot \mathbf{n}] \, d\sigma = \oint_C (\phi\nabla\psi) \cdot d\mathbf{r}$$

where $\mathbf{n} = \mathbf{n}(x, y, z)$ is a unit normal that varies continuously on S and the line integral is taken in the positive sense with respect to $\mathbf{n}$.

Answers to Chapter 17, Part C Questions

1. Place the origin at the center of the circular path C and use the time parameter t. Motion along C at constant speed is uniform circular motion

   $$\mathbf{r}(t) = r(\cos \omega t \mathbf{i} + \sin \omega t \mathbf{j}).$$

   Differentiation gives

   $$\mathbf{r}'(t) = r\omega(-\sin \omega t \mathbf{i} + \cos \omega t \mathbf{j}),$$
   $$\mathbf{r}''(t) = -r\omega^2(\cos \omega t \mathbf{i} + \sin \omega t \mathbf{j}).$$

   The force on the object is

   $$\mathbf{F}(\mathbf{r}(t)) = m\mathbf{r}''(t).$$

   Note that $\mathbf{F}(\mathbf{r}(t)) \cdot \mathbf{r}'(t) = 0$ for all t, and therefore W is 0 on every time integral.

   Physical explanation: At each instant the force on the object is perpendicular to the path of t object. Thus the component of force in the direction of the motion is always zero.

2. Follows from the linearity of the dot product and of ordinary integrals.

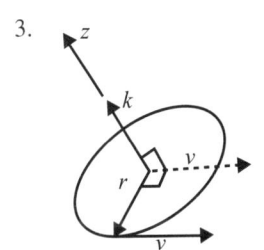

3.

   (a) $\mathbf{v} \perp \mathbf{k}$, $\mathbf{v} \perp \mathbf{r}$, $\|\mathbf{v}\| = \omega$ and $\omega \mathbf{k}, \mathbf{r}, \mathbf{v}$, form a right-handed triple

   (b) We can parametrize C counterclockwise by
   $$\mathbf{r}(t) = a\cos t\,\mathbf{i} + a\sin t\,\mathbf{j}, \quad 0 \le t \le 2\pi.$$
   Then
   $$\mathbf{r}'(t) = -a\sin t\,\mathbf{i} + a\cos t\,\mathbf{j}$$
   and
   $$\int_C (\omega\mathbf{k} \times \mathbf{r}) \cdot d\mathbf{r}$$
   $$= \int_0^{2\pi} (\omega\mathbf{k} \times \mathbf{r}(t)) \cdot \mathbf{r}'(t)\,dt.$$
   Now
   $$\mathbf{k} \times \mathbf{r}(t) = a\cos t\,\mathbf{j} - a\sin t\,\mathbf{i}.$$
   So
   $$(\mathbf{k} \times \mathbf{r}(t)) \cdot \mathbf{r}'(t) = a^2(\cos^2 t + \sin^2 t) = a^2.$$
   Thus
   $$\int_C (\omega\mathbf{k} \times \mathbf{r}) \cdot d\mathbf{r} + \int_0^{2\pi} \omega a^2\,dt = \omega a^2(2\pi)$$
   $$= 2\omega(\pi a^2) = 2\omega A.$$
   If C is parametrized clockwise, the circulation is $-2\omega A$.

4. $\mathbf{F}(x, y, z) = \dfrac{k}{(x^2 + y^2 + z^2)^{n/2}}(x\mathbf{i} + y\mathbf{j} + z\mathbf{j})$
   $$= \nabla f$$
   (a) $n = 2$: $f(\mathbf{r}) = K \ln r + C$
   (b) $n \ne 2$: $f(\mathbf{r}) = -\left(\dfrac{K}{n-2}\right)\dfrac{1}{r^{n-2}} + C$

5. (a) Since the denominator is never 0 in Ω, P and Q are continuously differentiable on Ω.
   $$\frac{\partial P}{\partial y} = \frac{x^2 - y^2}{(x^2 + y^2)^2} = \frac{\partial Q}{\partial x}.$$
   (b) Take $\mathbf{r}(u) = \dfrac{1}{2}\cos u\,\mathbf{i} + \dfrac{1}{2}\sin u\,\mathbf{j}$.
   $$\int_C \mathbf{h} \cdot d\mathbf{r} = \int_0^{2\pi}\left(\frac{\frac{1}{2}\sin u}{1/4}\mathbf{i} - \frac{\frac{1}{2}\cos u}{1/4}\mathbf{j}\right)$$
   $$\cdot \left(-\frac{1}{2}\sin u\,\mathbf{i} + \frac{1}{2}\cos u\,\mathbf{j}\right)du = \int_0^{2\pi} -du$$
   $$= -2\pi$$
   Therefore $\mathbf{h}$ is not a gradient since the integral over C (a closed curve) is not zero.
   (c) $\Omega: 0 < x^2 + y^2 < 1$ is an open plane region but is not simply connected.

6. $\mathbf{F}(\mathbf{r}) = \dfrac{k}{r^2}\mathbf{r} = \nabla f$ where $f(\mathbf{r}) = k\ln r$

7. $\mathbf{r}(u) = \left(2 - \dfrac{u^2}{2}\right)\mathbf{i} + \sqrt{1 - \dfrac{u^2}{4}}\,\mathbf{j} + u\,\mathbf{k}, u \in [0, 2]$
   $$\int_C y\,dx + yz\,dy + z(x-1)\,dz$$
   $$= \int_0^1 \left[-u^2\sqrt{1 - \frac{u^2}{4}} + \frac{u^2}{2}(2 - u^2)\right.$$
   $$\left. + u\left(1 - \frac{u^2}{2}\right)\right]du = -\frac{\pi}{2} + \frac{73}{120}$$

8. (a) $\mathbf{F} = \nabla f$ where $f(x, y, z) = x^2 y + xz^2 - y^2 z$
   (b) $\int_C (2xy + z^2)\,dx + (x^2 - 2yz)\,dy$
   $+ (2xz - y^2)\,dz = f(3, 2, -1) - f(1, 0, 1)$
   $= 25 - 1 = 24$
   (c) $\int_C (2xy + z^2)\,dx + (x^2 - 2yz)\,dy$
   $+ (2xz - y^2)\,dz = f(1, 0, 1) - f(3, 2, -1)$
   $= -24$

9. $\oint_C y^3\,dx + (3x - x^3)\,dy = \iint_\Omega (3 - 3x^2 - 3y^2)$
   $\times dx\,dy = 3\iint_\Omega (1 - x^2 - y^2)\,dx\,dy$

The double integral is maximized by
$$\Omega: 0 \le x^2 + y^2 \le 1.$$
(This is the maximal region on which the integral is nonnegative.) The line integral is maximized the unit circle traversed counterclockwise.

10. Let Ω be the region enclosed by C. Then
$$\int_C f(x)\,dx + g(y)\,dy = \pm \oint_C f(x)\,dx + g(y)\,dy$$
$$= \pm \iint_\Omega \overbrace{\left(\frac{\partial}{\partial x}[g(y)] - \frac{\partial}{\partial y}[f(x)]\right)}^{0} dx\,dy = 0$$

11. $\mathbf{r}(u) = [x_1 + (x_2 - x_1)u]\mathbf{i}$
$+ [y_1 + (y_2 - y_1)u]\mathbf{j}, u \in [0, 1]$

$$\int_C -y\,dx + x\,dy = \int_0^1 \{[-y_1 - (y_2 - y_1)u](x_2 - x_1)$$
$$+ [x_1 + (x_2 - x_1)u](y_2 - y_1)\}$$
$$= \int_0^1 (x_1 y_2 - x_2 y_1)\,du = x_1 y_2 - x_2 y_1.$$

12. (a) $\mathbf{r}_u' = -a\sin u\,\mathbf{i} + a\cos u\,\mathbf{j};\quad \mathbf{r}_v' = \mathbf{k}$
$\mathbf{N}(u,v) = \mathbf{r}_u' \times \mathbf{r}_v' = a\cos u\,\mathbf{i} + a\sin u\,\mathbf{j}$

(b) $A = \iint_\Omega \|\mathbf{N}(u,v)\|\,du\,dv = \iint_\Omega a\,du\,dv$
$= \int_0^{2\pi}\int_0^l a\,dv\,du = 2\pi l a$

13. (a) $I_x = \iint_\Omega k\sqrt{x^2+y^2}(y^2+z^2)\,d\sigma$
$= \iint_\Omega k\sqrt{x^2+y^2}(y^2+x^2+y^2)\,d\sigma$
$= k\sqrt{2}\iint_\Omega [y^2(x^2+y^2)^{1/2} + (x^2+y^2)^{3/2}]$
$\times dx\,dy$
$= k\sqrt{2}\int_0^{2\pi}\int_0^1 (r^4 \sin^2\theta + r^4)\,dr\,d\theta$
$= \frac{3\sqrt{2}}{5}\pi k$

(b) $I_y = I_x$ by symmetry

(c) $I_z = \iint_\Omega k\sqrt{x^2+y^2}(x^2+y^2)\,d\sigma$
$= \iint_S k(x^2+y^2)^{3/2}\,d\sigma$
$= k\sqrt{2}\iint_\Omega (x^2+y^2)^{3/2}\,dx\,dy$
$= k\sqrt{2}\int_0^{2\pi}\int_0^1 r^4\,dr\,d\theta = \frac{2}{5}\sqrt{2}\pi k$

14. $f(\mathbf{r}) = \frac{1}{r} \cdot \frac{\partial^2}{\partial x^2}\left(\frac{1}{r}\right) = \frac{\partial}{\partial x}\left(\frac{-x}{r^3}\right)$
$= \frac{-r^2 + 3x^2}{r^5},$ with similar formulas for y and z
Then $\nabla^2 f = \frac{-3r^2 + 3(x^2+y^2+z^2)}{r^5} = 0.$

15. Since $\nabla \cdot (\nabla f) = \nabla^2 f = 0$, the gradient field ∇f is solenoidal.
∇f is irrotational by Theorem 17.8.4

16. Since $\nabla \cdot \mathbf{r} = 3$, we can write
$V = \iiint_T dx\,dy\,dz = \iiint_T \left(\nabla \cdot \frac{\mathbf{r}}{3}\right) dx\,dy\,dz$
$= \iint_S \left(\frac{1}{3}\mathbf{r} \cdot \mathbf{n}\right) d\sigma$, by the divergence theorem.

17. $\nabla \times \mathbf{v} = \mathbf{i} - 2\mathbf{j} - 2\mathbf{k}$. Since the plane $x+2y+z=0$ passes through the origin, it intersects the sphere in a circle of radius a. The surface S bounded by this circle is a disc of radius a with upper unit normal
$$\mathbf{n} = \frac{1}{6}\sqrt{6}(\mathbf{i} + 2\mathbf{j} + \mathbf{k}).$$
The circulation of $\mathbf{v}$ with respect to $\mathbf{n}$ is given by
$\iint_S [(\nabla \times \mathbf{v}) \cdot \mathbf{n}]\,d\sigma = \iint_S \left(-\frac{5}{6}\sqrt{6}\right) d\sigma$
$= -\frac{5}{6}\sqrt{6}A = -\frac{5}{6}\sqrt{6}\pi a^2.$
If $-\mathbf{n}$ is used, the circulation is $\frac{5}{6}\sqrt{6}\pi a^2$.
Answer: $\pm\frac{5}{6}\sqrt{6}\pi a^2$.

18. $\nabla \times (\phi\nabla\psi) = (\nabla\phi \times \nabla\psi) + \phi[\nabla \times \nabla\psi]$
$= \nabla\phi \times \nabla\psi$
(17.8.7)
since the curl of a gradient is zero. Therefore the result follows from Stokes's Theorem.

CHAPTER 18 Elementary Differential Equations
Part B

18.1 Introduction: Review

1. Classify the differential equation $\dfrac{d^2y}{dx^2} + 3\dfrac{dy}{dx} + 2y = 0$ as ordinary or partial and give its order.

2. Classify the differential equation $\dfrac{\partial z}{\partial x} = z + x\dfrac{dz}{dy}$ as ordinary or partial and give its order.

3. Classify the differential equation $y''' + 2(y'')^2 + y' = \cos x$ as ordinary or partial and give its order.

4. Classify the differential equation $(y')^4 + 3xy - 2y'' = x^2$ as ordinary or partial and give its order.

5. Classify the differential equation $\dfrac{\partial^2 y}{\partial t^2} - 4\dfrac{d^2y}{dx^2} = 0$ as ordinary or partial and give its order.

6. Determine whether $y_1(x) = 3e^x$, $y_2(x) = 5e^{-x}$ are solutions of $y' + y = 0$.

7. Determine whether $y_1(x) = 1/x$, $y_2(x) = 2/x$ are solutions of $y' + y^2 = 0$.

8. Determine whether $u_1(x,t) = xe^t$, $u_2(x,t) = te^x$ are solutions of $x\dfrac{\partial u}{\partial x} = \dfrac{du}{dt}$.

9. For what values of C is $y = Ce^{-x}$ a solution of $y' + y = 0$ with side condition $y(3) = 2$?

10. For what values of C_1, C_2 is $y = C_1 \sin 2x + C_2 \cos 2x$ a solution of $y'' + 4y = 0$ with side conditions $y(0) = 0$ and $y'(0) = 1$?

11. For what values of r is $y = e^{rx}$ a solution of $y'' + y' - 6y = 0$?

12. For what values of r is $y = x^r$ a solution of $x^3 y'' - 2xy = 0$?

18.1 and 18.2 Bernoulli Equations; Homogeneous Equations; Numerical Methods

13. Find the general solution of $y' - 3y = 6$.

14. Find the general solution of $y' - 2xy = x$.

15. Find the general solution of $y' + \tfrac{4}{x}y = x^4$.

16. Find the general solution of $y' + \dfrac{2}{10 + 2x}y = 4$.

17. Find the general solution of $y' - y = -e^x$.

18. Find the general solution of $x \ln xy' + y = \ln x$.

19. Find the particular solution of $y' + 10y = 20$ determined by the side condition $y(0) = 2$.

20. Find the particular solution of $y' - y = -e^x$ determined by the side condition $y(0) = 3$.

21. Find the particular solution of $xy' - 2y = x^3 \cos 4x$ determined by the side condition $y(\pi) = 1$.

22. A 100-gallon mixing tank is full of brine containing 0.8 pounds of salt per gallon. Find the amount of salt present t minutes later if pure water is poured into the tank at the rate of 4 gallons per minute and the mixture is drawn off at the same rate.

23. Determine the velocity at time t and the terminal velocity of a 2 kg object dropped with a velocity 3 m/s, if the force due to air resistance is $-50v$ Newtons.

24. Use a suitable transformation to solve the Bernoulli equation $y' + xy = xy^2$.

25. Use a suitable transformation to solve the Bernoulli equation $xy' + y = x^3 y^6$.

26. Find the general solution of $y' = y^2 x^3$.

27. Find the general solution of $y' = \dfrac{x^2 + 7}{y^9 - 3y^4}$.

28. Find the general solution of $y' = y^2 + 1$.

29. Find the general solution of $x(y^2 + 1)y' + y^3 - 2y = 0$.

30. Find the particular solution of $e^x dx - y dy = 0$ determined by the side condition $y(0) = 1$.

31. Find the particular solution of $y' = y(x - 2)$ determined by the side condition $y(2) = 5$.

32. Verify that the equation $y' = \dfrac{y + x}{x}$ is homogeneous, then solve it.

33. Verify that the equation $y' = \dfrac{2y^4 + x^4}{xy^3}$ is homogeneous, then solve it.

34. Verify that the equation $y' = \dfrac{y}{x + \sqrt{xy}}$ is homogeneous, then solve it.

35. Verify that the equation of $[2x \sinh(y/x) + 3y \cosh(y/x)]dx - 3x \cosh(y/x) dy = 0$ is homogeneous, then solve it.

36. Find the orthogonal trajectories for the family of curves $x^2 + y^2 = C$.

37. Find the orthogonal trajectories for the family of curves $x^2 + y^2 = Cx$.

18.3 Exact Differential Equations

38. Verify that the equation $2xy + (1 + x^2)y' = 0$ is exact, then solve it.

39. Verify that the equation $ye^{xy} + xe^{xy}y' = 0$ is exact, then solve it.

40. Verify that the equation $(\cos y + y \cos x) + (\sin x - x \sin y)y' = 0$ is exact, then solve it.

41. Verify that the equation $e^{x^3}(3x^3 y - x^2) + e^{x^3} y' = 0$ is exact, then solve it.

42. Verify that the equation $(3 + 2x \ln xy + x) + \dfrac{x^2}{y} y' = 0$ is exact, then solve it.

43. Solve $(y + x^4) - xy' = 0$.

44. Solve $(x^2 + y^2 + x) + xyy' = 0$.

45. Solve $(2xy^4 e^y + 2xy^3 + y) + (x^2 y^4 e^y - x^2 y^2 - 3x)y' = 0$.

46. Solve $(y + \ln x) - xy' = 0$.

47. Find the integral curve of $(x + \sin y) + x(\cos y - 2y)y' = 0$ that passes through $(2, \pi)$.

48. Find the integral curve of $(y^2 - y) + xy' = 0$ that passes through $(-1, 2)$.

49. Find the integral curve of $(2y - 3x) + xy' = 0$ that passes through $(-5, 3)$.

18.4 The Equation $y'' + ay' + by = 0$

50. Find the general solution of $y'' - 5y = 0$.

51. Find the general solution of $y'' - 60y' + 900y = 0$.

52. Find the general solution of $y'' - 6y' + 25y = 0$.

53. Find the general solution of $16y'' + 8y' + y = 0$.

54. Find the general solution of $2y'' - 5y' + 2y = 0$.

55. Find the general solution of $y'' + 4y' + 5y = 0$.

56. Solve the initial value problem $y'' + 25y = 0$, $y(0) = 3$, $y'(0) = 10$.

57. Solve the initial value problem $y'' + y' - 6y = 0$, $y(0) = 4$, $y'(0) = 13$.

58. Solve the Euler equation $x^2 y'' + 5xy' + 4y = 0$.

59. Solve the Euler equation $x^2 y'' - 5xy' + 25y = 0$.

18.5 The Equation $y'' + ay' + by = \phi(x)$

60. Find a particular solution of $y'' - y' - 2y = 4x^2$.

61. Find a particular solution of $y'' + 5y' + 6y = 3e^{-2x}$.

62. Find a particular solution of $y'' + 4y' + 8y = 16\cos 4x$.

63. Find a particular solution of $y'' + 6y' + 9y = 16e^{-x} \cos 2x$.

64. Find the general solution of $y'' - y' - 2y = e^{2x}$.

65. Find the general solution of $y'' - 7y' = (3 - 36x)e^{4x}$.

66. Find the general solution of $y'' + 4y = 8x \sin 2x$.

67. Use variation of parameters to find a particular solution of $y'' - 2y' + y = e^x/x$.

68. Use variation of parameters to find a particular solution of $y'' + y = \sec x$.

69. Use variation of parameters to find a particular solution of $y'' + 4y = \sin^2 2x$.

18.6 Mechanical Vibrations

70. An object is in simple harmonic motion. Find an equation for the motion given that the period is $\pi/2$ and, at time $t = 0$, $x = 2$, and $v = 1$. What is the amplitude? What is the frequency?

71. An object is in simple harmonic motion. Find an equation for the motion given that the frequency is $4/\pi$ and, at time $t = 0$, $x = 3$, and $v = -3$. What is the amplitude? What is the period?

72. An object in simple harmonic motion passes through the central point $x = 0$ at time $t = 2$ and every 4 seconds thereafter. Find the equation of motion given that $v(0) = 3$.

73. Find an equation for the oscillatory motion given that the period is $3\pi/4$ and at time $t = 0$, $x = 2$, and $v = 5$.

74. Find an equation for the oscillatory motion given that the period is $5\pi/6$ and at time $t = 0$, $x = 1$, and $v = 4$.

Answers to Chapter 18, Part B Questions

1. ordinary, order 2
2. partial, order 1
3. ordinary, order 3
4. ordinary, order 2
5. partial, order 2
6. y_1 is not, y_2 is
7. y_1 is, y_2 is not
8. u_1 is, u_2 is not
9. $C = 2e^3$
10. $C_1 = 1/2, C_2 = 0$
11. $r = 2, -3$
12. $r = 2, -1$
13. -2
14. $y = Ce^{x^2} - 1/2$
15. $y = \dfrac{C}{x^4} + \dfrac{1}{9}x^5$
16. $y = \dfrac{40x + 4x^2 + C}{10 + 2x}$
17. $y = (C - x)e^x$
18. $y = \dfrac{\ln^2 x + C}{2 \ln x}$
19. $y = 2$ (identically)
20. $y = (3 - x)e^x$
21. $y = \dfrac{1}{4}x^2 \sin 4x + \left(\dfrac{x}{\pi}\right)^2$
22. $80e^{-0.04t}$ pounds
23. $v = 0.392 + 2.608e^{-25t}$;
    terminal velocity 0.392 m/s
24. $y = \dfrac{1}{1 + Ce^{\frac{1}{2}x^2}}$
25. $y = \left(\dfrac{5}{2}x^3 + Cx^5\right)^{-1/5}$
26. $y = \dfrac{-4}{x^4 + C}$
27. $\dfrac{1}{10}y^{10} + \dfrac{3}{5}y^5 - \dfrac{1}{3}y^3 + 7x = C$
28. $y = \tan(x + C)$
29. $(y^2 - 2)^3 x^4 = Cy^2$
30. $y = \sqrt{2e^x - 1}, x > \ln \dfrac{1}{2}$
31. $y = 5e^{(x-2)^2/2}$
32. $y = x \ln|Cx|$
33. $x^8 = C(y^4 + x^4)$
34. $-2\sqrt{x/y} + \ln|y| = C$
35. $x^2 = C \sinh^3(y/x)$
36. $y = Kx$
37. $x^2 + y^2 = Ky$
38. $y = \dfrac{C}{x^2 + 1}$
39. $e^{xy} = C$
40. $x \cos y + y \sin x = C$
41. $y = Ce^{-x^3} + \dfrac{1}{3}$
42. $3x + x^2 \ln xy = C$
43. $y = \dfrac{1}{3}x^4 - Cx$
44. $3x^4 + 4x^3 + 6x^2y^2 = C$
45. $x^2 e^y + \dfrac{x^2}{y} + \dfrac{x}{y^3} = C$
46. $y = Cx - \ln x - 1$
47. $1/2 x^2 + x \sin y - y^2 = 2 - \pi^2$
48. $y = \dfrac{x}{x + 1/2}$

49. $y = \dfrac{x^3 + 200}{x^2}$

50. $y = C_1 e^{\sqrt{5}x} + C_2 e^{-\sqrt{5}x}$

51. $y = (C_1 + C_2 x) e^{30x}$

52. $y = e^{3x}(C_1 \cos 4x + C_2 \sin 4x)$

53. $y = (C_1 + C_2 x) e^{-x/4}$

54. $y = C_1 e^{x/2} + C_2 e^{2x}$

55. $y = e^{-2x}(C_1 \cos x + C_2 \sin x)$

56. $y = 3 \cos 5x + 2 \sin 5x$

57. $y = 5 e^{2x} - e^{-3x}$

58. $y = \dfrac{C_1 + C_2 \ln x}{x^2}$

59. $y = x^3 [C_1 \cos(\ln x^4) + C_2 \sin(\ln x^4)]$

60. $y_p = -2x^2 + 2x - 3$

61. $y_p = 3x e^{-2x}$

62. $y_p = \dfrac{4}{5} \sin 4x - \dfrac{2}{5} \cos 4x$

63. $y_p = 2 e^{-x} \sin 2x$

64. $y = C_1 e^{-x} + C_2 e^{2x} + \dfrac{1}{3} x e^{2x}$

65. $y = C_1 + C_2 e^{7x} + 3x e^{4x}$

66. $y = C_1 \cos 2x + C_2 \sin 2x + \dfrac{1}{2} x \sin 2x - x^2 \cos 2x$

67. $y_p = -x e^x + x e^x \ln |x|$

68. $y_p = (\ln|\cos x|) \cos x + x \sin x$

69. $y_p = \dfrac{1}{6} \cos^2 2x - \dfrac{1}{12} \sin^2 2x$

70. $x(t) = 2 \cos 4t + \dfrac{1}{4} \sin 4t$

    amplitude $= \dfrac{\sqrt{65}}{4}$; frequency $= 4$

71. $x(t) = 3 \cos \dfrac{4}{\pi} t - \dfrac{3\pi}{4} \sin \dfrac{4}{\pi} t$

    amplitude $= \dfrac{\sqrt{144 + 9\pi^2}}{4}$; period $= \pi^2/2$

72. $x(t) = \dfrac{6}{\pi} \sin \dfrac{\pi}{2} t$

73. $x(t) = 2 \cos \dfrac{8}{3} t - \dfrac{15}{8} \sin \dfrac{8}{3} t$

74. $x(t) = \cos \dfrac{12}{5} t + \dfrac{5}{3} \sin \dfrac{12}{5} t$

Part C

1. For each real number r, each member of the two-parameter family $y = C_1 \sin rx + C_2 \cos rx$ is a solution of the differential equation $y'' + r^2 y = 0$.
   (a) Determine the numbers r such that the two-point boundary-value problem
   $$y'' + r^2 y = 0, \quad y(0) = 0, \quad y(\pi) = 0$$
   has a nonzero solution.
   (b) Determine the numbers r such that the two-point boundary-value problem
   $$y'' + r^2 y = 0, \quad y(0) = 0, \quad y(\pi/2) = 0$$
   has a nonzero solution. (18.1)

2. Verify that the equation is homogenous and then solve the equation.
   $$x\, dy = y\left[1 + \ln\left(\frac{y}{x}\right)\right] dx.$$ (18.2)

3. Find a particular solution determined by the initial condition.
   $$x \sin\left(\frac{y}{x}\right) dy = \left[x + y \sin\left(\frac{y}{x}\right)\right] dx, \quad y(1) = 0.$$ (18.2)

4. Show that the general solution of the differential equation $y'' + \omega^2 y = 0$ can be written $y = A \sin(\omega x + \phi_0)$, where A and ϕ_0 are constants with $A > 0$ and $\phi_0 \in [0, 2\pi]$. (18.4)

5. Show that the Euler equation $(x^2 y'' + \alpha xy' + \beta y = 0)$ can be transformed into an equation of the form (18.4)
   $$\frac{d^2 y}{dz^2} + a\frac{dy}{dz} + by = 0,$$
   where a and b are real numbers, by means c of variable $z = \ln x$. *Hint:* If $z = \ln x$, chain rule,
   $$\frac{dy}{dx} = \frac{dy}{dz}\frac{dz}{dx} = \frac{dy}{dz}\frac{1}{x}.$$
   Now calculate $d^2 y/dx^2$ and substitute into the equation.

6. In Section 8.8 we solved the linear differential equation $y' + p(x)y = q(x)$ by using the integrating factor $e^{\int p(x)\,dx}$. Show that this integrating factor is obtainable by the methods of this section. (18.3)

7. (a) Find functions f and g, not both identically zero, such that the differential equation (18.3)
   $$g(y)\sin x\, dx + y^2 f(x)\, dy = 0 \text{ is exact.}$$
   (b) Find all functions g such that the differential equation $g(y)e^y + xy\, y' = 0$ is exact.

8. Show that the change of variable $y = ve^{kx}$ transforms the equation (18.5)
   $$y'' + ay' + by = (c_n x^n + \cdots + c_1 x + c_0)\, e^{kx}$$
   into $v'' + (2k + a)v' + (k^2 + ak + b)v = c_n x^n + \cdots + c_1 x + c_0$.

9. (a) Show that $y_1 = x$, $y_2 = x \ln x$ are solutions of the Euler equation (see Exercises 18.3) (18.5)
   $x^2 y'' - xy' + y = 0$ and that their Wronskian is nonzero on $(0, \infty)$.
   (b) Use variation of parameters to find a particular solution of $x^2 y'' - xy' + y = 4x \ln x$.

10. Show that, if $\gamma \neq \omega$, then the method of undetermined coefficients applied to (18.6.8) gives (18.6)
    $$x_p = \frac{F_0/m}{\omega^2 - \gamma^2} \cos \gamma t.$$

11. Verify that the function $x_p = \dfrac{F_0/m}{(\omega^2 - \gamma^2)^2 + 4\alpha^2\gamma^2}[(\omega^2 - \gamma^2)\cos\gamma t + 2\alpha\gamma \sin\gamma t]$ is a particular solution of (*). (18.6)

Answers to Chapter 18, Part C Questions

1. Assume $r \neq 0$.
   (a) $0 = y(0) = C_2 \Longrightarrow y = C_1 \sin rx$.
   $0 = y(\pi) = C_1 \sin r\pi$. Since we want $C_1 \neq 0$, we need $\sin r\pi = 0$.
   Therefore, $r = n$, n any integer except 0.
   (b) Again, $y(0) = 0 \Longrightarrow y = C_1 \sin rx$
   $$0 = y\left(\frac{\pi}{2}\right) = C_1 \sin \frac{r\pi}{2}.$$ Since we want
   $C_1 \neq 0$, we need $\sin \frac{r\pi}{2} = 0$,
   so $r = 2n$, $n = \pm 1, \pm 2, \cdots$

2. $f(x, y) = \dfrac{y}{x}\left(1 + \ln\left(\dfrac{y}{x}\right)\right)$;
   $$f(tx, ty) = \frac{ty}{tx}\left(1 + \ln\left(\frac{ty}{tx}\right)\right)$$
   $$= \frac{y}{x}\left(1 + \ln\left(\frac{y}{x}\right)\right)$$
   $$= f(x, y)$$
   Set $vx = y$. Then, $v + xv' = y'$ and
   $$v + xv' = \frac{vx}{x}\left(1 + \ln\left(\frac{vx}{x}\right)\right) = v(1 + \ln v)$$
   $$xv' = v \ln v$$
   $$\frac{1}{v \ln v}dv = \frac{1}{x}dx$$
   $$\int \frac{1}{v \ln v}dv = \int \frac{1}{x}dx$$
   $$\ln|\ln v| = \ln|x| + K$$
   $$\ln\left(\frac{y}{x}\right) = Cx$$
   $$\frac{y}{x} = e^{Cx} \quad \text{or} \quad y = xe^{Cx}$$

3. $\dfrac{dy}{dx} = \dfrac{1}{\sin(y/x)} + \dfrac{y}{x}$. Set $y = vx$. Then
   $y' = v + xv'$ and
   $$v + xv' = \frac{1}{\sin v} + v$$
   $$\int \frac{dx}{x} - \int \sin v \, dv = C$$
   $$\ln|x| + \cos v = C$$
   $$\ln|x| + \cos\left(\frac{y}{x}\right) = C$$
   $y(1) = 0 \Longrightarrow 0 + \cos 0 = C \Longrightarrow C = 1$
   $$\Longrightarrow \ln|x| + \cos\left(\frac{y}{x}\right) = 1$$

4. $r^2 + \omega = 0 \Longrightarrow r = \pm \omega i$;
   $y = C_1 \cos \omega x + C_2 \sin \omega x$.
   Assuming that $C_1^2 + C_2^2 > 0$, we have
   $C_1 \cos \omega t + C_2 \sin \omega t = \sqrt{C_1^2 + C_2^2}$
   $$\times \left(\frac{C_1}{\sqrt{C_1^2 + C_2^2}} \cos \omega t - \frac{C_2}{\sqrt{C_1^2 + C_2^2}} \sin \omega t\right)$$
   $= A(\sin \phi_0 \cos \omega t + \cos \phi_0 \sin \omega t)$,
   where $A = \sqrt{C_1^2 + C_2^2}$ and
   ϕ_0, $\phi_0 \in [0, 2\pi]$, is the angle such that
   $$\sin \phi_0 = \frac{C_1}{\sqrt{C_1^2 + C_2^2}} \quad \text{and} \quad \cos \phi_0 = \frac{C_2}{\sqrt{C_1^2 + C_2^2}}$$

5. From the Hint, $\dfrac{dy}{dx} = \dfrac{dy}{dz}\dfrac{1}{x}$. Differentiating
   with respect to x again, we have
   $$\frac{d^2y}{dx^2} = \frac{d^2y}{dx^2}\frac{dz}{dx}\frac{1}{x} + \frac{dy}{dz}\left(-\frac{1}{x^2}\right) = \frac{1}{x^2}\left(\frac{d^2y}{dz^2} - \frac{dy}{dz}\right).$$
   Substituting into the differential equation
   $x^2 y'' + \alpha x y' + \beta y = 0$, we get
   $$\left(\frac{d^2y}{dz^2} - \frac{dy}{dz}\right) + \alpha \frac{dy}{dz} + \beta y = 0,$$
   or
   $$\frac{d^2y}{dz^2} + a\frac{dy}{dz} + by = 0,$$
   where $a = \alpha - 1$, $b = \beta$.

6. Write the linear equation as $p(x)y - q(x) + y' = 0$.
   Then $P(x) = p(x)y - q(x)$, $Q(x) = 1$
   and $v = \dfrac{1}{Q}\left(\dfrac{\partial P}{\partial y} - \dfrac{\partial Q}{\partial x}\right) = p(x)$
   depends only on x. Therefore $v = e^{\int p(x)dx}$ is an integrating factor.

7. (a) We need $g'(y) \sin x = y^2 f'(x)$.
   Take $g(y) = \dfrac{1}{3}y^3$ and $f(x) = -\cos x$
   (b) We need $g'(y)e^y + g(y)e^y = y$,
   that is $\dfrac{d}{dy}[g(y)e^y] = y$.
   It follows that $g(y)e^y = \dfrac{1}{2}y^2 + C, \Longrightarrow g(y)$
   $$= e^{-y}\left(\frac{1}{2}y^2 + C\right).$$

8. Follows directly from
   $(ve^{kx})'' + a(ve^{kx})' + b(ve^{kx}) = e^{kx}[v'' + (2k + a)v' + (k^2 + ak + b)v]$.

9. (a) $x^2 y_1'' - x y_1' + y_1 = x^2 \cdot 0 - x \cdot 1 + x = 0$:
y_1 is a solution.

$x^2 y_2'' - x y_2' + y_2 = x^2 \left(\dfrac{1}{x}\right) - x(\ln x + 1)$

$+ x \ln x = 0$: y_2 is a solution.

$W = y_1 y_2' - y_1' y_2 = x(\ln x + 1) - 1(x \ln x)$
$= x$ is nonzero on $(0, \infty)$.

(b) To use the method of variation of parameters as described in the text, we first rewrite the equation in the form

$$y'' - \dfrac{1}{x} y' + \dfrac{1}{x^2} y = \dfrac{4}{x} \ln x.$$

Then, a particular solution of the equation will have the form $y_p = y_1 z_1 + y_2 z_2$, where

$z_1 = -\displaystyle\int \dfrac{x \ln x \cdot [(4/x) \ln x]}{x} dx$

$= -4 \displaystyle\int \dfrac{1}{x} (\ln x)^2 dx = -\dfrac{4}{3}(\ln x)^3$

and

$z_2 = \displaystyle\int \dfrac{x \cdot [(4/x) \ln x]}{x} dx$

$= 4 \displaystyle\int \dfrac{\ln x}{x} dx = 2(\ln x)^2$

Thus, $y_p = -\dfrac{4}{3} x(\ln x)^3 + x \ln x \cdot 2(\ln x)^2$

which simplifies to: $y_p = \dfrac{2}{3} x (\ln x)^3$.

10. If $\gamma \neq \omega$, we try $x_p = A \cos \gamma t + B \sin \gamma t$ as a particular solution of $x'' + \omega^2 x = \dfrac{F_0}{m} \cos \gamma t$.

Substituting x_p into the equation, we get

$-\gamma^2 x_p + \omega^2 x_p = \dfrac{F_0}{m} \cos \gamma t,$

giving $x_p = \dfrac{F_0/m}{\omega^2 - \gamma^2} \cos \gamma t.$

11. Straightforward computation.